CD-AYF-687

# *United States Oil Policy and Diplomacy*

Recent Titles in
Contributions in Economics and Economic History
*Series Editor: Robert Sobel*

Who Owns the Wildlife?: The Political Economy of Conservation in Nineteenth-Century America
*James A. Tober*

The Rise of the American Electrochemicals Industry, 1880-1910: Studies in the American Technological Environment
*Martha Moore Trescott*

Southern Workers and Their Unions, 1880-1975: Selected Papers, The Second Southern Labor History Conference, 1978
*Merl E. Reed, Leslie S. Hough, and Gary M Fink, editors*

The Organization of Arab Petroleum Exporting Countries: History, Policies, and Prospects
*Mary Ann Tétreault*

Soviet Economy Today: With Guidelines for the Economic and Social Development of the U.S.S.R. for 1981-85 and for the Period Ending in 1990
*Novosti Press Agency Publishing House, Moscow*

The Era of the Joy Line: A Saga of Steamboating on Long Island Sound
*Edwin L. Dunbaugh*

The Role of Economic Advisers in Developing Countries
*Lauchlin Currie*

Perkins/Budd: Railway Statesmen of the Burlington
*Richard C. Overton*

The European Community and Latin America: A Case Study in Global Role Expansion
*A. Glenn Mower, Jr.*

Economics and Policymaking: The Tragic Illusion
*Eugene J. Meehan*

The Strategic Petroleum Reserve: Planning, Implementation, and Analysis
*David Leo Weimer*

A Portrait Cast in Steel: Buckeye International and Columbus Ohio, 1881-1980
*Mansel G. Blackford*

# *United States Oil Policy and Diplomacy*

## A TWENTIETH-CENTURY OVERVIEW

**Edward W. Chester**

Contributions in Economics and Economic History, Number 52

Greenwood Press
Westport, Connecticut • London, England

**Library of Congress Cataloging in Publication Data**

Chester, Edward W.
United States oil policy and diplomacy.

(Contributions in economics and economic history,
ISSN 0084-9235 ; no. 52)
Bibliography: p.
Includes index.
1. Petroleum industry and trade—Government policy—
United States. 2. Petroleum industry and trade—Political
aspects—United States. 3. United States—Foreign
relations—20th century. I. Title. II. Series.
HD9566.C52 1983 338.2'7282'0973 82-9379
ISBN 0-313-23174-5 (lib. bdg.) AACR2



Library of Congress Catalog Card Number: 82-9379
ISBN: 0-313-23174-5
ISSN: 0084-9235

First published in 1983

Greenwood Press
A division of Congressional Information Service, Inc.
88 Post Road West, Westport, Connecticut 06881

Printed in the United States of America

10 9 8 7 6 5 4 3 2 1

The University of Wisconsin Press has generously given permission
to use quotations from *Josephus Daniels in Mexico*, by E.
David Cronon. Copyright 1960, E. David Cronon, University of
Wisconsin Press.

In memory of my father

# Contents

# Preface

In recent years no strategically vital item for which the United States has to rely either in part or totally on foreign countries has assumed a greater importance than oil. The failure of this country to obtain sufficient quantities of petroleum at reasonable prices may well cripple, or at least deal a serious blow to, the national economy. When there were fewer expropriations of oil properties abroad and fewer state petroleum monopolies, American oil importers could operate with more freedom. Today, however, in many countries, it has become mandatory for foreign entrepreneurs to clear every business operation with the government of the host nation. Petroleum importing, therefore, has become more and more of a diplomatic concern, especially as groups of oil-producing nations periodically meet in an attempt to reach an agreement on setting the level of prices.

Since diplomacy is officially the concern of the American government, rather than of private firms or individual businessmen, authorities in Washington must base their actions upon a comprehensive understanding of the historical precedents. It is obviously impossible for them to negotiate intelligently if they have not assembled and analyzed all the background data relating to America's efforts to acquire petroleum from foreign oil-producing nations.

Unfortunately for modern-day U.S. diplomats, there never has been a comprehensive study of American oil policy and diplomacy in the twentieth century. Some of the leading textbooks on American diplomatic history devote only a few pages to petroleum and may not even cite oil in their indexes. Although scholarly monographs have better records of achievement, enormous gaps remain in the overall coverage, even with the glut of material on Mexico between the two world wars.

Much of the history of world petroleum during the twentieth century involves the so-called Third World, or developing nations. I have published two full-length books in this area: *Clash of Titans: Africa and U.S. Foreign Policy* (1974), and *The U.S. and Six Atlantic Outposts: The Military and Economic Considerations* (1980), which deals with Greenland, Iceland, the Azores, Bermuda, Jamaica, and the Bahamas. Thus in the present volume I have written about that type of country in which I have a long-established research interest.

In approaching the U.S. oil policy and diplomacy topic, I began my research by examining several major bodies of published materials, including the pertinent data from the Department of State's *Foreign Relations of the United States* series. Unfortunately, these volumes only extend up to the time of the Korean War. Various other government publications also proved to be of value. In addition, although there is a relatively limited output of retrospective scholarly articles, dating from the 1970s, there are thousands of contemporary pieces on various aspects of the petroleum industry. There are also several hundred books on various phases of U.S. foreign relations that mention oil at least briefly, as well as a far smaller number of volumes dealing with American petroleum operations overseas.

To examine these books and articles I used the University of Texas at Arlington Library and supplemented its holdings with the resources of other libraries in the Dallas-Fort Worth metropolitan area: the Dallas Public Library, the Southern Methodist University Library, the Fort Worth Public Library, the Texas Christian University Library, and the North Texas State University Library at Denton. Because a number of important titles were not available in this area, I paid an extended visit to Washington, D. C., to complete my research. I examined the holdings of the Library of Congress and those of the libraries of the American Petroleum Institute, the Joint Bank Fund, the Middle East Institute, the Organization of American States, and the Inter-American Development Bank.

Aided by a most generous grant from the Earhart Foundation of Ann Arbor, Michigan, I was able to visit five presidential libraries: those of Herbert Hoover (West Branch, Iowa), Franklin Roosevelt (Hyde Park, New York), Harry Truman (Independence, Missouri), Dwight Eisenhower (Abilene, Kansas), and Lyndon Johnson (Austin, Texas). It is impossible to trace the evolution of the oil policy and diplomacy of the American government since World War I without going through these archival holdings. Additional material for the years prior to 1950 came from the National Archives in Washington, D. C. and for the years since World War II from the *Declassified Documents Reference System* on microfiche. Finally, I would like to extend my sincere thanks to those individuals who most generously read and criticized portions of the manuscript: Harry N. Howard, Colbert C. Held, Irvine Anderson, Francis X. Murray, Thomas Hogarty, and Brian O. Lisle.

It is impossible to write a book of this type without referring to U.S. domestic oil policy, the histories of major oil firms, as well as independent ones, American economic foreign policy, and the various political developments in those nations in which U.S. oil interests have played an important role. A comprehensive treatment of each of these items, however, would destroy the continuity of this study through innumerable digressions. Cross-references, which are scattered throughout this book, as well as conveniently summarized in a single chronological table, also offer many opportunities for digressions. Developments in one country or region sometimes do have an unquestioned impact elsewhere—World War II, the Anglo-Iranian nationalization, and the Arab oil embargoes are cases in point—but it is not always easy to prove that what happens in one place significantly affects what occurs in another, either at that time or at some later date. Moreover, existing constitutions, laws, rules and regulations, and contracts frequently have slowed change.

Until the Arab oil embargo of 1973, American petroleum companies were able to maintain a more or less continuous flow of oil into the United States from abroad. They were able to do this despite the challenges posed at different times and in different places by nationalism, political reform and revolution, expropriation, the Israeli struggle for existence, and the long-standing global rivalry for oil involving both the USSR and Great Britain. This is a story that has never been told in its entirety from the standpoint of both U.S. petroleum policy and diplomacy. It is a topic, moreover, that needs to be explored in an objective, balanced manner without the emotional biases that so often distort conversations about oil, both here and abroad. Among the more important aspects of this subject are:

How much control—real or mythical—have the oil companies exercised over U.S. diplomacy?

What foreign nations have set up petroleum monopolies, and how has this affected the activities of American oil entrepreneurs?

Which types of diplomatic representations and weapons, including the imposition of embargoes and military shows of force, has the American government employed on behalf of U.S. petroleum firms, and how successful have these been?

What foreign countries have expropriated the properties of American oil companies, why did they do this, and what were the short- and long-range results of these expropriations?

What foreign countries have favored the petroleum firms of other nations over those of the United States?

Which branches of the U.S. executive branch have clashed over oil policy and diplomacy, and what positions did each take?

What has the U.S. government done at various times to restrict petroleum imports?

Which congressional committee investigated the oil giants before Idaho Senator Frank Church's Subcommittee on Multinational Corporations did so during the 1970s?

What have been the various suits that the Justice Department has brought against certain petroleum companies, and how have the latter fared in foreign courts?

How successful have attempts been to arbitrate oil disputes in foreign nations?

Under what circumstances have U.S. petroleum properties abroad been destroyed?

What has been the attitude of the American government to extending technical assistance concerning oil to other nations, and when has it attempted to make petroleum itself available to other countries?

Which diplomats have acted as agents abroad for U.S. oil companies, and which diplomats and oil men have advised foreign governments on petroleum matters?

Which American oil firms have had the most extensive paternalistic programs overseas?

Which U.S. petroleum companies have given political payoffs or bribes to foreign governments, parties, and individuals?

What role have American oil firms played in the recognition of regimes abroad and in foreign revolutions and wars?

How have groups such as labor, the military, and the press affected the oil policy of various foreign nations?

To what extent have the petroleum decisions made by foreign governments been the result of economic factors, and to what extent have they been influenced by political, social, and cultural concerns?

These questions only partially exhaust the list of possible inquiries that one might make about U.S. oil policy and diplomacy in the twentieth century. So complex is this topic that it is almost frightening to contemplate the fact that decisions have been made, or are being made, and opinions offered by individuals who have only a partial or fragmentary grasp of this immense subject. It is hoped that this volume will furnish a conceptual or structural framework that will enable those who read it to think and act in a more intelligent manner about questions relating to the international aspects of petroleum.

Dr. Edward W. Chester
Professor of History
University of Texas at Arlington

# Checklist of Name Changes and Abbreviations of Petroleum Companies

Standard Oil of New Jersey, or Jersey Standard, is in fact two separate companies: The parent firm is Standard Oil Company (New Jersey) Inc., today the Exxon Corporation. The operating company for most domestic activities is Standard Oil Company of New Jersey, Inc. In 1959 this merged with, and its name was changed to, Humble Oil and Refining Co. Inc., which later became Exxon USA Inc.

Esso—uniform trade name of Standard Oil of New Jersey abroad for several decades, with Exxon replacing it after 1973. Today it is the parent corporation only that operates outside the United States with worldwide holdings.

Standard Oil of New York, or New York Standard—abbreviated as Socony. The name changed to Socony-Mobil Oil Co. Inc. in 1955, then to Mobil Oil Corporation in 1966, now Mobil Corporation.

Socony-Vacuum—new firm resulting from the merger of New York Standard and the Vacuum Oil Company in 1931.

Standard-Vacuum Oil Company—abbreviated as Stanvac. New firm created by a merger of Jersey Standard and New York Standard to facilitate operations in the Eastern Hemisphere, that is "East of Suez," while the two companies remained competitors elsewhere. This joint endeavor lasted from 1933 to 1960 when the U.S. government outlawed it.

Standard Oil of California—abbreviated as Socal; sometimes as stancal.

California-Arabian Standard Oil Company—abbreviated as Casco. Became the Arabian-American Oil Company, abbreviated as Aramco, during 1944. A joint enterprise of California Standard Oil and Texaco.

California Texas Oil Company—abbreviated as Caltex. It is an equally owned subsidiary of California Standard and Texaco. It has operated in Indonesia as Caltex Pacific.

American Overseas Oil Corporation—abbreviated as Amoseas, a Libyan subsidiary of California Standard and Texaco.

Texas Company was incorporated in Texas as the Texas Company in 1902 and reincorporated in Delaware as the Texas Corporation in 1926. The name changed to the Texas Company Inc. in 1941, and then to Texaco Inc. in 1959.

Standard Oil of Indiana—abbreviated as Amoco, from American Oil Co., the name of its one-time leading domestic operating subsidiary.

Standard Oil of Ohio—abbreviated as Sohio.

American Independent Oil Company—abbreviated as Aminoil.

Atlantic Richfield Company—abbreviated as Arco. Originally this was the Atlantic Refining Company, or Arco, and remained known as Arco when the Richfield Oil Corporation was merged intoit in 1966.

Continental OilCompany—abbreviated as Conoco. This merged into DuPont in 1981.

Oasis—joint partnership of Conoco, Marathon, and Amerada.

Occidental—abbreviated as Oxy.

Anglo-Persian—later Anglo-Iranian, now British Petroleum, abbreviated as BP.

Royal Dutch-Shell Group—abbreviated as RD-S. In 1907 the Royal Dutch Petroleum Company of the Netherlands and the Shell Transport and Trading Company of Great Britain reached a still operative agreement under which the Dutch firm obtained a 60 percent interest in the British firm, and the British company a 40 percent share in the Dutch one. The Group frequently is referred to loosely, as "Shell."

Compagnie Française des Pétroles—French firm, abbreviated as CFP.

*United States Oil Policy and Diplomacy*

# 1.

# An Overview of the Historical Evolution of U.S. Oil Policy

## The Early Years, 1859-1914

The exploitation of petroleum seepages can be traced back to the early days of ancient history when the Sumerians, Assyrians, and Babylonians made use of bitumen, but it has only been during the last 125 years that the world has truly entered the Age of Petroleum. It was the development of the kerosene lamp around the middle of the nineteenth century that gave the initial impulse to the search for oil.

On December 30, 1854, a group of entrepreneurs organized the first American petroleum company, which was known as the Pennsylvania Rock Oil Company. Five years later, while deliberately searching for black gold, Edward L. Drake discovered petroleum at a depth of sixty-nine feet in a Titusville well on August 27, 1859. In 1860 the first oilfield refinery was built approximately a mile from the Titusville well, and in 1861 the first cargo of petroleum reached London from America in wooden barrels. At the end of the decade, in 1869, the first petroleum tanker to cross the Atlantic—the *Charles* of Antwerp, Belgium—began to carry oil from America to Europe.

Although in many people's minds the Standard Oil Company is synonymous with the petroleum industry itself, it was not until 1870 that John D. Rockefeller and others organized the company in Ohio, with a capitalization of $1 million. Through the efforts of Rockefeller and others, as early as 1874 crude oil output in the United States had surpassed 10 million barrels annually. By the early 1880s petroleum ranked fourth in value among American exports, behind cotton, breadstuffs, and provisions, but in terms of the percentage of the total national production sent abroad, it was second only to cotton.

As a result of this great surge in petroleum exports, Standard Oil's foreign representative, William Herbert Libby, declared that petroleum had "forced its way into more nooks and corners of civilized countries than any other product in business history emanating from a single source."[1] Leading the way for American operations overseas was William Rockefeller's Standard Oil Company of New York. After 1888, Standard Oil's marketing organizations began to proliferate in such diverse places as the Far East (an area that Libby personally favored), Great Britain, the Netherlands, Germany, Belgium, Italy, and Canada. The main petroleum rival of the United States at this time was Russia, which by 1897 furnished 45.3 percent of the world's crude oil and refined 23.3 percent of its kerosene. At the turn of the century, Standard Oil continued to dominate the American petroleum export market. But in 1902 the formation of the Texas Company (later known as Texaco) occurred; by 1905 the Texas Company had begun to establish overseas sales offices, and by the eve of World War I its products were available on five continents.

The year 1911 marked a major turning point for Standard Oil and the other monopolistic American businesses, since in that year the U.S. Supreme Court decreed the breakup of the petroleum colossus. Twenty-nine years earlier, in 1882, corporation lawyer Samuel Dodd had devised a scheme for Standard Oil, under which all owners of stock in allied firms surrendered their certificates to Standard Oil in exchange for "trust certificates." Although shareholders continued to receive their dividends, the management of the allied firms rested in the hands of nine trustees. Widespread criticism of the trust device led Congress to pass the Sherman Antitrust Act in 1890; the Ohio Supreme Court ordered the dissolution of Standard Oil two years later, but this did not become effective until 1899. At that time Rockefeller set up the Standard Oil Company of New Jersey, which he headed as president. This firm was a holding company.

Despite the precedent-making breakup of 1911, Rockefeller and his associates continued to control a number of petroleum firms. The largest of the survivors was a stripped-down Standard Oil of New Jersey; now known as Exxon, it is currently the largest petroleum company in the world. Another firm was William Rockefeller's Standard Oil of New York, abbreviated as Socony, which now operates as Mobil. Standard Oil of California, or Socal, was a giant that emerged from the wreckage. Even some of today's so-called "independents" have their roots in the old Rockefeller trust; these include not only Standard Oil of Indiana (Amoco) and Standard Oil of Ohio (Sohio) but also Continental (Conoco) and Atlantic Richfield (Arco). In the aftermath of the Supreme Court decree, no less than nine of these Standard-affiliated companies retained their foreign facilities, with Jersey Standard maintaining the most extensive empire.

As the United States moved into the World War I era, the American government became increasingly concerned about guaranteeing an "Open Door" for U.S. businessmen abroad, including petroleum entrepreneurs. Although Secretary of State John Hay had originally promulgated this policy of equal commercial opportunity with special reference to China at the turn of the twentieth century, the dream of a great China market for American merchants proved to be a myth rather than a reality. Thus, when the American government invoked the "Open Door" policy on behalf of petroleum entrepreneurs in such areas as the Middle East and Latin America after World War I, it was endorsing a concept set forth two decades earlier for another part of the world. The U.S. government has continued to support the "Open Door" policy down to the present day. This principle, moreover, has remained a constant in American economic foreign policy regardless of the existence of petroleum surpluses or shortages, or the imposition of protective tariffs or oil import quotas.

## World War I

By 1914, the year in which World War I broke out in Europe, the total direct American foreign investment in petroleum—including exploration, production, refining, and distribution—amounted to $343 million. Of this amount, $138 million was in Europe, $85 million in Mexico, $42 million in South America, $40 million in Asia, and $25 million in Canada. Of the $2.652 billion of direct American foreign investment, oil accounted for approximately 13 percent.

The outbreak of World War I made the possession of an adequate petroleum supply an absolute necessity to the belligerents. Although the United States did not declare war against the Central Powers officially until April 1917, the American military had been concerned about its oil supply as early as 1914. While the availability of petroleum directly affected the Navy, which had been converting its ships from coal to fuel oil since around 1890, the Army, too, needed petroleum products to operate its tanks, airplanes, and trucks. Even though in 1898 the U.S. Congress had appropriated $15,000 for the Navy to investigate the feasibility of oil powered vessels, Secretary of the Navy Paul Morton stated in 1904 that this was "a question that cannot by any means be regarded as settled,"[2] and other naval planners remained equally skeptical. Their hesitancy incurred the wrath of the *Oil Investor's Journal*, which attacked naval timidity intermittently during the years that followed.

In contrast, President Wilson's Secretary of the Navy, Josephus Daniels, showed a great concern for adequate petroleum supplies and in company with conservationist Gifford Pinchot favored the withdrawal of federal oil lands from public sale. Daniels and Pinchot were opposed by Franklin Lane, the Secretary of the Interior and a former California oil man, who advocated less restrictive leasing policies. This was only the first—or among the first—of innumerable disagreements over government petroleum policy at the cabinet level during the twentieth century. In 1914 Daniels did succeed in persuading Wilson to set aside the Teapot Dome reserves in Wyoming for naval use, while three years earlier President William Howard Taft had similarly withdrawn the Elk Hills and Buena Vista reserves in California.

With the advent of World War I, the United States did more than its share in supplying the Allied powers with petroleum, drawing significantly on domestic reserves in the process. As Lord Curzon observed, "the Allies floated to victory on a sea of oil."[3] Through the efforts of the American oil industry and the Petroleum War Service Committee (PWSC), the United States was able to furnish the Allies with 80 percent of their war requirements. Table 1 at the end of Chapter 1 illustrates the breakdown of the exports of refined petroleum oils for the years 1917 and 1918.

According to Alfred C. Bedford, President of Jersey Standard Oil and the

Chairman of the PWSC, the main objective of the PWSC was the filling of petroleum orders during the war emergency in a nondisruptive manner. Although the PWSC did advocate gasolineless Sundays, it did not fix prices until July 1918. In summarizing the work of the PWSC, the *Oil and Gas Journal* declared that "trade dissensions and competitions are forgotten and the idea prevailing is one of harmony and practical cooperation."[4] The Petroleum War Service Committee continued to function until 1919.

## The Post-World War I Years: Background

During the years immediately following World War I there probably were more talks and discussions about an official U.S. oil policy than there had been in the years between the first commercial exploitation of petroleum in America and the end of World War I. This generalization probably is as valid for the governments of other nations as it is for United States. Cordell Hull later noted in his *Memoirs* that "in many conferences after the last war the atmosphere and smell of oil was almost stifling."[5]

Official American interest in petroleum continued to intensify rather than lessen after World War I. In August 1945 John A. Loftus of the State Department observed that an "inspection of the *Papers Relating to the Foreign Relations of the United States* for the period subsequent to about 1910 and a review of the diplomatic history of the past 35 years will show that petroleum has historically played a larger part in the external relations of the United States than any other commodity."[6] Loftus later qualified his statement by observing that "until about 1920 the petroleum industry did not enter significantly into the foreign relations of the United States."[7] Prior to this date it was widely believed in America that the available domestic supply was more than adequate. Yet, according to an article published in 1973 by historian Norman Nordhauser, oil executives, far from opposing governmental regulation of their industry, led the movement to regulation. Nordhauser claims their basic objective was the stabilization of prices and profits.[8] Their enthusiasm for regulation became especially strong after 1927 when the discovery of the Seminole oil fields in Oklahoma and another great field in East Texas threatened to glut the market.

Although official American interest in petroleum did intensify after World War I, there was no unanimity among the executive branch, the U.S. Congress, and big business over what oil policy the American government should adopt. Not only did each group disagree with the other two on numerous occasions, but there was also considerable friction within each group no matter what political party occupied the White House. To complicate matters, petroleum firms frequently acted on their own. When a group of oil companies from the United States, Great Britain, the Netherlands, and France agreed to divide up a substantial portion of Middle Eastern petroleum under the Red Line Agreement of 1928 and three of the leading oil firms (Jersey Standard, Anglo-Persian, and Royal Dutch-Shell) decided to regulate world petroleum under the Achnacarry ("As Is") Agreement of the same year, they did so independently of the American government.

## The Last Years of Woodrow Wilson

In March 1919 Wesley Frost, the acting foreign trade adviser of the State Department, became chairman of a new economic liaison committee. The committee delegated the petroleum question to a subcommittee on mineral raw materials, and in July it issued a report recommending that U.S. consular and diplomatic officials abroad make special efforts in assisting those American interests that were seeking oil properties. This increasing concern for foreign oil reserves was not surprising in view of an opinion expressed by Secretary of State Robert Lansing to Viscount Grey, the British Ambassador to Washington. Lansing made known his fear that American petroleum reserves would be almost totally exhausted within a measurable period.

Another governmental agency to make a positive contribution to official U.S. oil policy at this time was the United States Shipping Board, which was in charge of the American Merchant Marine. Between 1919 and 1921 the Merchant Marine's total fuel oil requirements increased from 18 million to 40 million barrels. To meet these escalating needs, the service began to create fuel oil bunkering depots all over the world; among the first of these were the ones at St. Thomas, Honolulu, Manila, Brest, Ponta Delgada, and Bizerte. Although the prime objective of these bunkering stations was to service vessels of the U.S. Shipping Board, they were allowed to sell their surplus to any American steamer in need of petroleum. At this time the U.S. Shipping Board began to construct a fleet of government oil tank steamships whose main function was to transport petroleum to governmentally owned oil tanks.

While the executive branch took an active role in shaping an official oil policy, the legislative branch was by no means quiescent. In 1919 Senator James D. Phelan of California failed in his attempt to grant the President discretionary authority to embargo all oil shipments from the United States; the anti-British Democratic Senator's views clashed with those of Secretary of State Lansing, who preferred reciprocity to retaliation. That same year, moreover, Republican Senator Reed Smoot of Utah introduced a bill, which also did not pass, to ban all alien leases on public oil lands. But in the following year Congress enacted the Minerals Leasing Act, which President Wilson signed into law on February 25, 1920. This measure included the controversial provision that "citizens of another country, the laws, customs, or regulations of which, deny similar or like privileges to citizens or corporations of this country, shall not by stock ownership, stock holding, or stock control, own any interest in any lease acquired under the provisions of this Act."[9] During the years that followed, because of the discriminations that American oil firms faced in the Netherlands East Indies, the U.S. government directed this provision against the Netherlands more than any other country.

Congress persisted in its critical mood after the passage of the Minerals Leasing Act. In March 1920 a resolution adopted by the Senate required the State Department to draw up a report, summarizing the actions that various foreign governments had taken to block American entry into foreign petroleum fields. Two months later, on May 17, President Wilson submitted the completed report to the Senate. Despite later denials by Lord Curzon, the British Secretary for Foreign Affairs, this State Department document noted that as a rule Great Britain discriminated against aliens but not specifically against Americans. While admitting that the British had discouraged, or even prohibited, prospecting for minerals in its territories, especially in the former Ottoman Empire, the report pointed out that conditions were highly unsettled in occupied areas.

It was one day later, on May 18, that Senator Phelan of California introduced a bill into the Senate that would have established a United States Oil Corporation. Its foreign activities would have included exploration, refining, transportation, and storing with governmental sanction and backing. Controlled by a board of directors consisting of nine presidential appointees, the corporation would have been obligated to sell on demand any of the petroleum that it produced to the American government. Phelan, however, failed to gain the support of Woodrow Wilson's new Secretary of State, Bainbridge Colby, who believed that foreign countries might deny the corporation permission to operate within their borders.

Although the Republican party has long been portrayed as the party of big business and the Democratic party as the party of the people, a historical survey of the Democratic party reveals the existence of numerous ties between its politicians and business interests. The "petroleum plank" that appeared in the 1920 Democratic platform offers evidence that the Democrats were cooperating with oil interests following World War I, rather than denouncing them:

> The Democratic party recognizes the importance of the acquisition by Americans of additional sources of supply of petroleum and other minerals and declares that such acquisition both at home and abroad should be fostered and encouraged. We urge such action, legislative and executive, as may secure to American citizens the same rights in the acquirement of mining rights in foreign countries as are enjoyed by the citizens or subjects of any other nation.[10]

## Harding and Coolidge

In March 1921 a new Republican administration took over in Washington. For the next twelve years, the three GOP presidents showed a definite sympathy toward the operations of big business, both in the United States and abroad. Secretary of the Treasury Andrew W. Mellon, who served under Harding, Coolidge, and Hoover, favored tax reductions for the wealthy, arguing that this would enable them to build up industry and raise wages. Mellon also had played a key role in the founding of the Gulf Oil Company. Certain British newspapers referred to Harding's Secretary of State, Charles Evans Hughes, as the "Secretary for Oil," while the New York *World* observed: "Show Mr. Hughes an oil well and he will show you an oil policy."[11] And Herbert Hoover, as Commerce Secretary, also attempted to play a key role in shaping U.S. petroleum relations with foreign nations. Hoover, a former mining engineer, had extensive experience abroad prospecting for gold and other minerals in Australia and other countries.

By this time, the American government through its diplomatic representatives abroad was maintaining a constant vigil over world oil concessions to guarantee that U.S. petroleum interests received their fair share. In this connection the American legation at The Hague wrote the Netherlands Ministry of Foreign Affairs on April 19, 1921 that:

> . . .the real interest of the Government of the United States in these matters lies in the recognition of the principle of mutual or reciprocal access to vital and natural resources by the nationals of the United States and by those of foreign countries, and the belief that the recognition of the principle of equal opportunity is the only solution of the future oil problems throughout the world.[12]

At the same time, the U.S. government was keeping a wary eye on foreign investment in U.S. petroleum, a topic that the Federal Trade Commission (FTC) analyzed at length in a report released on February 12, 1923. Its leading target was the Royal Dutch-Shell group, which in 1922 had effected a merger of its leading American subsidiaries with the Union Oil Company of Delaware. By 1923 this new firm, the Shell Union Oil Corporation, held over 240,000 acres of petroleum lands in America, aside from its investments in refineries, pipelines, tank cars, and marketing equipment. California was one of those states where it was most active.

The Federal Trade Commission complained in its report about the discriminatory policies that the Dutch and British governments had implemented against U.S. oil companies in the Netherlands East Indies and India. The report went on to criticize the San Remo Agreement of 1920 between Great Britain and France, which had attempted to place a protective umbrella over the undeveloped petroleum resources of Mesopotamia (now Iraq), thus barring U.S. oil activity there. The FTC also attacked Italy and various African countries for their discriminatory policies.

During the summer of 1923 President Warren Harding died unexpectedly. The incoming chief executive, Calvin Coolidge, was hampered in his efforts to effect a rational oil policy by the emotional turmoil that followed the Teapot Dome disclosures. In December 1924, after he was elected President in his own right, Coolidge established the Federal Oil Conservation Board (FOCB), a cabinet-level committee of inquiry. This new body was to investigate and report on conditions in the petroleum industry; when the FOCB surveyed the oil companies during the following year, it found little sentiment for new laws.

Governmental officials in their attempts to formulate an oil policy quite naturally compared and contrasted what went on in the United States with developments in other nations. In 1924 Stanley Hornbeck of the State Department attempted to detail the differing conditions of competition among British, Dutch, and American petroleum firms. His summary is so perceptive that it should be cited in full:

(1) The American companies operate as strictly private enterprises. They compete with each other both at home and abroad. The United States Government makes no choice among them and has no connection with or financial or commercial interest in any of them.

(2) In Great Britain, however, some companies operate as a combination of private and governmental enterprise. Cooperation—particularly in foreign enterprises—is officially and privately encouraged and is to a considerable extent a fact. The Government owns stock, encourages pioneering and gives special assistance to certain companies at certain moments and in certain regions.

(3) In the Netherlands, substantially the same is true. There, in addition, the stock of the leading oil company, the Royal Dutch, is very widely owned—especially among the official class—which gives the company a very wide support as substantially a national enterprise, not, it should be noted, as a governmental enterprise but as a great national interest.[13]

Frederick Lee Moore concluded that from 1918 on British and Dutch petroleum policy remained constant. These two European governments continued to offer diplomatic assistance to British and British-Dutch oil firms abroad and to discourage foreign companies from operating in those areas of the globe that were under British and Dutch control. Similarly, Arthur Veatch was of the opinion that under the system in operation at that time, it was politically impossible for the State Department to match or to surpass the efforts which the British Foreign Office put forth on behalf of its nationals. The Chairman of the Board of Anglo-Persian (later Anglo-Iranian and then British Petroleum) declared in this connection that "the difficulties in which American interests find themselves arise from the uncertain diplomatic support they receive."[14]

When Frank Kellogg replaced Charles Evans Hughes as Secretary of State in 1925, American access to several rich petroleum areas was still blocked. On June 29, 1928, Kellogg advised Secretary of the Navy Curtis D. Wilbur to restrain the Navy Department from taking any discriminatory action against foreign oil interests operating in the United States, particularly those of the Netherlands. By this time U.S. petroleum firms had begun to enjoy greater access to the oil of the Netherlands East Indies, a development which will be examined in perspective when we discuss the NEI in the section on the Far East.

During the course of his memo to Wilbur, Kellogg offered a summary of government oil policy at this time, which proves with great succinctness that the policy had not changed significantly during the 1920s:

The Department of State, in connection with rendering assistance and support to American companies seeking or operating petroleum concessions abroad, is constantly seeking the recognition and practical application by foreign governments of the policy of the open door and equality of commercial opportunity. It is obvious that such a policy can be followed only as long as the United States accords to nationals of foreign countries treatment similar to that sought by this Government for its nationals abroad.

This Department frequently has had occasion to point out that the laws of the United States are very liberal as to access to private oil lands on the part of foreign interests, and that the laws of the United States appertaining to public oil lands do not discriminate against companies owned wholly or in part by foreign nationals as long as the foreign countries of which they are citizens grant reciprocal rights to American interests.[15]

## The Oil Firms: The Pre-Achnacarry Years

On March 14, 1919, representatives of the leading U.S. oil companies joined forces in New York City to organize the American Petroleum Institute (API) and to elect Thomas A. O'Donnell as its president. Stressing that it planned to cooperate with the government in all matters of national interest, the API selected as its first board of directors the members of the Petroleum War Service Committee. In July the new institute set up a foreign relations committee and appointed Walter C. Teagle of Jersey Standard as its chairman.

Still another organization that was active on behalf of U.S. oil interests was the American Institute of Mining and Metallurgical Engineers. In March 1920 a committee of its petroleum section, which was chaired by Mark Requa, drew up a petition on the "Imperative Need of Aggressive Foreign Policy as Regards the Oil Industry," which it then presented to both the President and the Congress. The petition declared that "the Government should at once make representation upon this matter under the alternative that the free entrance of foreign capital in the American development of natural resources will be reciprocally restrained."[16]

During April 1921 a dozen individuals from both the private and public sectors, including Mark Requa and Van H. Manning, met at the Engineers' Club in New York City to discuss foreign oil policy. The group adopted a number of suggestions or resolutions, in most cases by unanimous vote. While endorsing both equal commercial opportunity and the "Open Door" principle, those present called for the repeal of the Federal Leasing Act and "any other legislation whose object is retaliation for discrimination against our nationals in foreign oil fields."[17] In addition, they resolved that the State Department should take sides whenever there were disputes over—or even competition for—a foreign oil concession and so notify the foreign governments in question. However, according to the memorandum, "It was the sense of the meeting that only as a very last resort should the Government financially assist American foreign oil operations, and that existing conditions do not warrant any such assistance."[18]

Around this time the Foreign Relations Committee of the American Petroleum Institute selected R. L. Welch and Van H. Manning to draft a memorandum, suggesting a possible policy of the U.S. government for American business operations abroad. It was announced that the memorandum was necessary because of the growing tendency of foreign nations to discriminate against American businesses and nationals. While admitting that the efforts of the State Department to end this discrimination usually were conducted with conspicuous ability, Welch and Manning pointed out that it was the states and not Washington that possessed the right to determine which aliens could own property within their borders. But Welch and Manning felt that under certain circumstances the national government should have the right of retaliation. Accordingly, they recommended that Congress pass a law to this effect and suggested Elihu Root as a possible drafter of such a measure.

An international disarmament conference assembled during May in Washington, D.C. that has long overshadowed the great oil conference that met in the same city on May 16. The leading petroleum barons of the nation assembled in secret session at the request of Attorney General Harry Daugherty and Secretaries Hughes and Hoover. Unquestionably the most innovative idea that the participants discussed was the creation of a giant American petroleum corporation. Involving both producers and sellers, the corporation would operate abroad with governmental backing. Although there was the expected complaint that the establishment of such a corporation would violate the Sherman Antitrust Act of 1890, an equally serious objection was the possibility that it might alienate foreign governments more than the traditional U.S. policy of equal commercial opportunity.

For these and other reasons, the projected oil corporation failed to materialize. There was, however, a move in the direction of greater cooperation abroad on the part of American petroleum firms. Van H. Manning and Mark Requa had favored such a step under governmental auspices even before Herbert Hoover became Secretary of Commerce. On May 2, two weeks prior to the Washington oil conference, Requa had written Hoover and enclosed a ten-point "Memorandum Relating to a Corporation for Acquisition of Foreign Sources of Petroleum." Some of the key points set forth were:

(2) This is to be a cooperative effort on the part of the refining companies who are asked to participate on an equal basis, one company, one director, one share;

(3) Provision is made to retain control in the United States and only American companies may participate;

(5) This company is formed to meet a need that will become acute with the decline of Mexican production;

(8) The American companies agree with me that no one of them has sufficient capital to do what is necessary in the time available;

(9) No legislation is I think necessary at this time;

(10) I am asking no special privilege or Government financial aid for this corporation.[19]

In November 1921 seven U.S. oil companies, led by Jersey Standard, banded together to form the so-called American Group and shortly thereafter sought a share of the Mesopotamian oil fields from the Turkish Petroleum Company. Unfortunately, various attempts by the U.S. oil firms to cooperate abroad frequently ended in conflict. In Mesopotamia, for example, relations between the

American Group and the Chester Syndicate were not always harmonious, while in Persia Jersey Standard clashed with Sinclair. Nor was the record more positive in Latin America. In Costa Rica, for example, the United Fruit Company was at odds with both Sinclair and California Standard. Moreover, whenever such friction did occur, the State Department generally adopted a neutral stance, rather than backing one American firm against another.

The United States continued to dominate world petroleum during the mid-1920s. In 1925 the total American oil production was 764 million barrels, or 71.6 percent of the global total of 1.066 billion barrels. Its nearest competitors were Mexico (115 million barrels), Russia (52 million), Persia (34 million), the Netherlands East Indies (21 million), Venezuela (20 million), and Rumania (16 million). Significantly, oil imports into the United States had decreased from 128 million barrels in 1921 to 78 million barrels in 1925. Most of these at one time had come from Mexico, but more and more petroleum now was shipped in from Venezuela and Colombia.

American exports of petroleum and petroleum products at this time constituted over 9 percent of total U.S. exports and totaled $445,502,000. Of the 13 million barrels of crude petroleum that were shipped abroad, approximately two-thirds went to Canadian refineries. In contrast, approximately 65 percent of the 20 million barrels of exported gasoline was destined for Europe, especially Great Britain and France. American kerosene was transported to such diverse places as China, Japan, Italy, and Great Britain. Thus American oil dealings were not limited to those of a hemispheric, trans-Atlantic, or trans-Pacific nature but rather were truly global in scope.

## The Oil Firms: The Post-Achnacarry Years

In many respects modern oil competition reflects the old medieval doctrine of the two spheres: what is appropriate for one hemisphere is not suitable for the other. At the same time that American petroleum firms were lamenting the encroachments of Royal Dutch-Shell in the United States, that firm and Jersey Standard sought control jointly over the petroleum exports of the Soviet Union, which had nationalized all foreign oil properties. After Great Britain had recognized the Soviet Union in 1924, Walter Teagle of Jersey Standard and Sir Henri Deterding of Royal Dutch-Shell agreed to set up a company in Liechtenstein with the objective of obtaining a long-term contract from the Russians. Deterding, however, then began to purchase large quantities of oil from the Soviet Union, which he used in a drive to undercut the position of New York Standard in the Far East. In retaliation, New York Standard attempted to challenge Royal Dutch-Shell in India, only to have Deterding accuse the Standard Oil companies of collaborating with the Communists.

With the world petroleum industry thus torn by conflict, there was a summit conference of the global oil barons at Achnacarry Castle in Scotland, which was attended by Deterding (the host), Teagle, and Sir John Cadman, Chairman of the Board of Anglo-Persian. Two decades earlier Deterding had told Teagle that "cooperation means power." Remembering this sentiment, the petroleum kings drew up the so-called Achnacarry Accord of September 17, 1928, which attempted to end cutthroat oil competition throughout the world. The United States was excluded from this understanding on the grounds that such an accord might be in conflict with its antitrust legislation. At the heart of the Achnacarry Accord was the maintenance of U.S. prices. Using what became known as the "Gulf Plus System," it was agreed that all oil, no matter where it came from, would be charged for on the basis that it had originated from the Gulf of Mexico. Aside from the so-called Big Three, fifteen other American companies approved this concept, including the four other "sisters," New York Standard, Gulf, Texaco, and California Standard. Outside the United States, only the Soviet Union was exempted from coverage.

That Jersey Standard should represent the American oil industry in these negotiations was only appropriate. Its unparalleled power and influence can be found everywhere in the governmental documents of that era, such as the now forgotten letter that John E. Nelson of the Department of Commerce wrote to the department's Singapore office on June 26, 1928. In his letter Nelson advised those individuals then representing the American government in Malaya that:

> I have never yet seen an individual or small importer successfully defy the Standard Oil Company. Under similar conditions in other parts of the world the small importer usually imports a sizable shipment and just as it is about to be sold locally the market price is cut so low as to wipe out either the importer or local purchaser, and sometimes both, after which time the price goes back to its previous mark.[20]

At this time Jersey Standard submitted a plan to the Federal Trade Commission that called for the newly created Export Petroleum Association (EPA) to engage in operations abroad. As the stockholders of the EPA, which was set up in early 1929, consisted of Jersey Standard, over a dozen other American companies, and Royal Dutch-Shell, the association quickly endorsed the basing point pricing system established at Achnacarry; its members also agreed to follow the lead of the corporation's various committees with respect to prices and export quotas. Unfortunately, a Federal Trade Commission investigation revealed that the EPA had violated the Webb-Pomerene Export Trade Act, which applied only to those corporations that were set up solely to engage in export trade. The EPA collapsed in November 1930, after its members had proved unable to agree upon prices among themselves.

By the outbreak of the Great Depression in the fall of 1929, American investment in foreign oil-producing properties amounted to $854 million. Three-fourths of this amount was to be found in four Latin American nations, Venezuela, Mexico, Colombia, and Peru, with smaller investments in the Netherlands East Indies, Trinidad and Aruba, Canada and Newfoundland, and Rumania. In 1914 there had been only three American-owned multinational businesses with both market-oriented and supply-oriented direct investments abroad: Jersey Standard,

New York Standard, and the Texas Company (Texaco). By 1929 there were twelve. Nevertheless, in that year the British-Dutch group (Royal Dutch-Shell and Anglo-Persian) produced 41 percent of all foreign output, while American oil companies produced only 29.7 percent.

Despite U.S. investments in Venezuela, Royal Dutch-Shell at this time was importing large amounts of low-cost oil from that nation into America. As a result, it became necessary for Jersey Standard, Royal Dutch-Shell, and Anglo-Persian to set forth a "Memorandum for European Markets of 1930," following the demise of the Export Petroleum Association. Among other things, this memorandum established marketing quotas and barred outside concerns from participating in the quota system without the unanimous consent of the three companies.

Jersey Standard and New York Standard (now Socony-Vacuum) also agreed during 1933 to engage in integrated operations in the Orient as the Standard-Vacuum Oil Company. This important merger endured until 1960, when it fell victim to antitrust action. In 1936, in an equally important move, Texaco bought a half-interest in the California Standard holdings in Saudi Arabia and Bahrain. During World War II, this joint operation became known as Aramco in Saudi Arabia.

The change in political ideology resulting from the implementation of the New Deal did not significantly affect the principles agreed upon by Jersey Standard, Royal Dutch-Shell, and Anglo-Iranian at Achnacarry. Thus these firms drew up the "Draft Memorandum of Principles of 1934" using the "as is" concept, which assumed that future increases in business were to parallel the level of activity in 1928 with adjustments for changing patterns of consumption. But by 1938, Jersey Standard had decided to give verbal notice that it was withdrawing from the draft memorandum agreement, and the outbreak of World War II further undermined the understanding.

As war clouds began to collect over Europe during the late 1930s, the United States was not as concerned about a possible oil shortage as it had been following World War I. Not only had new oil fields been discovered in California, but during the 1920s and the 1930s they also had been found in East Texas and the midcontinent region. At this time the largely self-sufficient United States was supplying approximately two-thirds of the world's production from its own fields. The ascendancy of exports over imports was greater for the five-year period between 1936-40 than for any other; the leading petroleum importers during 1937-39 were Canada, Japan, France, and Great Britain. Writing in 1935, petroleum industry critic Samuel Guy Inman had observed: "The joke about all this tragic fight for oil is that after years of alarmist assertions that we were about to run out of this important fuel for our battleships and autos, experts have within the last month declared that this country has enough for at least another hundred years."[21]

### Hoover and Roosevelt

Although the Great Depression of 1929 was essentially a worldwide phenomenon, the official American reaction to it was one of economic nationalism. In

1930 Congress passed—and President Herbert Hoover signed into law—the highest tariff in U.S. history, the Hawley-Smoot Tariff. The tariff, however, did not halt the flow of imported oil. Although petroleum importers had begun to place quotas on oil voluntarily during 1931, angry American producers still deluged the White House with an avalanche of letters and telegrams, demanding legal restrictions on petroleum imports.

Defending the policies of the Hoover administration, Interior Secretary Ray Lyman Wilbur wrote to Republican Senator Hiram Johnson of California, the Chairman of the Senate Commerce Committee, that "there is general agreement that the notable efforts of the oil producers of the United States, and Federal and State officials concerned, to hold down output, have this past year prevented a disaster."[22] Irate U.S. petroleum producers ignored the fact that approximately one-fifth of the refinery products of this country were exported. Since over 76,000 workers held jobs in American refineries, these exports in effect provided work for approximately 15,000 laborers. Noting this, the Chairman of the Advisory Committee of the Federal Oil Conservation Board (FOCB) wrote Secretary Wilbur that "the danger of losing in exports far more than is gained through reduction of imports is obvious and should be considered in framing legislation."[23]

By March the White House had received several hundred telegrams, many proposing policy changes, from the West and the Southwest concerning petroleum imports. The governors of Texas, Kansas, and New Mexico recommended that Hoover hold a conference with the major oil-importing companies for the purpose of effecting limitations on imports. The Independent Petroleum Association of America, which had been organized in 1929, stated that pending congressional action the President should declare a temporary embargo against oil imports under Section 337 of the Hawley-Smoot Tariff of 1930. The administration, however, took the position that Hoover had never been granted this power by Congress.

In a letter to the chief executive on July 23, 1931, Attorney General William D. Mitchell wrote:

> . . . the question is whether the mere importation of oil into the United States in the ordinary course of business is an unfair method of competition or an unfair act within the meaning of the Tariff Act merely because the importation may have the effect of depressing an already glutted market.
>
> I find no basis for the conclusion that the importation of oil into the United States under these conditions constitutes an unfair method of competition or an unfair act within the meaning of the Tariff Act.
>
> It could not be seriously contended that Congress intended to give the President power to declare a general embargo on the importation of any articles of merchandise whenever, in his opinion, the domestic supply outruns the demand.[24]

Congress passed a Revenue Act in 1932 that imposed an import duty of 21 cents a barrel on crude and residual oil. In a desperate, last minute attempt to block passage of the tariff, opponents of the measure circulated a rumor that the

President had interests in oil wells either in South America or in other foreign nations. Hoover vigorously denied this claim on October 18, declaring that "I have not in fifteen years since I entered public service had a dime's interest outside of the United States of any kind including oil."[25]

While the President was wrestling with the worst economic collapse in American history, important developments were occurring at the state level as well. In the East Texas oil field there was constant overproduction, and private attempts at proration failed. On August 17, 1931, Democratic Governor Ross Sterling ordered out the state national guard, placing the field under martial law. The Texas legislature mandated the legal proration of the state's oil production in the aftermath of this spectacular gesture, and this regulatory system has remained in operation to the present day.

Although the New Deal of Franklin Roosevelt continued the Hoover administration's policy of discouraging foreign oil imports, the petroleum duty did not prove totally successful in barring the growing quantity of imported oil. Under the National Industrial Recovery Act of 1933 (NIRA), the petroleum industry drew up a code that restricted imports of oil to the level prevailing between July and December 1932, an average of 98,000 barrels daily. Even after the Supreme Court declared the NIRA unconstitutional in 1935, petroleum importers continued to follow these guidelines.

In 1935 Congress passed the Connally "Hot Oil" Act, which forbade shippers of petroleum in interstate commerce to exceed the quotas established by various states. In that same year Congress also placed its stamp of approval on the Interstate Oil Compact. By 1951 a total of twenty states had signed this agreement, which not only prorated petroleum production among the participating states, but also restricted the amount of oil that a state could ship in interstate commerce.

Even more radical was the proposal made in 1936 and again in 1940 by Secretary of the Interior Harold Ickes, to nationalize oil. While this scheme remained only a dream in the United States, it became a reality in Bolivia in 1937 and in Mexico in 1938. Both of these episodes were quite complex, and the final compromise settlements in both cases did not occur until after the Japanese attack on Pearl Harbor. Roosevelt believed, especially in the case of Mexico, that American petroleum firms were making inflated claims of compensation. His attitude mirrored the increasing hostility that the New Deal had begun to show toward big business by this time.

## World War II

On September 1, 1939, World War II broke out in Europe. Once again American entry into this conflict occurred long after the inception of hostilities, and once again there was a heightened interest in guaranteeing an adequate oil supply. But by 1939, the federal bureaucracy in the United States had vastly increased in size. It was during World War II that the continuing rivalry over petroleum among the departments and agencies of the national government first became a serious problem.

When President Franklin D. Roosevelt approved a bill on July 20, 1940, authorizing a two-ocean Navy, it became obvious that the military would need additional oil. But it was not until May 28, 1941—six months before Pearl Harbor—that Roosevelt appointed Secretary of the Interior Harold Ickes to serve as the Petroleum Coordinator. A day earlier FDR had proclaimed an unlimited national emergency.

Ickes, however, was not given unlimited power over oil matters. The Secretary of the Interior could only forbid petroleum exports or imports with the approval of Roosevelt or his Secretary of State, Cordell Hull. Explaining this particular restriction on Ickes' authority, the President wrote Ickes on June 18, 1941, that: "The reason for this is that exports of oil at this time are so much a part of our current foreign policy that this policy must not be affected in any shape, manner or form by anyone except the Secretary of State or the President."[26] Both Hull and Ickes were in agreement that there should be limitations imposed on oil exports from the Eastern seaboard, especially to the Far East, but FDR felt impelled to remind Ickes one week later to concern himself with oil conservation rather than with foreign policy.

Once he had assumed the office of Petroleum Coordinator, Ickes and other top officials of this new agency began to warn Roosevelt that the discovery of petroleum in America had begun to lag behind the actual consumption and that there was a need for a national oil policy. FDR, whose administrative techniques included the practice of playing one individual against another, on November 6 set up a coordinating committee composed of the Department of State, the Maritime Commission, and the Petroleum Coordinator to synchronize government petroleum efforts. This scheme was hardly pleasing to the outspoken Ickes, who wished to play the leading role in determining a national oil policy. On December 8, one day after Pearl Harbor, Ickes complained to the President that: "The fact is that we have no adequate national policy with respect to petroleum, and no international policy that I know of except to protect the interests of our nationals."[27] He added that to continue to sell oil abroad just as soon as it was produced might run counter to the national interest.

FDR, however, advised Vice President Henry Wallace on December 23 that it had become increasingly important to furnish both the United States and its allies with an adequate supply of oil and gasoline. In this connection the President instructed Wallace to set up a policy subcommittee under the Board of Economic Warfare. Highly displeased by this maneuver, Ickes sent letters to FDR on December 27, 1941, and January 5, 1942, in which he complained that a conflict had developed between his office and Wallace's Board of Economic Warfare. Rather than face the issue directly, the President sent a memorandum to Ickes on January 9 in which he stated that his oil policy was still in the formative stage.

The Japanese attack on Pearl Harbor and the German declaration of war immediately thereafter led to intensified Nazi submarine warfare against petroleum tankers in the Atlantic. In 1942 alone eighty-five American tankers were lost, twenty-three more than were constructed that year. With a fuel crisis confronting the northeastern states, the Roosevelt administration initiated a program of nationwide oil rationing on October 22, 1942. That December Ickes received a new title: Petroleum Administrator for War.

With the worldwide crisis deepening, it is perhaps surprising that FDR would pick this particular moment to have the antitrust Division of the Attorney General's Office challenge the overseas operations of Jersey Standard. Yet, it was during 1942 that the federal government pressured Jersey Standard into signing a consent decree in which it pledged not to violate U.S. antitrust laws abroad. From this point on, Jersey Standard could no longer divide up the overseas market with other oil firms.

The enlightened public, in addition to the executive and legislative branches of the national government, had begun to realize by 1943 that there was not enough petroleum in the United States to meet future needs. Accordingly, Secretary of State Cordell Hull set up an informal interdepartmental committee composed of members of the State, War, and Navy Departments, in addition to Ickes's office, to study how to increase supplies. Economic Advisor Herbert Feis, who served as Chairman, presented the committee's report on March 22; this painted a gloomy picture for the years ahead and suggested that a governmental agency should be created to supervise the activities of U.S. oil companies operating overseas.

That summer five Senators—Richard Russell of Georgia, James Mead of New York, Albert Chandler of Kentucky, Owen Brewster of Maine, and Henry Cabot Lodge of Massachusetts—visited American diplomatic outposts all over the globe. "It is time to utilize the petroleum deposits of other parts of the world," Senator Russell observed, "otherwise the end of the war will find our own deposits practically exhausted." [28] Their anti-British charges upon their return home did nothing to reassure the British government; Prime Minister Winston Churchill himself noted his irritation with this mission in a communication sent to Roosevelt in February 1944.

With U.S. involvement in World War II at midpoint, the struggle over the formulation of a governmental oil policy intensified. Not only were the Secretaries of State and the Interior involved but also the military. One of the most controversial issues was the acquisition of petroleum concessions abroad by the American government. Ickes was more favorably disposed to this idea than Hull, who preferred for the oil companies to develop those already in their possession. With this controversy still unsettled, on June 30, 1943, FDR established the Petroleum Reserves Corporation with Ickes as its head. Ickes' abortive scheme to involve the U.S. government in the ownership of Middle Eastern oil concessions will be discussed at length in the section on Saudi Arabia, but one might note here that it encountered widespread opposition in America both inside and outside the government.

Disturbed by the talk of an increased role for the government in the oil business, the Foreign Operations Committee, which was dominated by the private sector, released its own report, "A Foreign Oil Policy for the United States." In this document, dated November 5, a dozen American and two British petroleum executives urged the U.S. government to support diplomatically whatever deals private American interests might negotiate but opposed the assumption of any managerial role by the United States. "Private enterprise," they affirmed, "can operate with a minimum of political complications."[29]

The Foreign Operations Committee also endorsed still another abortive scheme, an international oil compact. Under this scheme, the nations involved would agree to produce petroleum at a rate that would both conserve reserves and meet consumers' needs; it also would establish the rate of output for various oil fields around the globe. The bureaucratic apparatus to implement this compact would consist of a general conference, a permanent commission, a technical institute, regional councils, and an arbitration tribunal. It is noteworthy that this report cited the Interstate Oil Compact of 1935 as a precedent, although that understanding was between states, rather than nations.

At the close of the Truman administration an unidentified official, or officials, drew up a memorandum on "The World Oil Cartel" that concluded the wartime allocation of supplies by the Foreign Operations Committee ". . . were identical in character with those found in the files of companies for operations prior to Pearl Harbor under the DMOP[Draft Memorandum of Principles of 1934]."[30] It is significant that the same individuals who had attempted to implement this DMOP were now members of the Foreign Operations Committee.

That the recommendations of the Foreign Operations Committee were not in line with the thinking of Harold Ickes is rather obvious. As *Time* magazine noted in 1943, "With a bang, the honeymoon between Ickes and the oil men was over."[31] Fortunately, State Department officials were able to reach a global understanding on petroleum policy with leading oil company executives in February 1944.

On April 11 the Department of State's Postwar Programs Committee approved the final version of an official memorandum, entitled "Foreign Petroleum Policy of the United States," that reflected the ideas of both that department and the Foreign Operations Committee. This document upheld the Atlantic Charter's equal access clause, as well as the "Open Door" principle of equal commercial opportunity for Americans abroad. It also opposed restrictions on the development and marketing of oil, such as the Red Line Agreement of 1928. The government was to see to it that there was no alienation of American petroleum concessions abroad and that the host nations received maximum economic and social gains from these. Eager to implement the policy set forth in the memorandum, the State Department by 1945 had begun to assign petroleum attachés to critical posts abroad.

As the war neared its end, it became increasingly difficult for American petroleum companies to tolerate the foreign oil that was imported into the country to meet emergency war requirements. On October 24, 1945, the Petroleum

Industry War Council, meeting in Washington, D.C., resolved that "it should be the policy of this Nation to so restrict amounts of imported oil so that such quantities will not disturb or depress the producing end of the domestic petroleum industry."[32] Oil exports at this time accounted for only 6 percent—or 2.3 billion dollars—of the 40.2 billion dollars worth of goods transferred to foreign governments between March 1941 and October 1945. Great Britain was the leading recipient and aviation gasoline was the primary petroleum product shipped abroad.

The Pacific phase of World War II ended in the fall of 1945, but Truman did not abolish the Petroleum Administration for War until May 1946. Deputy Petroleum Administrator Ralph K. Davies described the achievements of the agency as "unparalleled in the annals of government and industry."[33] Truman agreed that it had successfully discharged its wartime duties and then set up the National Petroleum Council (NPC) as its peacetime counterpart. The council was to serve as "a channel of communication between the federal government and the petroleum industry."[34] It was during this year that the Truman administration gave antitrust immunity to the NPC, which, among other things, was to prepare voluntary agreements for the allocation of scarce material and facilities.

Despite the heavy drain on its oil supplies during World War II, the United States still occupied a strong position with respect to petroleum. The optimistic mood regarding oil supplies that had been prevalent at the beginning of that conflict had been recaptured. The postwar era, however, saw the emergence of new oil controversies that the American government was hard pressed to resolve.

## The Truman Administration: Restricting Imports

The two most important disputes over oil during the Truman years involved imports and antitrust actions. Under the Marshall Plan of economic reconstruction (1948) the United States attempted to furnish Great Britain and Western Europe with needed oil, but official American support in the Middle East for the new Jewish state of Israel offended the oil-rich Arab states. In 1951 non-Arab Iran under the leadership of Mohammed Mossadegh nationalized the holdings of Anglo-Iranian. Although American firms were not directly affected, the expropriation caused significant dislocations in the world petroleum market of which the United States was a part.

After World War II foreign oil was imported into the United States in ever-increasing amounts, and in 1948 it became a net importer of petroleum for the first time since World War I. Among those who expressed dissatisfaction with the so-called oil invasion were the independent American producers whose lobby in Washington was—and is—among the most influential there. Other opponents of cheap foreign petroleum included the coal operators, the railroads, the United Mine Workers, the Railroad Brotherhoods, and the CIO Oil Worker's Union. Despite their differences of opinion in other areas, these groups united in support of an increase in petroleum import duties or the establishment of limited import quotas. Because it was committed to the reduction of trade restrictions around the globe, the Truman administration was not pleased with their campaign; it felt that if the campaign were successful it might prove damaging to U.S. foreign policy not only in Latin America but also in the Middle East. In 1947 the Truman administration negotiated a series of import duty reductions under the General Agreement on Tariffs and Trade, reductions that included tax cuts on gasoline and lubricating oil of 50 percent, from $.025 to $.0125 per barrel.

In Congress, however, representatives from petroleum producing states viewed the mounting imports with displeasure. Democratic Senator Elbert Thomas of Oklahoma offered an amendment to the Reciprocal Trade Act (GATT) that would have restricted oil imports to 5 percent of the national demand for the corresponding quarter of the previous year, as determined by the Bureau of Mines. The amendment failed passage by a single vote. In the years that followed there was to be a continuous struggle in Congress between senators and representatives from such oil-rich states as Texas and those from such oil-poor regions as New England over the federal petroleum importing policy.

During the first months of 1949 the Secretary of the Interior asked the National Petroleum Council, the successor to the Petroleum Administration for War, for advice about imports. As a result, the council set up a committee on petroleum imports in June. While the stormy debate over the federal petroleum import policy raged on, members of the oil industry, including domestic producers, adopted the slogan, "Imports to supplement but not to supplant."[35]

The committee made its initial report in January 1950. It took the position that it was essential to American national security for U.S. nationals to participate in the development of global oil resources. On July 24, the committee specifically addressed itself to the question of petroleum imports in a second report and sought to determine whether these were excessive or harmful. While rejecting a complete embargo as contrary to historical tradition, it concluded that:

1. Fair and equitable relationships should obtain between total imports of crude oil into the United States during periods of excess availability of domestic oil for United States consumption.

2. Imports of crude oil and its products, if increased beyond the limits of supplementing domestic production, will adversely affect the domestic industry, the national economy, and national security.[36]

Thanks to pressure from hundreds of independent oil producers, the Small Business Committee of the House of Representatives initiated an investigation of the second report, which had been entitled, "Effect of Oil Imports on Independent Domestic Producers." The investigation was begun in April 1949, and on January 26, 1950, Representative Wright Patman submitted a report to Truman pointing out that independent producers were responsible for three-quarters of the oil discovered in the United States. Patman recommended placing quantitative restrictions upon petroleum imports "at such level as will insure that independent producers do not suffer or are not threatened with serious injury."[37] He believed the authority for such a step existed in the "escape clauses" of various trade agreements.

Truman was sympathetic to the lot of the independents. "I can't see anything wrong," he observed, "with the importation of crude oil in equal amounts as our exports of the products, but when they begin flooding the market with fuel oil, diesel oil, greases and things of that sort, this is evidently a concerted effort on the part of the big companies to put the little refineries out of business."[38] Yet, two months later, noting that oil exploration was at unparalleled heights, he took the position that the U.S. government should not jeopardize the sale of millions of dollars in farm products abroad by cutting off foreign trade for the benefit of oil interests.

Meanwhile, the Oil and Gas Division of the Department of Interior had been studying the entire question and had concluded that petroleum imports to date had not seriously disadvantaged the independent producers. Joel D. Wolfsohn, an official of the Interior Department, had investigated the import situation and had decided that to reduce these by statute rather than by trade agreement might jeopardize American relations with other nations. Much of the oil that was being imported was not foreign petroleum but rather oil owned by Americans abroad, which for various reasons they could no longer sell in foreign markets.

When in 1952 the Truman administration decided to implement a new policy concerning oil imports, it was forced to cancel and renegotiate two bilateral international agreements. It cancelled its agreement with Mexico and in effect restored the tariff quota provisions of the Venezuelan understanding of 1939. The Truman administration also reached another agreement with Venezuela that reduced the tax on imported crude oil under certain conditions. At that time Venezuela, along with the Netherlands West Indies, Saudi Arabia, and Kuwait, supplied the United States with 94 percent of the total value of its imported petroleum.

## The Great Antitrust Oil Cartel Furor

The nationalization of Anglo-Iranian's oil holdings by the Mossadegh administration in 1951 triggered an international *cause célèbre*. In the aftermath of this seizure, the American government began to take a more active role in attempting to guarantee sufficient oil supplies to the free world, especially as the Korean War was going on at that time. Interior Secretary Oscar L. Chapman, who was also the Petroleum Administrator for Defense, suggested to Charles E. Wilson, Director of the Office of Defense Mobilization, that American oil firms operating abroad should join forces to advise the American government.

U.S. petroleum companies then began to suggest that in view of the impending world shortage they should be granted immunity from prosecution under the antitrust laws. Chapman was sympathetic to their pleas. He thus drew up a voluntary agreement under which the oil companies were to enjoy antitrust immunity for their overseas operations under Section 708 of the Defense Production Act. According to the summary of this found in the "World Oil Cartel" memorandum, the agreement permitted "the allocation of markets, the use of agreed upon supply schedules for the world outside the United States, the regulation of production of world oil in all foreign countries, [and] the regulation of imports and exports in the United States by agreement among the companies."[39] Immunity, however, did not extend to any past actions, nor to any future actions outside the scope of the agreement.

On June 25, 1951, the Truman administration gave its official approval to the agreement, which involved nineteen U.S. petroleum firms. At the same time, it set up the Foreign Petroleum Supply Committee (FPSC), and on August 2 it authorized the FPSC to proceed with a plan, drawn up under Chapman's guidance, to alleviate the oil shortage. The plan provided for increases in crude oil production involving eleven countries and in refined petroleum products affecting twenty-seven nations.

Despite this 1951 agreement, during the summer of 1952 the Federal Trade Commission presented a lengthy document to Congress entitled "The International Petroleum Cartel" later published by the Small Business Committee of the Senate. The report was basically factual in nature, lacking either conclusions or recommendations. While it made reference to an alleged cartel, apparently the 1928 Achnacarry Accord, in its preface, the focal points of its criticism were the so-called Seven Sisters, which had long dominated the world petroleum industry: Jersey Standard, Socony-Vacuum, California Standard, Texaco, Gulf, Anglo-Iranian, and Royal Dutch-Shell. Seemingly the FTC had been influenced by the Swedish parliament; one of its committees had complained in 1945 about the existence of a prewar petroleum cartel. The fact that in 1952 the seven major oil companies were producing 90 percent of the crude oil extracted outside of North America and the Communist world is evidence of the legitimate basis for the international petroleum cartel accusations. They also were marketing more than 75 percent of the petroleum products.

Although the FTC report did not go so far as to assert that the seven oil giants had damaged the world economy, the Mutual Security Agency (MSA) set up in 1952 to assist Western European rearmament took a more aggressive stance in a report of its own. Not only did it charge that the five American petroleum companies in this group had engaged in "exorbitant price discriminations," it also charged that they had added $50 million to the cost of the American foreign aid program by overcharging for oil. As a result of these two reports, on July 17, 1952, the Attorney General announced that there would be a grand jury investigation of the large firms. In addition, the government filed a suit in federal court against four of the oil companies, seeking to recover $67 million in alleged overcharges to the MSA and the Economic Cooperation Administration (ECA).

Among those members of Congress who were most critical of the petroleum firms was Senator Blair Moody of Michigan, a Democrat, who declared to the National Congress of Petroleum Retailers that the American government would smash the cartel. Equally displeased was Democratic Senator Thomas Hennings of Missouri. On July 18, 1952, he wrote Truman that: "The power of this oil industry to burden the American people through the charging of excessive prices abroad necessitating greater foreign aid appropriations, is clearly a matter of major concern to the American people."[40] (Hennings overlooked the fact that increased tax payments by U.S. petroleum firms to nations such as Saudi Arabia were in fact a substitute for foreign aid.) Still another critic of the oil companies was Democratic Senator John Sparkman of Alabama, the 1952 Democratic vice-presidential nominee, who as Chairman of the Senate Small Business Committee played a major role in convincing the FTC to release the report.

As expected, the petroleum firms denied the charges that had been made against them. They stated that their sales were open to everyone on a competitive basis and declared that "false" accusations such as these had caused Iran to nationalize Anglo-Iranian. Socony Chairman of the Board George V. Halton made a public statement, declaring that the MSA suit seemed "to be part of a deliberate program to make this and other oil companies engaged in international trade whipping boys during an election campaign."[41] James Terry Duce of the Arabian-American Oil Company (Aramco) claimed that the operations of the oil industry abroad were the outstanding example of the Point IV program in action and took the idealistic position that American petroleum firms overseas should be the first servant of the state.

Even the President of the National Petroleum Association, Earl M. Craig, attacked the oil cartel charges as unjust. As the NPA was composed of independent petroleum companies, it could hardly be regarded as a propaganda vehicle for the "Seven Sisters." It had been founded, in fact, fifty years earlier to assist smaller firms in fighting the Standard Oil trust.

An anonymous memorandum, dated December 1, 1952, and prepared for Interior Secretary Chapman, reveals that the cartel investigation did cause an uproar in certain quarters abroad. In Venezuela Dr. Jovito Villalba of the Union Republicana Democratica called for the nationalization of American oil properties, while in Australia Dr. Herbert Evatt, the leader of the opposition in the Australian Federal Parliament, concluded that the petroleum firms' chief desire was to increase prices. An inflammatory press report that appeared in the Greek paper *Elliniki Imera* on September 4 charged that the "Seven Sisters" had robbed all of the Marshall Plan nations, including Greece. Other reactions abroad, as chronicled by the New York *Journal of Commerce*, were equally negative. The South African Minister of Foreign Affairs declared that the charges had led him to deny the oil companies' request for a price increase; the prestigious Argentine newspaper *La Prensa* accused the State Department of having "covered up the oil fraud" for years. Terms such as "profiteers," "exploiters," and "plunderers," regularly appeared in German press accounts.

A week later Secretary Chapman in a report to the National Security Council stated that the reaction abroad to the Federal Trade Commission document could not have been predicted with accuracy. Nevertheless, Chapman observed, "The consequences remain catastrophic and compel re-examination of the acts taken from the standpoint of national security."[42] On December 17 the State Department's Office of Intelligence Research drew up a report of its own; it noted that Great Britain, the Netherlands, and Saudi Arabia were opposed to the submission of petroleum company records to American investigators armed with subpoenas. The French government also feared that the antitrust suit might have an adverse effect on those oil firms in which it held an interest.

In contrast, Federal Trade Commissioner Stephen J. Spingarn was quite pleased with the investigation. His collected papers at the Truman Library are an outstanding depository of anticartel materials for this period. As a young man he had attended a French university, and he had concluded on the basis of his first-hand knowledge that Adolf Hitler might never have risen to power in Germany, if cartels had been restricted. In any case, on October 29 Spingarn sent a memorandum to presidential advisor Richard Neustadt aboard the Truman campaign train, in which he noted that "the oil companies are screaming like stuck pigs over the F.T.C. oil report."[43] Perhaps fearful that he might be labelled as a radical, Spingarn released a statement on Christmas Eve affirming his belief in a competitive system of enlightened capitalism, and his opposition to cartels and monopolies.

Spingarn's anticartel zeal, however, did not find favor with the majority of the Federal Trade Commission, who voted three to two to release a statement on January 15, 1953, declaring that he was not speaking for the FTC as a whole. John Carson, another member of the FTC, had long been the spokesman for a tax privileged group of cooperatives, but his influence should not be unduly emphasized, for work on the FTC report had commenced prior to his appointment as a commissioner. The FTC, moreover, did not publish the report itself, nor did it ask the Justice Department to initiate the legal proceedings, or Congress to investigate the alleged cartel.

Outside the government, the petroleum firms had other defenders, many in the journalistic realm. One of the strongest of these was the *Saturday Evening Post*, which saw competition rather than cartel, when looking at the Middle East; it described the governmental campaign against the oil companies in terms of flogging a dead horse. *Newsweek* commented on the proximity of this investigation of big business to the presidential election campaign, and *The Oil Forum* lamented that "the gangrene will spread until America's sources of strategic minerals of all kinds are jeopardized."[44]

The most surprising attack of all was launched by the generally liberal Washington *Post*, which described the grand jury investigation on November 3 as "a huge fishing expedition." It also charged that "Counsel for the Government goes on from one piece of nonsense to another;"[45] a week later the *Post* declared that the suit was hampering both U.S. foreign policy and military planning. Obviously disturbed by the Department of Justice's demand for millions of documents from the petroleum firms, the *Post* bitterly observed, "This is not too different from the way in which Senator McCarthy has operated by demanding executive files in the hope that he might dig up something."[46] According to Stephen J. Spingarn, however, the law firm that acted as counsel for the Washington *Post* was also cocounsel for Jersey Standard.

So intense was the attack on the FTC report in certain journalistic quarters that Spingarn felt impelled to write to Truman on November 7, 1952, "It is being villified in editorials, articles and speeches all over the country in the most intemperate terms."[47] Spingarn also chastised Philip Graham, publisher of the Washington *Post*. In a letter to Graham, dated January 5, 1953, Spingarn alleged that the newspaper had not presented both sides of the controversy. Spingarn had been so upset by what he regarded as biased newspaper coverage that he had written a letter to the *New York Times* on January 4, stating that this newspaper was the *only* one in the United States to give fair coverage to the FTC report.

Even before Dwight Eisenhower took the oath of office as President on January 20, 1953, the Truman administration had been so pressured by the State and Defense departments, that it had agreed to drop any criminal indictments of the major oil companies on antitrust charges that might be forthcoming. In fact, the Secretary of State, Dean Acheson, had written a memorandum on April 16, 1952 opposing the entire proposed investigation.

On December 15 Federal District Judge James R. Kirkland directed the Justice Department to drop its cartel charges against the Anglo-Iranian Oil Company. Judge Kirkland argued that the firm was indistinguishable from the British government, which had ordered Anglo-Iranian not to release any information to the court without its approval. The Dutch and Indian governments were similarly uncooperative.

Four weeks later, on January 11, 1953, the Truman administration made public its decision not to press criminal charges against the petroleum companies. There were several reasons for this step. First of all, such a prosecution might endanger American petroleum supplies. Second, it might discourage U.S. private investment abroad in general. Third, it would give credence to Soviet propaganda. Fourth, it might jeopardize the position of the oil companies in such countries as Venezuela. Finally, the Korean War, which naturally had increased the demand for petroleum, was going on at this time. Both the State and Defense departments were opposed to a criminal suit and had complained to the National Security Council three days earlier that such a suit would violate U.S. foreign policy objectives.

Although the criminal investigation of the oil companies was called off, Attorney General James McGranery, following the recommendations of the State and Defense departments, attempted to substitute a civil suit in its place. This did not please the petroleum firms, which had received a series of proposals from McGranery with a one-day deadline to accept or reject them; there then ensued a stormy confrontation between McGranery and thirty-five petroleum firm lawyers. To Arthur H. Dean, Jersey Standard's attorney, the brief period of time that McGranery had allowed the oil companies to reach a unanimous agreement was "the most insulting statement I have ever heard in my many years at the bar."[48] It was at this time that Judge Kirkland decided to delay proceedings for two weeks in order to determine what the Eisenhower administration planned to do about the case.

Only three days after Eisenhower took the oath of office, he presided over a cabinet meeting at which the oil cartel case was discussed. The new Attorney General, Herbert Brownell, announced that he would ask for a delay because of the great controversy that this issue had aroused. Three months later the Eisenhower administration did decide to file a civil suit, but the criminal proceedings remained a dead letter.

During his remaining months on the Federal Trade Commission Stephen Spingarn continued to snipe at international petroleum firms with little or no success. As the Truman administration had drawn to a close, he had attempted to rally support for his crusade in the State Department, only to discover that such high-ranking officials as Dean Acheson, David Bruce, and Paul Nitze had disqualified themselves because of their assorted connections with the oil companies. On January 21 Spingarn wrote Democratic Senator Guy Gillette of Iowa that the members of the National Security Council had been "the unconscious victims of the world-wide propaganda campaign of the oil companies and their public relations experts in what I have described (borrowing the phrase from Army slang) as one of the biggest snow jobs in history."[49] In the final analysis Spingarn's "great crusade" was yet another of those abortive reform movements that clutter the landscape of American history. The steps that were eventually taken by the U.S. government to restrain the petroleum giants were a mere shadow of Spingarn's original vision.

## The Eisenhower Administration: Varying Concerns

The early years of the Eisenhower administration saw the government campaign against the so-called oil cartel so decline that in 1954 a group of American petroleum firms joined forces with participants from other countries to form a new Iranian oil consortium. With the nationalization of the Suez Canal by Egypt in 1956, the cartel question became even less important as it became necessary for the United States to oversee the supplying of petroleum to oil-hungry Europe. One question that did continually preoccupy Dwight Eisenhower during his tenure of office was that of oil imports. In a systematic attempt to restrict the movement of petroleum into the United States from abroad, the Eisenhower administration eventually was forced to implement a voluntary oil import program in 1957 and a mandatory one during 1959.

After three months of deliberation over the oil cartel issue the Eisenhower administration acted. On April 21, 1953, Attorney General Herbert Brownell filed a civil suit against five major petroleum firms: Jersey Standard, Socony-Vacuum, The Texas Company, California Standard, and Gulf Oil. This suit, according to the Washington *Post*, asked "for a sweeping perpetual injunction against future illegal monopolization in violation of antitrust and tariff laws."[50] Among other things, the suit accused the firms of monopolizing foreign production, dividing foreign markets, fixing global prices, excluding competition, and monopolizing refinery patents from 1928 to date. The petroleum companies were far from pleased with this action, as they had hoped that the government would refrain from filing a civil suit.

But in mid-June, with both the $67 million MSA suit and the conspiracy to violate the antitrust laws suit still pending, the Defense Mobilization Director, Arthur S. Flemming, asked fourteen U.S. oil companies to participate in a new voluntary agreement. Under the agreement they would supply oil to friendly countries abroad, furnish the U.S. government with data on foreign supply and demand, and draw up plans of action to meet possible future oil shortages should the Eisenhower administration request these. It was during this same month that a report by the House Foreign Affairs Committee praised the oil companies' good work abroad, a further indication that the tide of government opinion was turning in their favor.

With the oil policy of the American government in this state of flux, Prime Minister Mossadegh of Iran fell from power in August, perhaps in part as a result of Central Intelligence Agency intrigues and petroleum firm maneuverings. In the aftermath of this occurrence, Democratic Senator Thomas Hennings of Missouri demanded a congressional investigation for the purpose of determining whether the oil cartel "undermines the integrity of the ideals professed by Americans in their relations with friendly nations."[51] Eisenhower, however, did not proceed further against the petroleum firms but rather established an Advisory Committee on Energy Supplies and Resources Policy and appointed the Director of the Office of Defense Mobilization as chairman. Secretary of State John Foster Dulles questioned the committee's creation on a number of grounds, including the possibility that it might cause widespread ill-will in Latin America by recommending import duties or high tariffs.

Two years later, when Egypt seized the Suez Canal, Dulles held a private meeting, attended by representatives of the larger American oil firms doing business abroad. According to a memorandum kept by A. C. Ingraham of Socony-Vacuum, and never repudiated by the State Department, Dulles stated at this August 1956 conference that "nationalization of this kind of asset impressed with international interest goes far beyond the compensation of shareholders alone and should call for international intervention."[52] When he learned of Dulles' supposed statement, Democratic Senator Estes Kefauver of Tennessee asked on the floor of the U.S. Senate who was going to do the intervening. Shortly thereafter the Eisenhower Doctrine—which placed the Communist world on notice that the United States would permit no further Communist conquests in the Middle East—was before the Senate for debate; Kefauver sarcastically observed, "When the big oil companies take snuff, the State Department sneezes."[53]

At this time Eisenhower was considering direct U.S. military intervention in the Middle East during an emergency in order to safeguard the American supply of petroleum. During the summer of 1957, the President wrote to Dillon Anderson, who advised a special committee to investigate crude oil imports: "I think that you have proved that should a crisis arise threatening to cut the Western World off from Mid East oil, we would *have* to use force."[54] This statement, it must be remembered, dates from 1957, sixteen years prior to the great Arab petroleum embargo of 1973, after which schemes of military intervention became far more common.

As the Eisenhower administration drew to a close, however, the likelihood of a serious challenge to the American position in the Middle East became more and more remote. Instead, Eisenhower became increasingly preoccupied with implementing the voluntary and mandatory oil import programs of 1957 and 1959. Although a federal jury returned an indictment against a number of petroleum firms on May 29, 1958, charging that the Middle East Emergency Committee, set up in 1956 to supply oil to Western Europe and Great Britain after the seizure of the Suez Canal, had conspired to bring about an increase in petroleum prices, federal Judge Royce Savage dismissed the case in February 1960 for failing to rise above the level of mere suspicion.

The antitrust suit that the U.S. government had filed against several of the petroleum giants, however, finally bore fruit during 1960. Both Jersey Standard and Gulf Oil agreed to consent decrees that enjoined them from entering into any sort of price fixing and marketing arrangements that might hinder competition. In 1960 Standard-Vacuum also fell victim to a consent decree, thus ending a major international joint marketing enterprise; under the direction of Jersey Standard, however, Stanvac did continue to operate in Indonesia. Texaco agreed to a consent decree in 1963, but this was after the Antitrust Division of the Justice Department had decided in 1961 to emphasize the marketing aspects of the oil cartel case.

Nevertheless, a loophole in these decrees made foreign cartel arrangements possible, as U.S. petroleum firms were allowed to execute not only such acts required by foreign law but also acts "pursuant to request or official pronouncement of policy of the foreign nation."[55] In fact noncompliance would lead to a loss of business for the American oil companies. For example, in 1959 the West German government had pressured subsidiaries of Jersey Standard and New York Standard into participating in a price-fixing cartel. Such developments only prove the old French adage that the more things change, the more they are the same.

## Restricting Imports, 1953-1957

As soon as Eisenhower took office, high-ranking officials from oil-producing states began to exert pressure on the new chief executive to restrict excessive petroleum imports. Among those pressuring Eisenhower was Democratic Governor Allan Shivers of Texas, who had supported Ike over his Democratic opponent Adlai Stevenson in the presidential contest primarily because of Eisenhower's states' rights stand on the tidelands oil question. The Louisiana congressional delegation also took a firm position against the importation of foreign oil.

Two years after Eisenhower had established the Advisory Committee on Energy Supplies and Resources Policy, on February 26, 1955, the White House issued a press release setting forth its report. The fourth section of the report, which dealt with both crude oil and residual oil imports, declared:

The Committee believes that if the imports of crude and residual oils should exceed significantly the respective proportions that these imports of fuel bore to the production of domestic crude oil in 1954, the domestic fuels situation could be so impaired as to endanger the orderly industrial growth which assures the military and civilian supplies and reserves that are necessary to the national defense. There would be an inadequate incentive for exploration and the discovery of new sources of supply.

In view of the foregoing, the Committee concludes that in the interest of national defense imports should be kept in balance. . . . It is highly desirable that this be done by voluntary, individual action of those who are importing or those who become importers of crude or residual oil. The Committee believes that every effort should be made and will be made to avoid the necessity of governmental intervention.

The Committee recommends, however, that if in the future the imports of crude oil and residual fuel oils exceed significantly the respective proportions that such imported oils bore to domestic production of crude oil in 1954, appropriate action should be taken.[56]

In June, Congress passed the Reciprocal Trade Agreements Extension Act, which included the Byrd-Millikin Amendment. Under this act, the President received discretionary power to limit imports into the United States of any product, if the volume were great enough to threaten to impair the national security. Although Eisenhower technically did not invoke this amendment at the time, the Texas Independent Producers & Royalty Owners Association complained that imports of petroleum had already exceeded this critical level. That September Defense Mobilizer Arthur Flemming, who also had been pressured by independent oil producers, the coal industry, and certain U.S. Senators from coal and petroleum states, demanded that eighteen oil firms either cut petroleum imports voluntarily or face possible congressional legislation. Flemming gave the petroleum companies a week to formulate a workable solution. His action, however, did not satisfy the governors and members of Congress from the New England states, where there had been a chronic shortage of residual fuel oil.

At a Camp David cabinet meeting in November, Flemming stated that the voluntary program was working satisfactorily. Among those in favor of the petroleum industry policing itself were Secretary of the Treasury George Humphrey and UnderSecretary of State Herbert Hoover, Jr. According to a privileged cabinet paper, entitled "Summary Review of Developments with Respect to Oil Imports," the companies that were making a serious attempt to comply with the voluntary program were Atlantic Refining, California Standard, New Jersey Standard, Gulf, Sinclair, and New York Standard; those deviating included Cities Service, Eastern States, Sun, Tidewater and Texaco. Voluntary compliance obviously was far from uniform, which demonstrates that the petroleum firms did not continually maintain a united front on key issues.

Despite the existence of this program, by 1957 gross imports of petroleum into the United States amounted to 37.5 percent of the total American-owned crude production abroad. The percentage had increased steadily from 32.2 in 1954 to 33.2 in 1955 to 35.4 in 1956. As Eisenhower took office for the second time as

President, it became apparent that additional steps were needed to halt the mounting flow of oil into the United States.

### The Voluntary Limitation Program, 1957-1959

On April 23, 1957, Flemming officially notified Ike that crude oil imports had reached a level that was threatening national security. In his reply, dated two days later, the President stated that he would have the problem investigated but hoped that the importing companies would limit their oil imports voluntarily. That a time might come when additional imports would be needed was underscored in a letter that Deputy Under Secretary of State Robert Murphy sent to Budget Director Percival Brundage the following month. "In the event of emergency," wrote Murphy, "national security will require access to foreign petroleum supplies not only for direct importation, but also to supply United States armed forces and our allies abroad."[57]

Seeking further data, on June 26, Eisenhower created a special committee to investigate crude oil imports with Commerce Secretary Sinclair Weeks as its chairman. When the report of the special committee was presented on July 29, it emphasized the need for a reasonable limitation on petroleum imports, while noting that the policy of voluntary restrictions had been reasonably successful until the summer of 1956. The committee noted that it had considered and rejected three alternate plans: "1. Import foreign crude oil and store it in the country within depleted fields or elsewhere; 2. Enlarge government participation in exploring for oil reserves which, when discovered, would not be put into production; 3. Encourage increased importations in order that our own national resources might be conserved."[58] After endorsing its findings, President Eisenhower put a voluntary oil import program into operation, which remained in effect until March 10, 1959. Retroactive to July 1, the program at first only covered the area east of the Rockies, which was divided into Districts I, II, III, and IV.

In 1972, when Dillon Anderson, a committee advisor, was interviewed by the Oral History Research Office at Columbia University, he pointed out that the program proposed by the committee had not been perfect but that it had achieved its basic objectives and had kept the domestic industry both viable and healthy. He also noted that it wisely had attempted to protect and to stimulate the refineries of the interior, which were less vulnerable to attacks from without than those on the coast.

Despite his claim that "the program was not a very controversial thing at the time" and that no groups objected vigorously to it, an examination of the mail sent to the White House by various governors and members of Congress during 1957 reveals that there was conflicting advice over petroleum imports limitation.[59] Generally speaking, this White House mail followed a long-established geographic pattern. Public officials of the oil-rich Great Plains states called for restrictions on petroleum imports. These included Democratic Governors Price Daniel of Texas and Raymond Gary of Oklahoma, Democratic Senator Joseph

O'Mahoney of Wyoming, and Republican Senator Frank Carlson of Kansas. Another critic of the uninterrupted flow of foreign oil was Ernest O. Thompson of the Texas Railroad Commission, a misnamed body that regulates petroleum in Texas.

Restrictions on oil imports encountered stiff opposition throughout New England from members of both major political parties: Democratic Senator John F. Kennedy of Massachusetts, Republican Senator Norris Cotton of New Hampshire, and Republican Senators Margaret Chase Smith and Frederick Payne of Maine. Writing to the President on July 31, Payne complained that: "While this import limitation was investigated and executed within a few months, other industries such as textiles, plywood, and fisheries must wait years to even receive a hearing, and then only to have their appeals for protection rejected."[60]

Payne's appeal to the President fell on deaf ears. In endorsing the work of the special committee, Eisenhower pointed out that projected imports for the last half of 1957 would have discouraged petroleum exploration and drilling within the country. At a meeting of congressional leaders on August 27, however, Ike took the position that residual fuel imports which did not come under the new restrictions were not, in terms of price, competitive with coal.

On December 12 the special committee made its second report to Ike, which he immediately approved. This document dealt with District V—the area west of the Rocky Mountains—where the available domestic production of crude oil was insufficient to meet the demand. The committee believed that the projected future imports into District V were excessive and recommended limiting imports to approximately 220,000 barrels a day on a voluntary basis.

The debate over oil imports became even more intense in 1958 than it had been in past years. At a legislative leadership meeting held on February 25, Interior Secretary Fred Seaton declared that the voluntary program had worked well in terms of its original purpose. On March 7 at a cabinet meeting Commerce Secretary Sinclair Weeks claimed that it had been the independent oil refiners who had aggravated the situation by importing petroleum in excessive amounts. On March 14, however, Defense Mobilization Director Gordon Gray informed the cabinet he could not justify a certification that imports of residual fuel were endangering national security.

One of the most outspoken advocates of a tougher approach toward petroleum imports was Senate Majority Leader Lyndon Johnson of Texas, who wrote Eisenhower on March 8 that the domestic industry was "reeling" under their impact. LBJ recommended a mandatory reduction of 20 percent. In his reply the President claimed that established importers, with the exception of a few recalcitrants, had complied gladly with the voluntary program but that it still might be necessary to invoke mandatory controls. At a cabinet meeting held on March 21, Eisenhower, in citing the unusually low level of activity in the petroleum industry, again expressed his dissatisfaction with a situation in which a noncooperative firm could freely ignore the voluntary program.

An optimistic approach to this problem permeated the committee's supplementary report dated March 24, which favored a continuation of the voluntary plan on a more effective basis. According to the report, this would "involve a minimum of governmental regulation, the least possible restraint on our free enterprise system, and a lesser interference with normal trade relations."[61] Among other things, the special committee advocated a continuation of an import ratio of 12 percent in Districts I, II, III, and IV.

Rather than challenging its findings, Eisenhower recommended accepting the supplementary report with its suggested modifications. The President also issued an executive order (No. 10761) on March 27, in which he required the heads of all executive departments and agencies to apply the "Buy American" Act when they were issuing contracts for the purchase of oil. In a letter replying to a telegram from Republican Governor Milward Simpson of Wyoming, who favored a drastic reduction in petroleum imports, Eisenhower pointed out that he had "requested that the importers not only reduce their original allocation, but that the older importers further reduce their inputs to allow newcomers to participate in the program."[62]

During the twelve-month span from March 1958 to March 1959, the White House was bombarded with a series of letters and telegrams, primarily from outside of New England, that were highly critical of the voluntary program. Among the most impressive of these were a March 24, 1958, telegram signed by twenty-two governors and a June 26 telegram endorsed by thirty-three state chief executives. Critics of the voluntary program included spokesmen from such oil-producing states as Wyoming, Kansas, and Texas and such coal-producing ones as West Virginia and Kentucky. Republican Senator Edward Thye of Minnesota, who complained that small refiners in Minnesota and the Midwest were highly dependent on Canadian crude oil, was advised by Commerce Secretary Sinclair Weeks that it was inadvisable to exclude small refineries from the import program.

On December 22 the Special Committee to Investigate Crude Oil Imports made an interim report in which it recommended no changes in crude oil allocations until the end of February 1959 and that importers should attempt to keep their imports of unfinished gasoline and other unfinished oils to their present allocations. On February 27, however, Leo Hoegh, the Director of the Office of Civil and Defense Mobilization, reported to the President that crude oil and crude oil derivatives and products were being imported in amounts sufficient to impair the national security. This time there was not to be another extension of the voluntary program.

### The Mandatory Limitation Program, 1959-1961

During the extended discussion at the March 6, 1959, cabinet meeting, Commerce Secretary Lewis Strauss lamented over the unwillingness of a few companies to comply with the voluntary program and noted that the executive order supposedly restricting government purchases from noncompliers had proven ineffective. There followed an exchange of remarks, in which both Venezuela and

Canada figured prominently. In concluding the discussion of oil import restrictions, the President showed his irritation at "the tendencies of special interests in the United States to press almost irresistibly for special programs like this and wool and cotton, etc. in conflict with the basic requirement on the United States to promote increased trade in the world."[63]

The report of Chairman Strauss' Special Committee to Investigate Crude Oil Imports was released that same day. It called for a mandatory program in place of the voluntary program, except under certain conditions for refiners and pipeline firms, and for the control of liquefied petroleum gases, gasoline, kerosene, jet fuel, distillate fuel oil, lubricating oil, residual fuel oil, and asphalt in addition to crude oil. The report also recommended a different petroleum importing quota for Districts I, II, III, and IV, where oil production capacity exceeded actual production and District V, where production was declining.

On March 10 President Eisenhower, in implementing the recommendations set forth in the report, issued a proclamation that ended the voluntary era and ushered in the mandatory period. The new mandatory program established the level of imports for the area east of the Rockies at 9 percent of the total demand, not including that residual oil used as fuel. This figure was to be determined by the Bureau of Mines for each six-month quota period; at that time it would allow an import level of approximately 810,000 barrels a day. West of the Rockies the import level was to be approximately 195,000 barrels a day with unfinished imports making up no more than 10 percent of this figure. Although there was no attempt to include residual fuel oil imports in the 9 percent figure for Districts I, II, III, and IV, on a nationwide basis these were not to exceed the 1957 level of 480,000 barrels daily.

A month and a half later, on April 24, Secretary of State Christian Herter recommended to the President exempting Canadian petroleum from the mandatory program. Six days later the President issued another proclamation in which he freed from import restrictions crude oil and finished and unfinished products that entered the United States by pipeline, motor carrier, or rail from the nation that produced them. The two obvious beneficiaries of this measure were Canada and Mexico. As for Venezuela, Eisenhower had written President Rómulo Betancourt on April 28 that "the United States has been Venezuela's largest market and I am confident that it will continue to be so on an expanding scale."[64]

Still another aspect of this complex problem was whether or not U.S. flag tankers were carrying less and less of the restricted oil imports into America and consequently facing a bleak future. This fear was widespread among members of both the Committee of American Tanker Owners and the Joint Committee for American Flag Tankers, who talked with both the Department of Commerce and the Office of Civil and Defense Mobilization in January of 1960. Unfortunately for the tanker committees, Civil and Defense Mobilization Director Leo A. Hoegh concluded in December that national security considerations did not require a fixed share of petroleum imports to be carried in American tankers. Hoegh also offered the opinion that the U.S. flag fleet in conjunction with ships of Panamanian, Liberian, and Honduran registry was adequate to meet defense requirements.

That fall high-level officials of the Eisenhower administration again exchanged ideas about the mandatory program. Secretary of the Interior Fred Seaton sent a confidential memorandum to Eisenhower on September 12, 1959, in which he observed that residual imports were excessive and residual prices were weak. He concluded that "the petroleum industry would consider effective decontrol of residual as the first sign of weakness in [the] administration of the program."[65] But not every department was satisfied with the system of mandatory controls now in effect. Acting Secretary of State Douglas Dillon wrote to Maurice Stans, of the Bureau of the Budget, on December 1, "I have in mind also the serious concern, and even skepticism, with which the entire oil import program is viewed by friendly countries which have legitimate interests in our petroleum market and our trade policies in general."[66] The State Department was especially concerned about maintaining the friendship of Venezuela and with this in mind favored the liberalization of imports of residual petroleum.

But rather than innovate boldly, President Eisenhower released a proclamation on December 24 that required the Secretary of the Interior to make corrective adjustments in the oil import quotas to prevent overestimates and underestimates of total demand from having an unintended impact on import levels. It was not until January 18, 1961, two days before he left office, that Ike issued a more far-reaching document, which not only permitted the entry of new importers into the residual fuel oil markets of the East Coast on the basis of their deepwater terminal inputs but also attempted to set up more equitable fuel oil import allocations for established importers. Since fuel oil imports into the other four districts were negligible, their importing practices were not altered at this time. This, then, was the amended mandatory program that confronted John F. Kennedy as he took office as President. The program was to remain in operation with some additional changes until the penultimate year of the Nixon presidency.

## The Birth of OPEC

During the post-World War II era the petroleum-producing nations of the world began to think in terms of joint action in order to deal more effectively with the increasingly voracious demands of such oil-consuming countries as the United States. As early as 1947 Venezuela had approached various Middle Eastern governments with the idea of a common cause vis-à-vis petroleum; six years later Iraq and Saudi Arabia agreed to exchange information on oil. A year after that, in 1954, President Gamal Abdel Nasser of Egypt in his book *The Philosophy of a Revolution* noted that petroleum was one of the chief sources of Arab power.

Five years later, in 1959, an Arab petroleum congress met in Cairo. Although the U.S. oil companies sent representatives, the State Department did not, for it had not received an official invitation. At this period there was a substantial oversupply of crude oil, and at the urging of Venezuela, the Cairo congress came out in favor of a producers' cartel. While the delegates did reject setting aside for regional development 5 percent of the profits from petroleum going to both the oil companies and oil nations, it recommended the integration of the global petroleum industry.

The following year in Baghdad, representatives from the oil-producing nations assembled to establish the Organization of Petroleum Exporting Countries. OPEC, the brainchild of the Saudi Arabian Minister of Oil, Sheikh Abdullah Tariki, counted among its members Iraq, Iran, Kuwait, Saudi Arabia, and Venezuela. In fact, the Venezuelan Minister of Mines and Hydrocarbons, Juan Pérez Alfonzo, was a leading figure at the time of OPEC's inception. Although the average American inevitably thinks of oil-rich Arab sheikhs when his attention turns to OPEC, such nations as Indonesia, Nigeria, and Ecuador also belong to it.

Ironically, it was the giant petroleum firms that inadvertently contributed to the establishment of OPEC by reducing the posted price for Middle Eastern oil, a step that irritated both the countries of the Middle East and Venezuela. As Edith Penrose has pointed out, following the creation of OPEC, "the companies effectively lost their freedom to alter posted prices unilaterally, even when these became increasingly out of line with realized prices."[67]

In the years that followed, OPEC directors sometimes copied American practices and sometimes relied strictly upon Third World ideas, such as those of President Nasser. Anthony Sampson has observed that "OPEC was to look back to Texas as their model for controlling Middle East production."[68] Certainly it would appear that the cartel policies of the major oil companies in the Middle East during the 1950s and 1960s served as models for the oil-producing countries. By encouraging these cartel policies, the Truman and Eisenhower administrations, as well as the Kennedy and Johnson administrations, may have contributed to the conceptualization, creation, and continuation of OPEC. The evidence brought to light in 1972 by a subcommittee of the Senate Foreign Relations Committee that was investigating multinational corporations would seem to substantiate this interpretation.

### JFK: Mandatory Controls

On January 20, 1961, John F. Kennedy took the oath of office as President. Although he was more liberal and less favorably disposed toward big business than his Republican predecessor, he did not significantly modify the Eisenhower administration's foreign oil policy during his abbreviated term of office. But the Kennedy administration did initiate a review of the mandatory control program in May 1961 by holding public hearings in Los Angeles and Washington, D.C. Most witnesses believed that import controls were necessary and that the present level of imports was close to the optimum. That fall a subcommittee of the House of Representatives' Small Business Committee held its own public hearings. While the May witnesses primarily came from the oil and coal industry, those who appeared before the subcommittee were largely government witnesses. With the exception of the Department of Interior, which favored a further cut in petroleum imports, spokesmen for other departments generally opposed any limitation on petroleum imports. These witnesses included representatives from the Defense, State, and Commerce departments.

On December 2, 1961, Kennedy announced that the Director of the Office of Emergency Planning (OEP), Edward A. McDermott, would oversee a thorough investigation of petroleum requirements and supplies from the standpoint of national security. Curiously, McDermott's office released a statement on February 16, 1962, declaring that "there is no immediate need for submission of views and material by the public," since "a number of recent studies and investigations have been made, and hearings held, by the Government departments and agencies concerned."[69] Nevertheless, representatives of the oil industry did make thirty-two written submissions to the Office of Emergency Planning.

The OEP report appeared on September 4. With the Interior Department dissenting on various grounds, McDermott and the seven governmental officials who signed this report concluded that there would be enough oil available through 1965 from the domestic supply to cover both defense and civilian needs. This, they claimed, would be the case even if a foreign government, or governments, cut off other sources of supply in an act of political or economic reprisal. In assessing the system of controls then in effect, the committee observed that the abandonment of controls would lead to a reduction of approximately $1 a barrel in the price of domestic crude oil. Even so, the signers of the OEP report were of the opinion that there should be no rapid alteration in the price of crude petroleum under the present system of supports, nor should there be a drastic change in the level of imports. They did feel, however, that a gradual shift toward liberalization should take place and that domestic petroleum costs then should correspondingly decline. As for a tariff on oil, the committee believed that the approach in theory might be preferable but in practice it would not be feasible at that time as it might have an adverse effect within the Western Hemisphere. The report failed to show any concern over the possible decrease in the number of U.S. flag tankers under the oil import control program.

Approximately three months after this report, JFK released another presidential proclamation adjusting the imports of petroleum and petroleum products into the United States. This document contained a number of important provisions: it related the total level of imports for the area east of the Rockies (Districts I, II, III, and IV) to domestic production rather than demand; it exempted overland oil imports from Canada and Mexico from the total level in advance of making allocations to individual refiners; it directed the Secretary of the Interior to provide for the gradual reduction in the employment of historical quotas, which had tended to favor established importers; and it mandated that allocations to individual companies be made on the basis of a graduated scale so that smaller refiners would receive larger proportionate quotas. There were those small companies, however, that already had received quotas.

Then on April 22, 1963, Kennedy set up an appeals board to consider petitions concerning the revised mandatory program. The board was composed of representatives from the Interior, Defense, and Commerce departments. Finally, on June 11, the President issued still another presidential proclamation, altering the oil importing formula for Districts I, II, III, and IV from 12.2 percent of past production to 12.2 percent of estimated production. This seemingly minor revision offended many independent domestic producers since it eliminated an anticipated reduction in imports for the last half of 1963. Having done this, President Kennedy did nothing of major consequence to alter the international oil policy of the United States in the months prior to his assassination on November 22, 1963.

## The Johnson Administration: Varying Concerns

Unlike Kennedy, who was from oil-poor Massachusetts, the new chief executive, Lyndon Johnson, hailed from petroleum-rich Texas. Because of his close ties with home-state oil interests, LBJ deferred many decisions in this area to his Interior Secretary, Stewart Udall, thus allowing Udall a great deal of influence over government oil policy. During Johnson's administration the debate over oil policy intensified not only at the highest levels of the executive branch but also in Congress. Even so the mandatory program limiting oil imports was still in operation when Lyndon Johnson left office in January of 1969.

The leading congressional critic of Johnson's oil policies was a fellow Texan, Democratic Representative Wright Patman, the Chairman of the House Banking and Currency Committee. In May 1965, Patman announced that his committee would undertake a thorough investigation of the international operations of American petroleum firms; he stated that government policies toward these firms had contributed to the national balance-of-payments deficit. This suggested probe had the backing of the Texas Independent Producers and Royalty Owners Association. Patman also suggested that LBJ should stop delegating oil policy matters to Stewart Udall's Department of Interior. "The present program," he concluded, "can be described as successful only if it is being evaluated by meaningless criteria."[70]

As for developments overseas, when the third Arab-Israeli war broke out in June 1967, the United States was using large quantities of petroleum, much of which was imported from abroad, in the Vietnamese conflict. While the supply situation worsened temporarily after the Arab countries placed an embargo on oil shipments to the United States and Great Britain and Egypt closed the Suez Canal, it was not just developments in the Middle East that troubled the American government. In 1968 Peru seized the holdings of the International Petroleum Company. The expropriation proved to be only the first of several that U.S. firms would have to endure in Latin American countries in the years ahead. Furthermore, throughout the 1960s the newly created OPEC attempted to build up its power, hoping eventually to pose a real threat to petroleum-consuming nations.

At the time of the outbreak of the Arab-Israeli war, the Foreign Petroleum Supply Committee held a meeting on June 8, 1967, at which representatives of twenty-one petroleum firms, most with foreign holdings, were present. There was an extensive discussion of the consequences of the disruption in the oil supply and of the type of action that would be appropriate under the circumstances. Two days later, on June 10, Assistant Secretary of the Interior J. Cordell Moore declared, with the concurrence of State and Defense departments and the Office of Emergency Planning, that a petroleum emergency did exist.

Confronted with the necessity of working out the specifics of pooling both ships and supplies, representatives of nine U.S. oil companies assembled in Washington, D.C., on June 13. On June 15 Interior Secretary Udall ordered refiners of petroleum to follow "contingency contracts" in meeting Defense Department demands for oil products. Despite this step, on the following day Udall observed that the Middle Eastern petroleum situation appeared to be returning to normal.

On June 20, the major American petroleum firms that had assembled in Washington agreed on a plan that provided for a tanker fleet to transport oil to Western Europe and elsewhere. The situation was aggravated, however, by the failure of the six-year-old, twenty-one nation Organization for Economic Cooperation and Development, which was meeting in Paris at this time, to endorse the American suggestion that it declare an oil emergency. The OECD, created to promote economic cooperation between America and Europe, instead set up an information-gathering advisory body. Faced with this setback, the Office of Emergency Planning approved the Foreign Petroleum Supply Committee's plan of action on June 30, after consultations with governmental officials. On July 6 Assistant Secretary Moore created an Emergency Petroleum Supply Committee, which was to take the place of the FPSC; this body was to include representatives from twenty-six U.S. oil firms. By October 6, John Ricca of the Emergency Petroleum Supply Committee was able to write to Moore that "tanker tonnage will be tight, but should be adequate to meet demands."[71]

As the Johnson administration entered its final year, there was some talk at the White House of limiting direct foreign investment as a part of a balance of payments package. Nothing, however, came of this. The question of restricting oil imports remained the continuing preoccupation of the latter days of LBJ's administration.

## The Mandatory Limitation Program, 1963-1969

During January 1964 Interior Secretary Stewart Udall openly took the position that the governmental program of mandatory controls on oil imports had his support and needed only a minimum amount of change. Udall, moreover, wrote approximately twenty members of Congress defending his position. In a memorandum dated April 28, 1964, he stated that "the [Interior] Department has consistently adhered to the line established in White House discussions last spring and made no move either to tighten the program to meet the demand of the coal people or to relax it to meet the demands of the consumers groups."[72] Another key Johnson administration official, J. Cordell Moore, at this time was the U.S. oil imports administrator. Moore publicly stated that American petroleum restrictions were helpful to the Canadian oil industry, since they prevented other foreign competitors from grabbing a share of the U.S. market. Moore also observed that the mandatory control program was geared to national security considerations and probably would remain in operation for a long time.

Not everyone, however, was so complimentary about the petroleum policies of the Johnson administration. Critics of excessive oil imports included the Republican Senator John Tower of Texas and state Senator Charles F. Herring of Austin, chairman of the Texas delegation to the Sixth World Petroleum Congress in Frankfurt, Germany. On June 8 Tower wrote to LBJ, complaining about "the deteriorating conditions of the domestic oil and gas industry."[73] Although Herring and Tower had overlooked the fact that petroleum production was up in neighboring Louisiana, further to the east, the Charleston, West Virginia, *Daily Mail* and Walter J. Tuohy, the chief executive officer of the Chesapeake and Ohio Railway and chairman of the Baltimore and Ohio Railroad, complained about the threat of foreign residual oil to the American coal industry.

Diametrically opposed to this position were two Republican congressmen from Maine who were of the opinion that residual fuel imports posed no threat to domestic coal producers. Republican Senator Kenneth Keating of New York entertained similar views. An editorial entitled "Throw Us A 'Home Run Ball,' Mr. President," which appeared in the Taunton, Massachusetts *Gazette*, also embraced this position.

In November 1964 a task force on natural resources, appointed by LBJ, released its report, after a year of working on a comprehensive study. This document recommended that pending the establishment of an energy policy commission: "1. Import quotas on residual fuel oil should be removed; 2. Crude oil import quotas should not be reduced; 3. Imports from Canada and Mexico should not reduce the quantity of oil that is permitted to enter under quota from other countries."[74] The first of these recommendations paralleled a suggestion made by John G. Winger of the petroleum department of the Chase Manhattan Bank on November 27; Winger had stated that controls over residual petroleum did more harm than good and that as end-use controls they were incompatible with the free enterprise system.

Defense Secretary Robert McNamara sent a memorandum to President Johnson on March 17, 1965, in which he set forth the findings of a cabinet committee that Johnson had created the previous summer to investigate the problems facing the coal industry. Rather than endorsing the position of the National Coal Policy Conference, the McNamara committee members with one exception believed that a reduction in residual oil quotas would neither increase jobs nor stimulate investment. The dissenting voice was that of the Secretary of Labor, W. Willard Wirtz, who felt that the Appalachian economy would be adversely affected by the elimination of these quotas and that the government's oil policy was offsetting the benefits derived from the region's redevelopment program.

When Interior Secretary Stewart Udall attempted to inaugurate a new residual oil import program at the end of March, he encountered White House opposition.

Udall had planned to place Florida and the five New England states under "open end" controls, while maintaining strict quotas for the Middle Atlantic states. In summing up the probable reaction to this new program, Udall noted that "1. The Florida and New England Congressional people will applaud long and loud. 2. The coal people and the two West Virginia Senators will undoubtedly be highly critical. 3. Congressmen from the Middle Atlantic area will be mildly disappointed."[75]

But at a press conference on the following day, Udall declared that certain legal questions had been raised about the new program and that it had been concluded from the national security aspect that it was not a viable solution legally. The Interior Secretary, therefore, took the alternate step of increasing the import quota for residual fuel oil on the East Coast by 75,000 barrels a day. Udall failed to mention publicly that both the West Virginia congressional delegation and Republican Senator Jacob Javits of New York had complained bitterly to the White House after news of the new program had leaked out.

Republican Representative Joseph Martin of Massachusetts denounced Udall's abrupt reversal on the import question as "a cruel April Fool's joke on New England."[76] The Providence *Journal* editorialized on April 2: "The only man who can blow a whistle on a member of the President's Cabinet is the President himself, and the situation cries out for further Congressional investigation."[77] Caught in the middle by this switch in policy was Venezuela, the source of most of the residual fuel oil imported into the United States; the Interior Department already had notified the Venezuelan Embassy that the American government was going to eliminate the quotas on residual fuel oil.

Shortly before the end of the year, on December 10, Lyndon Johnson issued a presidential proclamation, adjusting the import level of both petroleum and petroleum products. The proclamation, among other things, granted import allocations for petrochemical plant feedstocks. Udall had earlier advised the chief executive that the time was ripe to announce a new program, since there would be a minimum of criticism from both Congress and the oil industry. Although there was general agreement on the new program among the departments and bureaus of the executive branch, the Bureau of the Budget did not feel the program should be so extended that a new petrochemical industry could be created in Puerto Rico.

During the winter months a revision of the residual fuel oil program enjoyed top priority at the White House. Several key Johnson administration officials unsuccessfully advocated the total elimination of these quotas, but those favoring the substantial relaxation of quotas eventually emerged triumphant. On December 23, Udall sent a memorandum to Joseph Califano, a presidential aide, in which he noted, "The reaction to the press release on the liberalization of the residual fuel import program has been more favorable than could reasonably have been expected."[78] Udall then quoted Democratic Senator John Pastore of Rhode Island to the effect that liberalization was the next best thing to decontrol. The expected objections from the National Coal Policy Conference, the National

Coal Association, and the United Mine Workers were rather mild in comparison with their past attacks on the government's residual oil policy.

Joseph Califano obtained presidential approval for a new program in March, 1966, under which the Secretary of the Interior would maintain the policy that he had been following for the past several months of setting an annual residual import level of 30 million barrels, while only releasing 15 million. By implementing this in the months ahead, Udall hoped not only to demonstrate to skeptics and critics that unlimited residual oil imports would not threaten the coal industry but also to place in operation a program that would benefit small and medium consumers. In addition, the government would grant individual allocations on the basis of business done, rather than on an arbitrary formula that was not related to sales. When Udall's office announced the new program officially on March 25, it pointed out that it was consistent with advice received from both the Office of Emergency Planning and the Department of Defense. Yet, during the months that followed, the Johnson administration apparently began to have second thoughts about its policy of relaxing import quotas. Thus, at the end of the year Udall announced a program that decreased these quotas slightly for 1967 rather than increasing them significantly. While expressing relief that the influx of oil imports had been checked, at least for the moment, F. Allen Calvert, President of the Independent Petroleum Association of America, complained that the administration had not dealt adequately with the question of foreign trade zones and the increasing imports from Puerto Rico and the Virgin Islands. The independent oil companies also attacked the alleged favoritism shown to the petrochemical firms.

The debate in Congress over the government's petroleum policy again intensified during the summer of 1967. The usual regional attitudes again manifested themselves. Democratic Representative Omar Burleson of Texas came out in favor of legislation that would terminate the Canadian and Mexican overland exemptions, "plug up" all of the other loopholes, and impose an across-the-board restriction on importations of foreign crude oil to 12.2 percent of the domestic production. In contrast, Democratic Senator Edward Kennedy of Massachusetts lamented the fact that price rises in fuel oil supplies had adversely affected New England.

Having considered these opinions and others, including those of Udall, on July 17 Lyndon Johnson issued a presidential proclamation. He stated that "it is necessary to enhance the ability of the petroleum industry to provide adequate supplies of low sulphur residual fuel oil to be used as fuel" and added "it is necessary to permit the entrance of new importers and to provide for allocations which will assure that adequate supplies of low sulfur residual fuel oil to be used as fuel will be distributed to users of such products."[79] While reaffirming the power of Udall, LBJ's proclamation also revealed an awareness of the various criticisms of Udall made by Ronald F. Hornig, Special Assistant to the President for Science and Technology.

As the Johnson administration entered its last year, dissension over the petroleum policy increased rather than lessened among the officials of the executive branch. On May 1, 1968, De Vier Pierson, a White House aide, complained to LBJ that "the oil import program is a mess. While 'scandal' is too strong a term, evidence of shoddy administration, poor coordination and questionable policy decisions is growing."[80] The heart of the problem, concluded Pierson, was the exercise of final authority on petroleum matters by the Secretary of the Interior rather than the President. But instead of offering an alternate program, the White House aide merely proposed a review of the government's oil policy.

In August, the House of Representatives' Committee on Interior and Insular Affairs issued a report on the mandatory oil import control program. The report made six major recommendations:

1. The mandatory oil import program must not be weakened by use for purposes unrelated to the preservation of national security, regardless of the merits of the alien objectives.
2. All inequities in the program should be removed so that the Government is administering a plan that is equitable and fair for all who are governed by it.
3. The program should be simplified to the greatest extent possible to eliminate the present chaotic condition which causes serious inefficiencies in the planning and operations of the entire petroleum industry.
4. Rules of procedure should be strengthened and observed so as to eliminate decisions made without benefit of adequate notice and public hearings.
5. Canadian sources of crude should continue to be considered within the scope of our national security planning and therefore should receive special treatment. However, participation of Canadian crude oil in our growing U.S. market must not be disproportionately greater than the growth achieved by domestic producers.
6. It is the view of the committee that any program authorized under the national security provisions of the Trade Agreements Extension Act should be administered strictly in accordance with the purposes of that act and not extended to unrelated matters, notwithstanding the merit of such other programs.[81]

But it remained impossible to separate petroleum as an international concern from its domestic political implications. Thus, a minority report setting forth dissenting views complained that the majority report reflected "accurately the views, interests, and aspirations of the Southwestern oil producing and refining States," but failed to concern itself with "the wider impact of the oil import control program on all Americans, not just those who produce or refine oil."[82] This report was signed by Democratic Representatives John Tunney of California, Bob Kastenmeier of Wisconsin, Lloyd Meeds of Washington, Patsy Mink of Hawaii, Thomas Foley of Washington, and Hugh Carey of New York.

The official view of the executive branch was set forth in the findings that appeared in the Department of the Interior's publication, *United States Petroleum Through 1980*. The document concluded that from 1959 through 1966 the stated objectives of the mandatory oil import program had been thoroughly fulfilled. Production of both crude oil and natural gas in the United States had risen during the 1960s, and there had been only a modest increase in the percentage of imported petroleum in the total American supply. The growth rate, in fact, was much lower than that which had been operative during the 1950s. Finally, the nations of the Western Hemisphere had been responsible for 90 percent of the increase in total oil imports during the 1960s, thus reducing U.S. reliance on the more politically unstable Middle East. Despite this statistic, however, it was the withholding of petroleum exports from the Middle East to the United States and other consumers that ushered in a new era in the government's energy policy only half a decade later.

## Richard Nixon: Petroleum Imports and Offshore Oil

When Richard Nixon took office as President in March 1969, the United States was on the verge of becoming a net importer of petroleum. In addition, independent companies were playing a larger role than ever before in foreign oil operations, and because of their dependence on foreign supplies, many of these companies, especially Armand Hammer's Occidental, were in a vulnerable position. Still another negative development from the standpoint of the American government was the growth of militant regimes in Third World nations dedicated to socialist and nationalist principles.

The new President soon was confronted with still other troublesome developments relating to petroleum. In the fall Bolivia seized the holdings of the Gulf Oil Company, and in 1973 the Libyan government expropriated Nelson Bunker Hunt. Throughout the Nixon years, the OPEC nations exerted an increasingly vigorous role in global petroleum matters. Offshore oil also became a growing international concern; the United States eventually renounced its claim to much of the petroleum and gas off its coasts.

Despite these and other happenings abroad, which were making it increasingly difficult for the United States to guarantee petroleum supplies, American oil firms continued to maintain an imposing global stance in the period immediately preceding the 1973 Arab oil boycott. The five major U.S. petroleum companies, for example, controlled no less than 70 percent of the oil outside the Communist bloc; not only were they making most of their profits overseas, they were escaping taxation on most of these profits in the United States. At this time petroleum was responsible for more than a third of all direct American investment abroad, accounting for more than three-fifths of the total earnings of U.S. firms in underdeveloped countries.

The decade-long debate over the merits of the mandatory program restricting petroleum imports continued to divide the country. On January 14, 1969, the Special Representative for Trade Negotiations had recommended to President Johnson a reexamination of the oil quota program to determine whether it needed to be retained or improved. Two months later, on March 26, President Nixon appointed a task force, headed by Labor Secretary George Shultz, to study the oil import policy.

A bitter debate now ensued between the Justice Department, which wished to abolish quotas, and the Interior Department, which wished to keep them. While Assistant Attorney General Richard McLaren felt that the quota system was not only expensive but also unnecessary for national security, Interior Secretary Walter Hickel (whom Nixon later dismissed) believed it was reasonable to predicate the quota system on national security grounds.

In February 1970 a majority of the cabinet task force gave the President its report. The report advocated phasing out the quota system over a three-year period and implementing in its place a system of variable duties to become effective no later than January 1, 1971. According to George Shultz, the oil import restrictions were no longer justified on national security grounds: "Besides costing consumers an estimated $5 billion each year, the quotas have caused inefficiencies in the market place."[83]

Nixon, however, opted for a continuation of the review of oil import policy, rather than immediately abandoning the quota system. On February 20, 1970 he established an interdepartmental oil policy committee "to consider both interim and long-term adjustments that will increase the effectiveness and enhance the equity of the oil import program."[84] While the American Petroleum Institute was encouraged by the President's action and Republican Senators Clifford P. Hansen of Wyoming and John Tower of Texas were pleased, a number of members of Congress expressed their dissatisfaction in a letter to Nixon.

Although Nixon retained the oil import quota system for the time being, the decision to set the quotas on a "gap" basis (the difference between domestic demand and domestic production) led to a 40 percent rise in petroleum imports between 1971 and 1972. Another decisive factor in the rise of imports was the administration's imposition of price controls in 1971, a decision that substantially increased Middle Eastern exports of petroleum to the United States between then and 1973.

Between 1965 and 1971 the Canadian share of American petroleum imports had increased sharply, while because of increased Nigerian production, the West African share also had risen although more modestly. During the same period the Caribbean share declined. Although the import level from Venezuela had not increased greatly, that nation remained America's largest supplier. As for the Middle East, there was a considerable annual variance in imports from the various countries there. As both Middle Eastern and African crude oil was processed in either Europe or the Caribbean for shipment to the United States, the total is somewhat confused.

The State Department's Office of Systems Analysis issued a study in August 1972 calling for a new oil import policy which emphasized the national security factor. The OSA took the position that low-cost U.S. reserves were nearing exhaustion. It suggested that "import tickets for Middle Eastern oil should be given to importers in proportion to their holdings of stored oil or their development of spare productive capacity and shut-in production (unpumped excess capacity)."[85] This scheme, the OSA hoped, would enable the United States to circumvent any cutoff in its foreign petroleum supply.

On December 8, 1972, the Office of Emergency Preparedness announced that it was suspending import curbs through May 1973 on crude oil used mainly for home heating. But on April 18, while the suspension technically was still in effect, President Nixon officially terminated the mandatory controls that had been in effect for fourteen years. By the spring of 1973 the price of the delivered foreign crude oil had reached approximately the same level as that of domestic crude oil. The quota system no longer had any function other than that of distributing the foreign crude oil among U.S. importers.

A new system of license fees replaced the mandatory controls, and it was hoped that they would stimulate domestic production and refining. Actually, these license fees were disguised tariffs. Not only were there to be different rates on crude oil and gasoline, but also these rates were to increase gradually in the months ahead; moreover, there were to be no licensing fees on ethane, propane, butane, and asphalt. The vast majority of petroleum imports entered America under fee-exempt licenses during 1973, and the licenses were programmed to decrease to the vanishing point by 1980. This, then, was the oil policy of the government on the eve of the great Arab petroleum embargo, which rendered all such plans obsolete.

Turning to a different question, in 1969 the issue of offshore petroleum was before the Senate Commerce Committee's Special Subcommittee to Study U.N. Subocean Lands Policy. While Democratic Senator Claiborne Pell of Rhode Island wanted to limit American rights offshore to 50 nautical miles or a 550 m water depth, the Interior Department and the petroleum industry wanted these rights to extend to 2,500 m, the abyssal ocean floor. The 1958 Geneva convention had defined the continental shelf of each nation as including the submerged lands to 200 m, the then current technological limit of exploitation.

By 1969, however, the National Petroleum Council was predicting that within ten years it would be possible to drill and produce in water as deep as 4,000 to 6,000 feet. It thus issued a report, which was supplemented by another one two years later, declaring that "the United States, in common with other coastal nations, now has exclusive jurisdiction over the natural resources of the submerged continental mass seaward to where the submerged portion of that mass meets the abyssal ocean floor and that it should declare its rights accordingly."[86] The Interior Department, traditionally more assertive on such matters than the State Department, by this time had begun to issue leases in deeper and deeper water.

Yet, in a decision that came as a surprise to many people, in May of 1970 Nixon broke with the offshore exploiters and proposed a treaty by which the United States agreed to surrender its mineral rights in waters deeper than 200 m, or 660 feet. As a result, America in effect renounced its claim to approximately half the oil and gas off the U.S. coast, an action that greatly displeased the National Petroleum Council. Nixon's decision was a defeat for the Interior Department and a victory for the State and Defense departments which wished to enjoy the greatest possible freedom in operating off the shores of other countries. The terms of existing leases, however, remained inviolate under this treaty, including those in the Santa Barbara Channel off California, where drillers had already found oil. Given the increasing reliance of the United States on imported foreign petroleum during the 1970s, the wisdom of a policy that surrendered control over so much oil and gas must be questioned.

## The Rise of OPEC

The years from 1961 to 1973 witnessed the oil-producing nations of the world adopting a more aggressive stance toward their petroleum reserves and constructing a more elaborate organizational network. Qatar joined OPEC soon after its founding, Indonesia and Libya in 1962, Abu Dhabi in 1967, and Algeria in 1969. OPEC was to enjoy its first—if partial—success as a result of several resolutions endorsed by the delegates at its fourth conference in mid-1962. The conference declared that oil prices should be elevated to their pre-1960 levels, and that royalty payments should not be treated as credits against tax liabilities.

Five years later, in late August and early September of 1967 there was an Arab summit conference at Khartoum. The assembled delegates voted to remove the selective petroleum embargo that had been imposed on the United States, Great Britain, and West Germany. At this time the delegates reasoned that the Arab countries would only injure themselves economically by continuing the embargo, a quite different attitude from that which characterized their actions in 1973.

Largely ignored by both the Western press and the oil companies, the sixteenth OPEC conference, in June 1968, approved a number of important principles that had a significant impact on the evolution of OPEC strategy during the 1970s. Point one, for example, affirmed that "member Governments shall endeavor, as far as feasible, to explore and develop their hydrocarbon resources directly."[87] The fourth point stated that the governments of the oil-producing nations, and not the petroleum companies, should determine the posted price, and the ninth point gave to these governments the authority to hand over any dispute they might have with these petroleum firms to competent national courts.

During this same year Kuwait, Libya, and Saudi Arabia combined to establish the Organization of Arab Petroleum Exporting Countries, or OAPEC. Although Algeria, Qatar, Abu Dhabi, Bahrain, and Dubai joined this group in 1970, it remained a less important bloc than OPEC in the years that followed.

OPEC had begun to step up its offensive against the oil companies by 1970. During this year every member of the organization, aside from Indonesia, advised foreign petroleum firms that the production of oil would cease, if OPEC demands were not met. In December OPEC approved a resolution that gave the petroleum companies a mere fifteen days in which to accept its demands.

In January 1971, after holding meetings for several days in New York City, a group of twenty-three petroleum firms approached the Secretary of OPEC with a demand for joint negotiations. This bloc included both internationals and independents, as well the Compagnie Française des Pétroles, the Arabian Oil Company (Japanese), Petrofina (Belgian), and Elwerath (German). In their statement they noted, "we cannot further negotiate. . .on any other basis than one which reaches a settlement simultaneously with all the producing governments concerned."[88] The oil companies agreed to assist each other, should a petroleum-producing nation or nations attack one of them. The State Department, however, gave only lukewarm support to the oil firms' demands; the head of the Office of Fuels and Energy, James Akins, felt that this was not the time to risk a confrontation with OPEC.

As a result, it was the oil-producing countries that took the offensive, as the Tehran and Tripoli agreements of 1971 demonstrate. The first of these was signed by Iran, Saudi Arabia, Iraq, Kuwait, Qatar, Abu Dhabi, and a group of sixteen petroleum firms. Among its key provisions were a tax rate of 55 percent—a rise of 38 cents per barrel in the posted price of oil—upward price adjustments each year to compensate for the inflationary spiral, and assurances against "leap-frogging" and embargoes. "Leapfrogging" was especially distasteful, because it allowed various host countries to obtain better terms than others. The Tehran agreement lasted for only five years and only covered Gulf-exported petroleum.

The Tripoli agreement also was to last until December 31, 1975. Although it was between Libya alone and a group of oil firms, it too provided for a 55 percent tax rate, but the increase in the posted price was to be 90 cents, rather than 38 cents, per barrel. Not only did Libyan oil have a low sulphur content, but it also was nearer to Western markets than Gulf petroleum, and thus logically commanded a higher price. Other agreements in this same year were the one between Aramco and Saudi Arabia and the one between the Iraq Petroleum Company and Iraq.

This series of tax and price understandings culminated in two Geneva agreements. The first of these, reached in January 1972, raised posted prices 25 cents per barrel; the second, that of June 1973, increased them an additional 39 cents. By this time, too, the OPEC members had begun to think in terms of participation, a sharing in the operations and assets of the international oil firms. A controversy now ensued as to whether or not the Tehran agreements had postponed action on this issue until 1976; in the opinion of the State Department, any attempt to renegotiate these contracts as circumstances changed was contrary to both Western and Islamic jurisprudence. Nevertheless, in December 1972, Minister Sheikh Yamani of Saudi Arabia drew up a prototype participation agreement that was acceptable to Saudi Arabia, Qatar, Abu Dhabi, and tentatively Kuwait, calling for at least half-ownership in the petroleum firms as of 1982.

Even though it was the petroleum companies that had made the concessions to the oil-producing countries, many petroleum firm officials placed the blame for the OPEC victory on the governments of the oil-consuming nations. To quote one company negotiator:

> The Italians, French and Japanese on the whole saw the weakness of the companies, and planned separate initiatives vis-à-vis OPEC; the Americans, British, Germans and Dutch preferred to hope that the hurricane would pass without causing much more damage.
>
> But none of them saw what was really needed—a solidarity among all of them—to be a practical policy aim. The French were too anti-American, the British too bemused by their

debt to Pompidou, the Italians too much governed by Mattei's legacy of dislike for the Seven Sisters, and the Japanese too scared of the OPEC reaction for any discussion to get started toward what was needed.[89]

By this time there was a growing awareness throughout the United States that its petroleum supply from abroad was in jeopardy, with possible dire consequences for the future of the nation. As Gene Kinney observed in *The Oil and Gas Journal*, "oil supply is the jugular vein of America's greatness."[90] The most famous journalistic response, however, was a contemporary cover of *Newsweek*, which pictured a young man in desert garb holding a gasoline hose in his hand; the headline above him proclaimed "The Arab Oil Squeeze."[91] (Ironically, this "Arab" was in real life a Jewish boy from Brooklyn.)

That OPEC took such a long time to confront petroleum-consuming nations is perhaps surprising, given the fact that both Western Europe and Japan had long been importing most of their oil. OPEC, though, was set up in 1960 to stabilize rather than to raise prices and before the participating Third World nations had gained sufficient experience in mastering the petroleum trade. Moreover, with the exception of Iraq, most of these oil-producing nations were pro-United States and anti-Communist in an era during which America and Russia had yet to make a major effort to achieve détente.

Today, neither the United States nor the Soviet Union enjoy the prestige throughout the Third World that they once did. The American image was tarnished by its failure to win the Vietnam War, and Russian forces were expelled from once friendly Egypt by President Anwar Sadat in 1972. Latin American, African, and Far Eastern countries now dominate the General Assembly of the United Nations. What was unthinkable for Third World nations in 1960, or even 1965 or 1970, is not only the subject of contemplation today, it is the object of policy.

***Table 1.***
***Refined Petroleum Oil Exports, 1917–1918***

| | | **1917** | | | |
|---|---|---|---|---|---|
| *Products* | *England* | *France* | *Italy* | *Others* | *Total* |
| Gasoline | 60,443 | 68,009 | 10,721 | 4,851 | 153,024 |
| Napthas | 89,675 | 50,585 | 18,253 | 1,751 | 160,164 |
| Kerosene | 167,512 | 85,294 | 35,714 | 15,109 | 303,629 |
| Lubricating Oil | 113,384 | 60,025 | 14,812 | 892 | 189,113 |
| Gas & Fuel Oil | 89,388 | 1,675 | 31,841 | 2 | 122,906 |
| | | **1918** | | | |
| Gasoline | 78,965 | 66,726 | 36,314 | 1,074 | 183,079 |
| Napthas | 70,376 | 81,887 | 19,615 | 0 | 171,878 |
| Kerosene | 178,052 | 81,939 | 38,926 | 214 | 299,131 |
| Lubricating Oil | 106,127 | 65,601 | 20,406 | 673 | 192,807 |
| Gas & Fuel Oil | 787,909 | 4,735 | 26,553 | 1,132 | 820,329 |

SOURCE E.G. Hanson to Ida Tarbell, Washington, D.C., June 29, 1934, General Correspondence Secretary of the Navy (1930-1942), Record Group No. 80, FileNo. JSM-11/L14-2 (340626), National Archives, Washington, D.C.

# 2.

# Western and Eastern Europe

## Great Britain to World War II

Outside the Western Hemisphere, it has been Great Britain more than any other nation that has played a key role in shaping global oil policy. This is largely due to the control, either complete or partial, of two of the "Seven Sisters"—Royal Dutch-Shell and Anglo-Persian (later Anglo-Iranian and today British Petroleum). Because of the special relationship that exists between these firms and the government of Great Britain, it is impossible to separate the activities of private British oil firms from those of the British government.

Great Britain, prior to World War II, was a different England from the Great Britain of the post-World War II era. The British empire was still intact; the breakup of the colonies had yet to commence. Nevertheless the petroleum that Great Britain sought generally lay outside its colonial empire, which is one reason why the British made a major effort to strengthen their position in the oil-rich Middle East and elsewhere between the two world wars. No such search was necessary for coal, in which Great Britain is rich.

As early as 1882 Admiral Lord Fisher suggested that the British navy convert from coal to oil: "The use of fuel oil adds 50 percent to the value of any fleet that uses it."[1] Six years later Standard Oil organized the Anglo-American Oil Company, and its trust monopolized £500,000 in Anglo-American capital stock. John D. Rockefeller, however, was soon challenged by the Samuel brothers, British merchants who long had been involved in commerce with the Far East. In 1890 Marcus Samuel completed the negotiations that allowed him to market petroleum in bulk throughout the Orient; he also obtained permission for his tankers to pass through the Suez Canal, a privilege that Standard Oil did not enjoy. By 1897 Samuel had established the Shell Transport and Trading Company, so named because Samuel had once traded in seashells.

During the early years of the twentieth century, the two most important developments relating to petroleum in Great Britain were the merger of Royal Dutch with Shell as Royal Dutch-Shell in 1906 and the purchase by the British government of a controlling interest in the Anglo-Persian Oil Company in 1914. Royal

Dutch was smaller than Shell, but it possessed valuable reserves in the Netherlands East Indies, and Shell was facing increasing competition in Europe at that time. Under the terms of the merger Royal Dutch obtained 60 percent control and Shell 40 percent. Because Henri Deterding of Royal Dutch rather than Samuel became the managing director, the merger disturbed the British government, which feared German pressure on the Dutch as conditions worsened in Europe. So suspicious were many Britishers of Royal Dutch-Shell's foreign connections that the India Office prohibited Royal Dutch-Shell from obtaining petroleum concessions in Burma after oil was discovered there. A firm, Burmah Oil, directed by a group of Scots, not only won the right to develop the Burmese oil fields but also after 1910 the right to supply the British navy with petroleum. Founded in 1902, Burmah Oil also had been active in Persia.

Winston Churchill, then the First Lord of the Admiralty in a Liberal government, unleashed a bitter attack on Royal Dutch-Shell before Parliament in 1914. Earlier the British government had announced plans to buy 51 percent of the stock of Anglo-Persian, a company formed in 1909 through an alliance between William D'Arcy, who had supervised oil exploration in Persia, and Burmah Oil. Stung by this obvious favoritism toward Anglo-Persian and the equally obvious discrimination against it, Royal Dutch-Shell voluntarily furnished the British navy with petroleum during World War I at moderate prices and even outsupplied Anglo-Persian. Although the Netherlands technically adopted a neutral stance during the war, Henri Deterding made it clear that his firm was on the side of the Allies in this instance, although he later ended his life as a Nazi, living in Germany.

Neutral maritime rights also were a bone of contention during World War I. Standard Oil of New Jersey took the position that illuminating oil should not be regarded as contraband. At the start of the war the British government held up the American vessel *John D. Rockefeller*, but released it after learning that the petroleum was destined for the Danish Petroleum Company, for use in Denmark rather than for reexportation. Later in 1914, however, the British government seized other American vessels that were bound for Denmark.

To complicate matters further, Great Britain balked at releasing "innocent" goods found in the cargo of these and other captured vessels until the shippers paid freight and other charges. This attitude irritated American authorities even further.

While the lack of sufficient petroleum reserves in the British Isles following the close of World War I disturbed many people, there were those who were convinced that it was the United States which was running out of oil. Thus, Sir Edward Mackay Edgar of the British Controlled Oilfields observed in the September 1919 issue of *Sperling's Journal* that America soon would have to turn to foreign sources and that all of the known oil fields outside the United States were in British hands. "What it comes to, therefore," observed Edgar, "is that with [the] exception of Mexico, and to a lesser extent Central America, the outer world is securely barred against an American invasion in force."[2]

Edgar was by no means the only British spokesman to take such an outspoken position. First Lord of the Admiralty Walter Hume Long boldly proclaimed in the following year that "if we secure the supplies of oil now available in the world we can do what we like."[3] And A. Beeby Thompson, a noted petroleum specialist, asserted in 1921 that Great Britain had preserved its oil resources unlike the United States, which had squandered its petroleum riches.

Making a major effort to strengthen its access to overseas oil, the British government signed the San Remo Agreement in 1920 with its French counterpart. The two countries hoped to divide the Middle Eastern petroleum supply between themselves. A year earlier the French oil expert Henri Berenger had prepared a memorandum for Georges Clemenceau in which he had stated that the British policy in Asia Minor was based primarily on its quest for oil there. After the agreement was signed, Lord Curzon, the British Secretary of State for Foreign Affairs, pointed out that the current British petroleum interests in Persia had been the key factor underlying Great Britain's approval of the San Remo Agreement.

In a related development, the British Parliament enacted a law that barred foreigners from owning any petroleum lands in British territory or from even owning stock in a British oil concern. Although technically each dominion and colony had its own legislation pertaining to oil, there was little, if any, deviation from the guidelines laid down by London. Because at this time only the Board of Trade or the Minister of Munitions or persons authorized by them could prospect for petroleum, the only active driller was S. Pearson and Company, the petroleum development managers for the British government.

On the floor of the United States Senate, Republican Senator Henry Cabot Lodge of Massachusetts charged that the British government controlled Royal Dutch-Shell as well as Anglo-Persian and that "England is taking possession of the oil supply of the world.[4] Although the British government categorically denied this accusation, a number of U.S. petroleum men, irritated by developments in the Middle East and elsewhere, had begun to adopt a hard-core anti-British position. As a result of this agitation, the British oil secretariat and Lord Curzon sent a memorandum to the British Ambassador on April 21, 1921, asserting that Great Britain had an "open" rather than a "closed" door policy toward foreign petroleum interests, and that the assumption that the British controlled the oil of the world was mythical rather than factual.

A comparative examination of developments in India, the Middle East, the Far East, and Latin America, developments that are examined at greater length in subsequent chapters, clearly indicate, however, that the British petroleum policy throughout the world was hardly uniform toward American oil interests at this time. As early as 1884, five U.S. oil companies had experienced difficulties when they attempted to do business in either Burma or India. Between 1902 and 1905 the British government blocked an attempt by American petroleum firms to obtain prospecting licenses in Burma; in 1917 Standard Oil of New Jersey failed in its efforts to lease petroleum lands in India. The British government again stated that prospecting for petroleum in India was a British monopoly in 1921, and two decades later it still was discouraging American firms from obtaining oil concessions there.

Conversely, British petroleum interests were less strongly entrenched in Latin America than in the Middle East or Far East. At the time of the outbreak of World War I, U.S. protests led the Pearson group to withdraw from both Colombia and Costa Rica. President Woodrow Wilson and Secretary of State William Jennings Bryan also criticized the Huerta administration in Mexico for adopting a too favorable stance toward Great Britain and the British petroleum interests; American hostility contributed significantly to Victoriano Huerta's fall from power during the summer of 1914. At the end of World War I, in 1918, the American government again objected strongly when the short-lived Tinoco regime awarded a legally questionable concession in Costa Rica to Amory and Sons, a New York City based firm with British backing. In 1919 the State Department conducted an investigation of British petroleum operations in Argentina; in 1920 the U.S. government successfully blocked the awarding of a Royal Dutch-Shell concession in Peru; and in 1921 the State Department refrained from supporting Great Britain when British oil firms active in Venezuela encountered problems in their dealings with that country's government.

But in March 1921, the Harding administration took over, and under its new Republican leadership, the U.S. government began to take a more sympathetic position toward British oil companies in Latin America. Therefore, the State Department supported the British and the Dutch position in Venezuela during 1922 in a dispute over the Caribbean Petroleum Company concession, and in 1923 it backed a Royal Dutch-Shell subsidiary, British Controlled Oilfields, when this firm experienced difficulties in its dealings with the Venezuelan government.

It was during the 1920s, too, that the British began to moderate their hard-line approach to American petroleum companies in the Middle East. An agreement between Great Britain and Russia in 1907 had allowed the former to enjoy a sphere of interest in southern Iran and Russia a similar sphere in northern Iran. The British also had assumed an ascendant position in both Bahrain and Kuwait by the end of the nineteenth century, and on the eve of World War I they had obtained from the rulers of both countries written promises to give the British due notice before they began to develop any oil finds. By the end of the war, British troops had occupied both Mosul and Baghdad in Mesopotamia (present-day Iraq), much to the displeasure of the American government, which believed that Great Britain was using its military presence to advance its petroleum interests. Under the World War I peace treaties, the British government obtained a mandate over both Iraq and Palestine where they immediately began to interfere with American oil operations.

In part because of the more favorable American stance toward British oil activities in Latin America and in part because of continuing U.S. pressure for a role for America firms in the Middle East, the British announced at the Lausanne Conference in 1923 that they were ready to support an "Open Door" policy that would allow U.S. oil companies to operate in the Middle East more freely than before. The American government countered by announcing its willingness to recognize the Turkish Petroleum Company, which had emerged victorious in the protracted war for concessions. Even with this concession, it took five more years before the American firms and the Turkish Petroleum Company could work out the terms of U.S. participation.

Anglo-American petroleum rivalry had lessened drastically by 1928. It was in that year that Sir Henri Deterding of Royal Dutch-Shell met with Walter Teagle of Jersey Standard and Sir John Cadman of Anglo-Persian at Achnacarry Castle in Scotland. At this point the oil barons made a serious effort to eliminate the cutthroat petroleum warfare that had been erupting intermittently around the globe. This was the year of the so-called Red Line Agreement, which was signed by a group of American companies and three European firms that were already exploiting Middle Eastern oil: Royal Dutch-Shell, Anglo-Persian, and the Compagnie Française des Pétroles. Unlike the earlier San Remo Agreement, the Red Line Agreement allowed U.S. participation in developing Middle Eastern petroleum. With the Great Depression at hand, competition for world oil resources gave way to cooperation.

During this period, the main concern of the American government remained the opportunities, or lack thereof, for U.S. oil entrepreneurs in the British empire. The British Petroleum (Production) Act of 1934 not only vested in the Crown ownership of oil-bearing strata in Great Britain, but also empowered the Board of Trade to grant licenses to look for and to obtain petroleum. On December 30, 1935, the British Ambassador, Sir Ronald Lindsay, asked the Secretary of State, Cordell Hull, whether or not the American government now regarded Great Britain as a reciprocal country under the Mineral Leasing Act of 1920. Hull did not reply until June 3, 1936, but he did affirm on that occasion that: "An opinion establishing the determination of this status has now been rendered by the Attorney General indicating that under certain conditions Great Britain is to be regarded as a reciprocal country under the Act referred to."[5] The question of the eligibility of foreign nations had only arisen on a limited number of occasions since 1920, and apparently never with respect to Great Britain. On July 20 Hull terminated this series of diplomatic exchanges and informed the British Ambassador that the Interior Department was now ready to recognize Great Britain as a reciprocating country.

This type of cooperation between the American and British governments continued to occur. Prior to the outbreak of World War II, on November 17, 1938, the American government concluded a reciprocal trade agreement with Great Britain that provided for tax exemptions on petroleum products imported for use as supplies on American war vessels. In 1938 Congress began to loosen some of the restrictions which it had incorporated into the neutrality legislation it had begun passing in 1935. Its conciliatory spirit paved the way not only for this agreement, but also for the later destroyers-for-bases agreement and the Lend-Lease Act.

Elsewhere in the world, there had emerged greater Anglo-American solidarity on oil matters in the decade following the formulation of the Achnacarry Accord and Red Line Agreement. During the mid-1930s both the British and American governments protested the establishment of a petroleum monopoly in Manchukuo and the Japanese government's requirement that foreign oil firms keep a six-month supply of petroleum on hand. In 1938 when the Mexican government expropriated both British and American oil properties, both the United States and Great Britain protested.

On the other hand, during the late 1920s and the early 1930s, the British government temporarily blocked attempts by U.S. oil firms to operate in Bahrain and Kuwait. With the outbreak of World War II, Great Britain wavered in its Far Eastern petroleum policy. At first it protested the 1940 U.S. embargo on the exportation of aviation gasoline and scrap iron to Japan, but by winter it was advocating restrictions on the flow of Netherlands East Indies petroleum to the Japanese. The oil-rich Middle East remained a trouble spot throughout World War II. Great Britain feared that the United States might undermine its position in Iran and Iraq, and the United States was suspicious of expanding British influence in Saudi Arabia. Fortunately by 1944 President Franklin Roosevelt and Prime Minister Winston Churchill were able to reach an understanding on the Middle East.

**Northwest Europe: The Netherlands and France**

Recognizing that any scheme of categorization has its imperfections, the following six divisions—Northwest Europe (the Netherlands and France), Southwest Europe (Spain), World War II Axis Powers (Germany and Italy), Eastern Europe (Rumania and Albania), and the Soviet Union—offer appropriate groupings for the purpose of examining American petroleum relations with continental Europe up to the outbreak of World War II. These multinational groupings draw together nations having one or more key similarities, although obviously one might cite various differences as well.

Both the Netherlands and France played a major role in the development of Middle Eastern oil. The Dutch held part ownership in Royal Dutch-Shell, and the French totally controlled the Compagnie Française des Pétroles. Both these firms were among the participants in the Red Line Agreement of 1928.

Historically the Netherlands has been dependent on imported raw materials for its industries, since its only significant mineral resource is coal. Thus, the Netherlands have felt it necessary to maintain both a large ocean-going fleet and a large colonial empire as embodied in the Netherlands East Indies, today the independent nation of Indonesia.

Beginning in 1913, a controversy arose between the United States and the Netherlands over the opening of the Netherlands East Indies' petroleum fields to American entrepreneurs. Americans were able to obtain concessions there only after fifteen years of acrimonious bickering. The government of the Netherlands suspended the issuance of new concessions in the Netherlands East Indies in 1913. This drastic action was followed by restrictive mining legislation in 1917 and 1918 and the granting of a lucrative concession in the Netherlands East Indies to Royal Dutch-Shell in 1921.

The confrontation over petroleum concessions in the Netherlands East Indies had its counterpart in the United States. When the Shell Company of California sought a public lands lease in Utah on September 12, 1922, Secretary of the Interior Albert Fall gave California Shell sixty days to offer proof that the Dutch government would grant similar oil concessions to U.S. petroleum entrepreneurs. The application was withdrawn in the face of this challenge.

Despite this rebuff, by 1926 Royal Dutch-Shell was extracting 5,793,267 metric tons of oil annually in the United States. This figure was twice its production total for the Netherlands East Indies and approximately 37 percent of its worldwide output. Commenting on these statistics, Secretary of State Frank Kellogg observed, "The Government of the United States is quite certain that the Netherlands Government will admit that Netherlands interests have received most liberal treatment in the United States."[6]

Two years later, however, there was an exchange of correspondence between the American and Dutch governments over the treatment of U.S. oil men in the Netherlands East Indies and the treatment of Dutch petroleum firms in America. In a memorandum dated July 10, 1928, the Ministry of Foreign Affairs noted that U.S. citizens were free to exploit the petroleum fields of the Netherlands East Indies provided that they complied with Article Four of the Netherlands East Indies mining law. A week later the Dutch Minister of Colonies signed a contract with a Jersey Standard subsidiary awarding it a concession there. But as another Dutch memorandum of the same date notes, "after the granting of the concession in question the company concerned will be able to exploit its fields under complete legal security, whereas the situation of [sic] American companies with Netherlands capital which desire to obtain and exploit fields in the public lands run grave risks by the absence of legal security."[7] The Dutch government had two concerns: not only might state legislation block a claim, but also a U.S. court might invalidate a lease. There was also the possibility that an incoming President might interpret the 1920 law differently from his predecessor.

Officials in the Netherlands sought either a formal treaty with the United States regarding this matter or various modifications in the American law. U.S. Minister to the Netherlands Richard M. Tobin countered with a declaration that the Netherlands was a reciprocating nation under the Minerals Leasing Act of February 25, 1920. The State Department asserted that Dutch fears were groundless for three reasons: first, a state could not interfere with oil leases on public lands of the United States; second, challenges to concessions rarely reached the courts; and third, regardless of a change in presidential administration, a twenty-year lease of federal public mineral lands would only become inoperative because of noncompliance with its terms. With the air thus cleared, this issue did not attract the attention during the Hoover and Roosevelt administrations that it had during the Harding and Coolidge ones.

A decade later the outbreak of World War II raised the possibility that the petroleum fields of the Netherlands East Indies might fall into the hands of the Japanese. By 1940 the American government had come to fear that the Dutch might destroy the oil properties there to prevent this from happening. The State Department also worried about the level of the Netherlands East Indies petroleum shipments to Japan, but it refrained from guaranteeing that the United States would defend the Netherlands East Indies should oil-hungry Japan attack it. Two months after Pearl Harbor, the Japanese occupied the Netherlands East Indies only to find the Standard-Vacuum refinery at Palembang, Sumatra, in ruins.

The American government became interested in the petroleum policy of its French counterpart far sooner than it had in that of the Netherlands. In 1890 the U.S. government became highly upset when the Budget Committee of the French Chamber of Deputies recommended an increase in duties on mineral lubricating oils. It was because of the French tax laws that Standard Oil decided to keep Bedford et Compagnie, a partnership registered in France during the summer of 1893, separate from the American firm; the first action of Bedford et Compagnie was to establish a modern refinery at Rouen. During the following year Standard Oil and fourteen French refiners reached an agreement under which Standard Oil would furnish the refineries with crude petroleum.

France, however, produced only a small amount of oil, and her coal fields did not match those of Great Britain, Poland, or Germany. Thus, it is not surprising that the French attempted to expand their influence throughout the petroleum-rich Middle East following World War I. They willingly signed such government understandings as the San Remo Agreement of 1920 with Great Britain and such private understandings as the Red Line Agreement of 1928. Although the French empire was a large one, it was not until after World War II that oil was discovered in Algeria. Prior to that time, France had to look elsewhere for its petroleum.

The French government approved a new law in 1919 that nationalized all mineral resources but made no specific mention of petroleum. During the following year the State Department in a report to the Senate concluded that: "The French Government exercises wide discretionary powers in the granting of concessions. This wide discretion makes possible discriminatory action but there is no evidence of its operation against citizens of the United States."[8] Six years later, in 1927, France asked the assistance of Jersey Standard in establishing a national marketing cartel, but the American firm rejected the offer. In the opinion of its President, Walter Teagle, the Russians were spending money freely in France and an established oil monopoly there might become an outlet for Russian petroleum.

In 1928 the State Department made informal representations over possible discrimination against American oil imports into France. A bill then before the French Parliament would have based future importation licenses on a firm's average annual imports in the last five years rather than on the maximum figures. Such an act would give preference to French refineries and French importing companies in supplying surplus allotments to meet increased consumption. Fortunately for American petroleum interests, these provisions did not appear in the final version of the bill, which was known as the Loi Poincaré. Prime Minister Raymond Poincaré and Foreign Minister Aristide Briand had opposed what the United States had found objectionable in this bill. Briand informed American Ambassador Myron Herrick that the general policies of the U.S. and French government were not likely to conflict. However, Loi Poincaré did impose a tariff on refined products.

Teagle's fears over a possible French petroleum monopoly gained additional credence early in 1933, when the socialists presented a proposal to the French Parliament, calling for the establishment of a monopoly over oil imports into France. As early as 1919, in response to inquiries from the American embassy, the French government had stated that the establishment of a monopoly was an internal matter. The only exceptions were those cases in which the rights of Americans or other nationals were jeopardized or preferential treatment was given to exporters of certain nationalities. The U.S. oil companies then operating in France were especially disturbed by the proposed monopoly because working under twenty-year leases, they already had spent approximately $50 million on import facilities and refineries.

Protests from both American and British diplomats temporarily halted action on the 1933 proposal. The French press itself, aside from the far Left, had not been favorably disposed toward government monopolies. When the Chamber of Deputies approved a measure creating a commission to investigate the feasibility of a petroleum monopoly, Paul Reynaud protested that this plan, if adopted, would cost the government several billion francs in indemnities for broken contracts. The bill encountered greater opposition in the Senate, but a compromise measure, calling for the establishment of a commission to make a report on oil, eventually became a law. The commission finally rejected a plan less drastic than the originally proposed monopoly, and American petroleum entrepreneurs were able to operate in France down to the time of the German invasion.

**South West Europe: Spain**

Unlike France, Spain remained neutral during both World War I and World War II. In 1923 Primo de Rivera seized dictatorial power in Spain, but after the bitter civil war of 1936-39 General Francisco Franco emerged with the backing of the Falangists as the leader of the Spanish nation for the next generation. Only in the mid-1970s did Spain move back again toward democracy.

Although Spain does possess certain important natural resources, its coal deposits are of low quality and difficult to extract. Over the years, the country has been forced to import its oil from abroad, as its empire also lacked petroleum deposits.

During the middle years of the Rivera dictatorship American interest in Spanish petroleum caused a flurry of diplomatic activity. On June 30, 1927, the the government of Spain established a petroleum monopoly under the Spanish abbreviated title of Campsa. It also took over certain oil properties belonging to such American firms as the Vacuum Oil Company and Jersey Standard, which at that time held an ascendant position in Spain over Royal Dutch-Shell. The initial reaction of the U.S. Department of State emphasized three major points: (1) Spain or any other country had the right to establish such a monopoly; (2) the U.S. government would continue to attempt to protect American property rights abroad and would demand complete and fair compensation in the event of expropriation; and (3) the Spanish oil monopoly might prove unworkable. The State Department reasoned that it would be premature for the American goverment to discuss U.S. property rights with Spanish officials in case the latter should adopt a hard-line approach. The French government was especially disturbed by the creation of the Spanish monopoly; it held that the Spanish action was in violation of existing Franco-Spanish treaty rights.

In attempting to explain the reasons underlying the Spanish decision to establish a monopoly, Philadelphia District Manager Edwin B. George of the Bureau of Foreign and Domestic Commerce wrote to John H. Nelson of the Minerals Section on July 28, 1927, that:

> [there was a] conviction on the part of a number of people in responsible positions that the entire Spanish attitude has been provoked by the squabbling of the Standard Oil of New Jersey throughout the past two years. The alleged attempt of the latter to defeat the Spanish tariff by a sort of mock synthetic refining process in Spain itself has apparently aroused keen and lasting resentment on the part of the Spanish Government.[9]

Both Jersey Standard and Royal Dutch-Shell were reluctant to deal with the Spanish oil monopoly, which they feared would offer high prices and inferior service. During an interview with U.S. Ambassador to Spain Ogden Hammond in November 1927, Primo de Rivera asserted that not only would the American firms receive fair compensation, their properties to be valued at $30 million, but also that he always had favored U.S. interests. Later that month the Spanish government seized the Alicante installations of the Standard Oil subsidiary in Spain, as well as those of two Royal Dutch-Shell subsidiaries, presumably for use in conjunction with the storage of monopoly-held petroleum supplies.

By the beginning of 1928 the Vacuum Oil Company and another U.S. firm, the Atlantic Refining Company, had expressed a willingness to deal with the Spanish petroleum monopoly. The activities of the Valuation Commission, however, proved disturbing to all oil companies because of their fear that it would offer inadequate compensation. By May American and French diplomatic pressure had forced the Valuation Commission to agree to allow practically in full the claims for physical property asked by the companies, but it remained reluctant to offer goodwill compensation in addition to this.

Ambassador Hammond then complained about the award the Valuation Commission had made before adopting a new set of rules in February. These earlier awards included onc involving a Jersey Standard subsidiary, Babel and Nervion. Hammond requested that awards be set aside until they conformed with the new guidelines. During the summer, however, Babel and Nervion, as well as Royal

Dutch-Shell, settled their physical property claims with the Spanish government. France now asked for international arbitration of the amount due French owners for their oil properties in Spain, since the local courts, in the opinion of the French, were too subject to political pressure. The U.S. government offered only moral support to France's moves. American firms were favorable to arbitration, since there was a minority of French stockholders in Babel and Nervion; thus an international arbitral ruling would affect American petroleum claims in Spain as well.

On December 8, 1928, Royal Dutch-Shell received its final payment from the Spanish government of £1.1 million for its properties. Yet, at the same time it sent a legally dubious petition to Primo de Rivera, claiming that the settlement had been made under duress and that it thus reserved all of its rights in the event that other foreign concerns obtained more favorable treatment. In May 1929 Babel and Nervion accepted a Spanish offer of approximately 25 percent for goodwill value over the physical worth of its properties in Spain. Two French oil companies, however, decided to hold out for 30 percent goodwill compensation, and the French Foreign Office informed Standard Oil that the claims of the three companies had to be considered jointly. During the following month the Spanish government agreed to increase the overall financial settlement, and since Standard Oil was willing to give the French companies part of its own *pro rata* claim, the matter came to a close.

Oil did not again become during the decade that followed a focal point in diplomatic relations between America and Spain. The Spanish Civil War failed to cause the furor over petroleum supplies in the United States that the Italo-Ethiopian War did. In 1939 the Franco regime decided to limit the foreign participation in the Spanish petroleum industry to 25 percent. It was only after World War II had broken out that the American government began to show concern over supplying oil to neutralist Spain and Portugal for possible reshipment to the Axis Powers.

## Two Emerging Totalitarian States: Germany and Italy

Although Germany and Italy fought on the same side during World War II, during World War I they were on opposite sides. While Germany was one of the Central Powers, Italy remained neutral until 1915, when it joined the Allies. The postwar peace treaties, however, proved to be as unacceptable to expansionist Italy as they were to defeated Germany. The frustrations experienced by both nations were among the reasons why extremist political groups assumed power. In 1922 Benito Mussolini and the Fascists took over in Italy, while Adolf Hitler and the National Socialists (Nazis) followed suit in Germany in 1933.

The end of World War I found Germany still an industrial nation but badly crippled by its loss of an overseas empire and by the forced cession of territory in Europe to other countries. There already was considerable state participation in German industry, and it was this long-standing existence of a "mixed" economy that facilitated the transition to the Hitler policies after 1933. The German government was much more sympathetic toward cartels than the American government. I.G. Farben, which controlled 400 firms in Germany and 500 abroad, monopolized the production of both synthetic gasoline and rubber without encountering governmental difficulties. After the Nazis came to power, they subjected the economy of the country to rigid experimentation and assumed increasing control over German industry. Germany enjoyed the advantage of rich coal deposits, but not of petroleum.

American oil entrepreneurs were concerned about developments in Nazi Germany, but their activities in that country, especially the involvement of Standard Oil of New Jersey, predated World War I. On the eve of that conflict there was talk of ousting the giant U.S. petroleum firm and establishing a government petroleum monopoly similar to the tobacco one then operating in France. That such a move could prove quite risky for Germany was pointed out by the *Hamburger Nachricten*, which warned, "It is certain that the German Government will feel the cudgel on their shins if they dare to show their teeth to Mr. Rockefeller and his Oil Trust."[10] In February 1917, with American entry into World War I only two months away, Jersey Standard agreed to sell its shares in its German marketing and refining affiliate, Deutsche-Amerikanische Petroleum Gesellschaft, to its German partner. In return the American firm obtained as collateral securities held by its German partner in the United States.

After the end of World War I, a number of American oil men decided to invest in the Weimar Republic, despite the massive problems confronting it. Henry L. Doherty was one American oil baron who did not care to take advantage of its petroleum potentialities; in 1930 he elected not to exercise his option there. Nevertheless, the oil resources of Northwest Germany had considerable appeal for many Wall Street promoters.

A year later the German government awarded the contract for building the first large refinery in the country to Arthur G. McKee and Company of Cleveland. The refinery's capacity was to be 60,000 to 70,000 tons, and its estimated cost was $1 million. The Prussian government, however, also was a participant in the emerging German petroleum industry. Thus the Preussische Bergwerks und Huetten A. G., which already owned coal and potash mines, acquired an oil unit. The federal post office (Deutsche Reichpost) signed a gasoline contract with the oil companies, and, pending the completion of the new refinery, the Cologne Sinclair subsidiary was to meet the post office's needs.

German interest in both U.S. and other foreign oil firms continued to grow. In July 1934 the Secretary of State for Economy summoned to Berlin representatives of Jersey Standard, Royal Dutch-Shell, and Anglo-Persian. The German government had earlier required the maintenance of a four month supply of petroleum; it now asked that proceeds from all sales remain in the country for five years. Jersey Standard and Anglo-Persian opposed this new scheme, but the Shell interests were somewhat more receptive. To complicate matters, Shell made its counterproposal to the German government contingent upon acceptance of its terms by Jersey Standard and Anglo-Persian, but the two firms were in total disagreement over what should be done with the fuel made from coal.

The German Dye Trust, obviously acting as the spokesman for the German government, suggested that the three petroleum companies deliver to it oil products worth £4 million within five years, a quantity of approximately one million tons. The advice of the State Department on the proper course of action for Jersey Standard, as formulated in a memorandum written by Economic Advisor Herbert Feis, was that "this Government does not wish to intervene in its business relationships."[11] If the question of protection arose, Jersey Standard probably would be dependent upon the British government's goodwill.

In a later memorandum, Assistant Economic Advisor Frederick Livesey noted that the specific items involved in the proposed transaction were aviation-quality gasoline and lubricating oil. Livesey pointed out that the three petroleum firms would slow the German drive for self-sufficiency in oil by supplying these items and at the same time prevent Rumania and the Soviet Union from acting as suppliers. Yet, when the German contract with the three petroleum companies was settled in December 1934, it specified the delivery on a cash basis of only 400,000 tons of oil over the next eighteen months with further discussions over 600,000 additional tons to take place in 1935. This procedure greatly reduced the possibility that Germany might obtain large quantities of petroleum without paying for it just before a crisis.

Whereas in 1934 American petroleum companies sold to Germany $12 million worth of petroleum products, by 1938 this figure had nearly tripled to $34 million. Motor fuel and lubricating oil were particularly in demand. As a result, the United States joined Russia, Mexico, and Rumania as one of the leading suppliers of petroleum to the German navy. Aside from Jersey Standard and its German subsidiary, the Deutsche-Amerikanische Petroleum Gesellschaft, the other U.S. firms active in Germany included Socony-Vacuum, the Atlantic Refining Company, and the Pure Oil Company.

Germany attacked Poland in September 1939, and World War II began. It was during that month that William Rhodes Davis, a wealthy independent American petroleum entrepreneur, sought Roosevelt's approval for a mission to Berlin and Rome, which he hoped would lead to American mediation of the conflict. Davis, though, was hardly a neutral observer; the war was threatening his large-scale oil sales to Germany, which had been arranged with the assistance of President Lázaro Cárdenas of Mexico and Dr. Hertslet, the representative of the German banking and industrial interests. Davis did receive permission at a White House conference on September 15 to go to Berlin. But in his talks with Hermann Goering, the German leader stressed the need for a world conference to liquidate the Versailles Treaty and to allow Germany to reassume its rightful place in the world order. To arrange such a gathering obviously was beyond the capacity of Davis.

The president of the Texas Company, the Norwegian-born Torkild Rieber, was another American oil entrepreneur who suffered widespread criticism for his dealings with Nazi Germany. Throughout 1937 Rieber had petroleum shipments diverted from Belgium to Spain, which was then torn by civil war. After the U.S. Attorney General, Homer Cummings, complained about this violation of the neutrality legislation, Rieber had the shipments continued through Italy. Rieber also agreed to supply Germany with petroleum from Colombia, routing the tankers through neutral ports; these shipments continued even after the outbreak of war in Europe in September 1939. When Rieber asked the Germans for three tankers, Goering requested diplomatic assistance in return . Rieber went so far as to approach FDR with a "peace plan" designed to undercut the British, but the President advised him to sever his German contacts. Shortly thereafter, the New York *Herald Tribune* published a story exposing Texaco's financial support through Reiber of Dr. Gerhardt Westrick, a German agent in New York City, and angry Texaco stockholders forced Rieber to resign.

Surveying the two interwar decades as a whole, one of the most controversial aspects of this era was a 1926 agreement between Jersey Standard and I.G. Farben that provided for an exchange of patents and information. Although this exchange allowed the American firm to develop later a number of products critical to the war effort, including synthetic rubber, it also permitted Farben to obtain the patent for tetraethyl lead, a vital item in the manufacture of hundred-octane aviation fuel. This scientific dialogue continued after the German attack on Poland.

Twenty-six months after the United States had entered World War II as a participant, Jersey Standard became the object of widespread criticism. Because of its reliance on the German firm, Jersey Standard had held back the development of synthetic rubber in the United States. To Democratic Senator Harry Truman of Missouri, who chaired a Senate committee investigating national defense, this example of trans-Atlantic cooperation was treasonous. By this time the Justice Department had brought suit against Jersey Standard for making restrictive agreements with I. G. Farben. The government eventually agreed to a settlement under which Jersey Standard would pay a $50,000 fine and release its patents. Obviously the safest course of action for the American firm would have been to repudiate its Farben agreement once war had broken out in Europe.

On the eve of World War II in 1938, Germany was consuming 44 million barrels of oil each year, a total amount dwarfed by Great Britain's 76 million barrels, the Soviet Union's 183 million barrels, and the United States' 1 billion barrels. In that year Germany imported 28 million barrels of petroleum from overseas and 3.8 million barrels overland from Europe, mostly from Rumania. In addition, it produced 3.8 million barrels of oil domestically and manufactured 9 million barrels synthetically. During 1940 and the first half of 1941 the Germans seized 5 million barrels of petroleum when they occupied Norway, Belgium, the Netherlands, and France and imported an additional 5.6 million barrels from the Soviet Union. Even so, the impending exhaustion of Nazi oil stocks was a major factor in the German decision to invade Russia in the summer of 1941.

Unfortunately for the Germans, they only were able to capture the Maikop petroleum fields of the Soviet Union, rather than the larger fields and refineries of Grozny and Baku. As a result, Germany was forced to rely more and more on Rumania, but Allied bombing raids eventually crippled the Ploesti oil fields and refineries. Germany, however, still had its synthetic fuel output, which increased fourfold from 9 million barrels in 1938 to 36 million in 1943, approximately one-half of the total petroleum supplies available to Germany in 1943. Allied air attacks on Nazi hydrogenation plans began in May 1944 and into the spring of 1945 with devastating results. After the war, however, the Americans discovered that the Germans had been engaged in building complete oil refineries underground in huge man-made (and hence bomb-proof) caverns carved out of mountain interiors.

Turning to Italy, the focal point of Italian-American petroleum relations during the interwar decade was the Italo-Ethiopian War, although the United States remained officially neutral. As early as 1871 the American Minister at Florence, George P. Marsh, complained about the imposition of import duties on U.S. petroleum at Italian ports. The Italian government apparently levied these duties in retaliation for the high duties imposed by America on marble, wines, and various other products from Italy. This little-known episode marks the first extended reference to oil in the *Foreign Relations* series.

The end of World War I found Italy physically exhausted with a badly strained economy. In October 1922, when the Italian government failed to provide needed leadership, Benito Mussolini and the Fascists came to power, following their so-called March on Rome. Under the direction of Mussolini employers and employees organized separate syndicates; representatives on employee syndicates served on corporation councils along with representatives of the Fascists. Mussolini also set up a state petroleum monopoly.

Prior to to World War II there was little governmental control over petroleum firms from abroad, although these enterprises controlled approximately three-fifths of the Italian market. After the onset of the Great Depression, however, intervention in the private sector increased. By 1934 Mussolini, operating through two dozen key corporation councils, had fully regimented the Italian economy. It was during this year that a licensing act took effect, encouraging both Italian and foreign oil companies to build refineries there.

For four decades Italy had been brooding over its 1896 defeat at Adowa by Ethiopia. By the mid-1930s Italy was girding for a war of honor against the East African state. In early 1935 Mussolini attacked Ethiopia and by May 1936 had revenged Adowa with a military victory.

Until the mid-1930s the question of Ethiopian petroleum did not loom large in American diplomacy. In 1915 the Anglo-American Oil Company, which was under the control of Standard Oil, had obtained a concession there. But one of the conditions of the grant was that work would commence within five years; unfortunately, World War I delayed the commencement of operations. After World War I Anglo-American discovered that its concession was relatively worthless, and a rival concern backed by British capital eventually obtained half of the territory originally granted to Anglo-American.

But then in August 1935 a furor erupted when Ethiopian Emperor Haile Selassie made available a petroleum concession to a subsidiary of the Standard-Vacuum Oil Company. (Because the British Oil promoter Francis M. Rickett was active in the negotiations, it is sometimes referred to as the Rickett oil concession; previously he had offered it to British petroleum firms without finding a taker.) It was quite apparent that Haile Selassie hoped to gain official U.S. assistance in his war against Italy by granting the concession. The reaction of the State Department, though, was negative: "The granting of this concession had been the cause of great embarrassment, not only to this Government but to other governments who are making strenuous and sincere efforts for the preservation of peace."[12] Standard-Vacuum officials were then quickly persuaded to withdraw immediately and unconditionally from the concession.

Selassie's reaction was one of frustration and rage. He complained to the American Minister there, Cornelius Van H. Engert, that Ethiopia had granted the concession as a gesture of friendship toward the United States, not just to take advantage of U.S. technology. Secretary of State Cordell Hull instructed Engert that the pressure the American government had placed on Standard-Vacuum was not a departure from established policy but rather designed to be helpful to the cause of peace. Critics of the State Department pointed out that the only way in which the department had made peace more possible by its interference was by speeding up the date of the Italian victory.

An even more explosive issue was the question of American oil shipments to Italy, which had signed a treaty of commerce and navigation with the United States in 1871. On September 20, 1935, Secretary of State Hull asserted that the American government "would not join in the imposition of sanctions upon any nation involved in the pending controversy between Italy and Ethiopia."[13] Later, however, Hull did advocate a moral embargo on certain exports. Then on October 6, President Franklin Roosevelt officially recognized that a state of war existed between Italy and Ethiopia and informed the American people that they would trade at their own risk with belligerents. Three days later the Export-Import Bank announced that credits or loans would no longer be extended for trade with belligerents. Finally, on October 11 the League of Nations indicted Italy as an aggressor.

Even with these actions, American trade with Italy increased rather than decreased during October. Shipments of petroleum, copper, trucks, tractors, scrap iron, and scrap steel grew and continued during November. Direct U.S. exports of gasoline to Italian Africa, for example, escalated from 25,714 barrels in October to 110,109 barrels in the following month; American gasoline exports to Italy itself rose from none in October to 78,284 barrels in November. During this two-month period exports of lubricating oil direct to the war zone and exports of gas oil to Italy also increased.

One of the more outspoken critics of American petroleum shipments to the

Mussolini regime was Secretary of the Interior Harold Ickes. On November 21 he suggested that U.S. producers keep "the spirit and letter" of the 1935 Neutrality Act by refraining from selling oil to Italy, but as the Neutrality Act did not prohibit such shipments, Ickes had to withdraw his protest. Slightly more successful in its protest was the Shipping Board, which blocked the *Ulysses* from sailing for Italy with a cargo of oil. Companies owing money to the Shipping Board were at the mercy of this government agency; but, since only two tanker companies were in arrears, most U.S. firms could sell petroleum to Italy without interference.

Although Italy was winning its war against Ethiopia, the Italian government was not pleased with the actions of the United States. Mussolini himself observed that "those whom we helped in the World War are conspiring against us."[14] The Italian Ambassador to the United States, Augusto Rosso, complained to an American official that the moral embargo was only injuring Italy. The United States, he claimed, had thrown in its lot with Great Britain and the League of Nations against his country. Italian-Americans began to pressure Washington; newspaper editorials attacked the American neutrality policy, while a deluge of letters inundated the U.S. government from such organizations as the First Ward Italo-American Democratic Club of Philadelphia; the Italian-American Federation of Societies and Clubs of Waterbury, Connecticut; and Cincinnatus Lodge No. 1191, Order Sons of Italy in America. Most Italian-Americans, it must be remembered, were Democrats.

If Washington did not make a total effort to embargo oil shipments to Italy, neither did the League of Nations, which was meeting in Geneva, Switzerland. Nevertheless, on January 12, 1936, a League committee of experts reported that the American policies were an obstacle to the imposition of oil sanctions against Italy; it pointed to the fact that during 1935 the percentage increase in American petroleum exports to Italy was greater than those of the other leading oil suppliers, such as the Soviet Union and Great Britain. Yet the League itself postponed consideration of oil sanctions against Italy on four occasions: November 2 and December 12, 1935, and January 20 and March 2, 1936.

Representative Samuel McReynolds of Tennessee and Senator Key Pittman of Nevada, both Democrats, were two advocates of a more vigorous anti-Italian policy by the United States. On January 3, 1936, they introduced bills in Congress that would have allowed the President to hold the trade in important raw materials with both Italy and Ethiopia to prewar averages. Unfortunately for McReynolds and Pittman, such powerful isolationist Senators as William E. Borah of Idaho and Hiram Johnson of California felt that these bills would surrender the right of freedom of the seas; in addition, various export-import organizations and manufacturers were not enthusiastic. As a result, on February 12 the Senate Foreign Relations Committee voted to extend the 1935 neutrality legislation.

On February 29, 1936, President Roosevelt announced that the American government was now assuming that a state of war existed between Ethiopia and Italy. It was perhaps a rather late date to take such a step, and less than four months later, on June 20, FDR declared that the state of war no longer existed. In so doing he ended the arms embargo and withdrew the earlier warning to American citizens concerning travel on belligerent ships. At the same time the rather ineffective moral embargo also was terminated. During this June, too, the State Department finally ended the 1871 treaty of commerce between the United States and Italy. The League of Nations was slower to recognize that hostilities between Italy and Ethiopia had halted; it was not until July 4 that it voted to lift sanctions against Italy.

Having won the war, the Italian government apparently grew less critical of the American version of neutrality. When Democratic Representative Fred J. Sisson of New York asked the White House on October 14 for a copy of the letter that Mussolini had sent FDR, congratulating him on refusing to apply oil sanctions against Italy, Assistant Secretary to the President Marvin H. McIntyre replied that there was no such letter. McIntyre, though did advise Sisson, who wanted a copy of the letter to reassure Italian voters in his district, that:

> Signor Mussolini in a conversation in May with a member of our Embassy staff, expressed appreciation of the position of neutrality taken by this Government during the Ethiopian crisis and, on June 23, the leading Rome newspaper, the *MESSAGGERO*, in an inspired editorial, acknowledged with gratitude the American attitude of strict neutrality and pointed out that the United States had resisted all pressure to subscribe to embargoes on metals and oils.[15]

While it is true that British and Russian petroleum sales to Italy did decline between 1934 and 1935, the Italian government obtained so much oil from countries other than the United States that an American petroleum embargo would not have had a significant impact as a "lone wolf" gesture. A universal oil embargo against Italy, however, might have been quite effective, as Mussolini admitted to Hitler when they met in the fall of 1938.

## Rumania, Albania, and Eastern Europe

Aside from various activities in Russia during prior decades the European oil industry began in Rumania in 1857, two years before Edward L. Drake discovered petroleum in Pennsylvania. During that year hand-dug wells in Rumania produced 275 tons. But it was not until 1900 that Standard Oil tried to obtain a concession there; on this occasion critics of the Rockefeller firm claimed that the Rumanians were "selling their heritage for a mess of potage and would give themselves over into the hands of a monopolistic combination which would endanger their economic independence."[16] Nevertheless, by 1902 Jersey Standard was purchasing large amounts of Rumanian petroleum for resale elsewhere, and in 1904 it organized Romano-Americana, a wholly owned subsidiary.

Halfway through World War I, in 1916, Rumania entered the war on the Allied side. The three leading oil companies then in operation there were the Astra Romana (Royal Dutch-Shell), the Steaua Romana (German), and the Romano Americana (Jersey Standard). Between 1914 and 1916, the U.S. firm had entered into several large contracts with the German government. At the end of 1916, however, the new British Prime Minister, David Lloyd George, sent Major John Norton-Griffiths of the Allied War Council to Rumania. Norton-Griffiths persuaded the Americans to plug their wells with scrap iron, to dismantle their tank "farms," and to destroy the machinery in the Ploesti refinery. In return Griffiths offered a written guaranty that Romano Americana would be reimbursed for its losses.

Following the close of World War I, the Germans withdrew from Rumanian oil, leaving the United States and Great Britain in a dominant position. But in 1920 Great Britain and France signed the San Remo Agreement, pledging to act together in diplomatic support of the claims of their nationals to the petroleum fields of Rumania. While the United States protested this as a violation of the "Open Door" principle, the British and Rumanian governments reached an agreement to the effect that the compensation Great Britain owed to Rumania for the destruction of its oil properties should be subtracted from the debt the Rumanians owed to the British.

Between the two world wars, there was an intense struggle over foreign petroleum rights in Rumania. In 1921 the Rumanian Parliament passed an agrarian law that gave embatic (perpetual lease) owners full ownership of the land covered by their leases. The government soon began to advance the theory that embatic owners were only entitled to surface rights; subsoil rights, the government theorized, belonged to the state. In a major defeat for the state, however,the Rumanian Supreme Court ruled in its Decision No. 81, dated February 29, 1924, that the embatic holder possessed the subsoil of embatic lands. Nationalistic elements now mounted a counterattack against the Jersey Standard subsidiary; a leading oil periodical of Rumania, the *Romana Petrolifera*, spearheaded the offensive in a series of articles entitled "The American Peril." The paper complained that "The Romano-Americana excluded Rumanians from the staff of its employees notwithstanding the fact that it is living on Rumanian soil. Its officers and most of its employees are Americans who have no knowledge of the Rumanian language."[17]

In such a supercharged atmosphere the Parliament passed a new mining law on June 30 that provided for the nationalization of the subsoil of the nation. The state received the right to grant oil concessions even on privately held lands, but no concession was to last more than fifty years. Concessions granted before the passage of the new mining law were to expire in 1957, in the event that the original grant extended to that date or beyond. Within ten years Rumanian citizens were to hold at least 55 percent of the stock of every oil company; two-thirds of the board of directors of each petroleum firm were to be Rumanian nationals, and the government of Rumania possessed the right to fire the manager of any petroleum company.

The new mining law brought an immediate protest from Peter Jay, the American Minister to Rumania, who then left Bucharest for a trip home. British,

Dutch, Belgian, and French diplomats were equally irritated. The United States, moreover, had a formidable weapon against this law in the unpaid $44 million war loan that it had made to Rumania. Acting Secretary of State Joseph Grew tartly observed in a letter to Chargé Riggs dated October 15, 1924: "It is difficult to understand how in an economic situation in which it is said that funds are not available for the payment of existing debts it is yet possible that there will be available the substantial sums necessary for the purchase of important American holdings as apparently contemplated under the law."[18]

Even before the Rumanian Parliament had passed the new mining law, there had been complications over the claims Jersey Standard held against Great Britain for the destruction of its properties in Rumania in 1916. That fall Jersey Standard recommended subjecting the matter to arbitration, but the suggestion proved most distasteful to the British. British Secretary of State for Foreign Affairs J. Ramsay McDonald stated that he did not see how his government could accept arbitration unless the French and Russian governments also participated. McDonald went on to say that were Great Britain to pay off Jersey Standard, it would have to make similar payments to other foreign interests as well.

During subsequent months the American government continued to emphasize the British responsibility for the destruction of the U.S. oil properties in Rumania, and the British continued to deny that they were in any way liable. Great Britain's position in this matter was strengthened by a statement that the Rumanian Finance Minister made to the Chamber of Deputies on February 11, 1925: "From the very beginning we laid down the rule that we did the destruction in our territory and that we make compensation. We pledged ourselves to those who suffered damages and hence, the sums that become due under that head you must deliver to us."[19] For Rumania to allow Great Britain to negotiate with foreign firms, he continued, would be a violation of the country's sovereign rights.

Having thus placed full responsibility for the destruction of the U.S. oil properties on the Rumanian government, the Minister of Finance then completely undercut the American claims. He observed that "the principle that 'war losses' do not give rise to a legal right to compensation is not limited to war losses in the sense of loss or damage inflicted by the enemy, but covers also loss and damage which the commander in the field is himself obliged to inflict upon the owner of property in the area under his authority."[20] The American response to this claim was that the British military had brought about the property destruction in Rumania against the wishes of the Rumanian authorities.

The Rumanian government, far from being intimidated by the U.S. challenge, then promulgated a new law in April, 1926, which took the position that the subsoil rights to embatic lands belonged to the state. Furthermore, this measure was made retroactive to June 1921. Given these circumstances, it is difficult to understand why the Rumanian government made an agreement with the Eldorado Company early in 1928, that recognized the rights of this firm over embatic lands acquired prior to 1926. In return, however, the Eldorado Company was to pay 4 percent royalty to the state in addition to the 2 percent government tax and to the sum paid to the peasants.

When Rumania approached Romano-Americana with a similar contract, the American firm balked, complaining that it was paying higher royalties to the lessors and to the peasants than Eldorado, apart from the 2 percent tax to the state. It was not interested in an agreement that would force it to pay an additional 4 percent to the state. The Rumanian government replied that it would not budge on the 4 percent royalty in the case of Romano-Americana but that it would pressure the peasants to reduce the payments due to them from the American firm.

By the early part of 1928 relations between the Rumanian government and Romano-Americana had become more harmonious. Officials in Bucharest notified the American legation that they would restore to the American firm its subsoil rights in embatic lands acquired prior to the interpretive law of 1926. The question of the royalties, though, remained unsettled. Unfortunately, the friendly atmosphere evaporated in June with the publication of extensive mining law regulations concerning the methods of exploitation of petroleum and gas. The focus of the controversy was a provision to the effect that 75 percent of the staffs of all oil companies, whether or not they held concessions from the state, were to be Rumanian citizens. In addition, the Ministry of Industry would determine whether or not the employees of these firms would continue in their service and in what capacity.

Mr. Hughes, the Director of Romano-Americana, replied that 95 percent of the employees of foreign companies in the country were indeed Rumanian subjects. For that nation to demand that the staffs of these firms be made up of 75 percent Rumanian nationals would wreck the oil industry, as only foreign specialists had the technological know-how to bring about modernization. Significantly, within a month the Rumanian government announced another, and less severe, set of mining law regulations.

Unfortunately, Romano-Americana was bargaining from a position of growing weakness rather than increasing strength. Its share of the domestic crude production in Rumania had decreased from 22 percent in 1921 to 7 percent in 1926. This decline, according to Jersey Standard, was the result of the discriminatory restrictions the Rumanian government had imposed on foreign companies. In his 1930 volume *America Conquers Britain*, Ludwell Denny made the claim that other foreign oil companies had suffered less from the Mining Law of 1924 than Romano-Americana. The Rumanian and German governments finally agreed in 1929 to pay the Standard subsidiary approximately $16.5 million for wartime damage to its refining installations, but the sum probably did not completely cover the cost of the damages.

In 1930 Jersey Standard and Royal Dutch-Shell joined forces in an attempt to curtail production in Rumania. The effort proved quite effective during the months of August, September, and October, but in the fall the Rumanian-controlled Steaua Romana entered into a contract with the Compañia Española de Petróleos, a Madrid-based firm, to supply Spain with nearly 75 percent of its annual petroleum requirements. In the months that followed, the Anglo-American curtailment agreement collapsed. Even so, the Rumanian government faced declining prices; crude oil, which had sold for $15 a ton in 1924, now only brought $2.50. As a result, the Rumanian government imposed increased taxes on gasoline, lamp oil, and crude oil.

As in other countries, the Great Depression intensified the economic problems Rumania experienced between the two world wars, but the petroleum industry reached an interwar production peak of 8 million tons in 1936. It was Rumania's oil that especially appealed to the Nazis, who were eager to acquire the means by which to implement their dream of dominating the European continent. During World War II, control over Rumanian petroleum once again came to be a major American concern.

The political and economic dislocation in Eastern Europe following World War I was highlighted by the breakup of the Austro-Hungarian empire in the south and the creation of Poland and the Baltic states in the north. Curiously, during the interwar years U.S. petroleum relations were perhaps the most extensive with the least-known state of Eastern Europe, Albania, which did not gain its independence from Turkey until 1912. Following World War I Albania experienced constant pressure from other countries, as American, British, French, and Italian interests sought Albanian oil. When Italy finally overran Albania in 1939, Albania at last began to develop its petroleum deposits.

The Sinclair Exploration Company learned in 1921 that the Albanian government preferred American capitalists, especially petroleum entrepreneurs, to those of other nations. Although Albania had discussed an oil concession with Anglo-Persian, it was somewhat hesitant to award the concession to European nationals. According to the president of the Albanian Parliament, the country was "filled with gratitude to America for all that the American Red Cross had done for her and. . . she would never forget that or the fact that President Wilson had stood for the unity and independence of Albania when every other nation wished to see the country divided up."[21]

In April 1922 Secretary of Commerce Herbert Hoover asked that the American government recognize Albania diplomatically to make it easier for the Sinclair interests to obtain a petroleum concession. At this time both British and Italian nationals were seeking such concessions from the Albanian government, and both Great Britain and Italy had recognized Albania diplomatically. U.S. recognition finally came in July.

By the fall of 1922 Sinclair was facing a challenge in the race for concessions from both Royal Dutch-Shell and Anglo-Persian. At this point it began to debate the feasibility of a loan to Albania. Anglo-Persian was hostile to such a loan, while Royal Dutch-Shell was inactive. For this and other reasons, in late October, Sinclair reached an agreement with the Albanian government for an oil concession lasting seventy-five years. The royalty rate was around 13 percent, and the right of exploitation covered 1,500 square miles; Albanian citizens, moreover, were offered the right to subscribe for shares totalling up to 24 percent. But the loan discussions failed to lead to a similar understanding. By the end of the year no one in the American financial market had come forth with the offer of a loan, with the result that Sinclair sought a delay in the granting of the concession until the opening of the National Assembly in March 1923.

Anglo-Persian then moved in, attempting to obtain a monopolistic petroleum concession in Albania. Early in 1923 the American government again raised the issue of an "Open Door" policy. The Italian Ambassador to America, who also was disturbed by this turn of events, suggested that the United States complain about the activities of Anglo-Persian in conjunction with Italy and France. But in May a joint committee of the Albanian Parliament notified the government that the terms that Anglo-Persian offered were not acceptable. That summer New York Standard proposed a joint undertaking with the British firm, but Anglo-Persian rejected the proposal.

Anglo-Persian now proceeded on its own, and in February 1925 it ratified a contract with the Albanian government. The final concession encompassed 50,000 hectares, and the royalty rate on the gross production of crude oil was 13.5 percent. Earlier in the month the Italian Chargé d'Affaires had informed the Albanian Minister for Foreign Affairs that the Italian government would consider the granting of a petroleum concession to British interests as an act of hostility. Perhaps because of this diplomatic pressure, the Mussolini regime, over the objections of Standard Oil, obtained second choice for a petroleum concession, involving 30,000 plus hectares of oil territory. Among the other contingent agreements of the Italian settlement were the creation of a state bank in Albania and the contracting of a £2 million sterling loan in the Italian market.

On March 17 Secretary of State Frank Kellogg complained to the American Chargé in Albania, Trojan Kodding, that "the Albanian Government, by inducements which are still unfulfilled, has led American companies to expend considerable time and effort in Albania, and that the Government has reaped considerable benefit from the geological and other work carried on by these companies."[22] The reaction of the Albanian government was to assert that the "Open Door" policy was still being respected. By July New York Standard had reached an agreement that was acceptable to the Albanian Parliament for an oil concession of its own. One of the tracts extended along the northern shore of the Bay of Valona and north up the Adriatic coast, while the other was on the north side of the Bay of Durazzo. That October Anglo-Persian began drilling its first well, but neither the Americans nor the British were to commence large-scale petroleum operations before the outbreak of World War II. In contrast, by that time the Italians had discovered and developed small oil fields in Albania, built a pipeline, and begun exporting low grade crude oil to their own refineries.

## Russia: From Czar to Communism

At the outbreak of World War I Russia was still a backward country, only fifty years past the abolition of serfdom in 1861. It was just beginning the transformation from a rural, semifeudal society into an urban, industrialized one. The result was increasing political, economic, and social dislocation, which was to lead to the establishment of a central parliament, or Duma, in 1905 and then to the Communist takeover and the abolition of the monarchy in 1917.

Although the full-scale development of Russia's natural resources had not begun in 1914, the nation, like the United States, possessed great mineral riches, including petroleum. Commercial oil production, in fact, began in Russia during the middle decades of the nineteenth century. Output had increased to over 24,000 barrels by 1840, with the 130 hand-dug pit wells gradually increasing their production. Output reached 38,000 barrels by 1863 and peaked at 182,000 barrels in 1870. Shortly thereafter, petroleum development was opened to competitive enterprise, and a large number of wells were sunk and refineries built in the Baku area. Unfortunately, however, Russian entrepreneurs were so inefficient that it was necessary to import large amounts of oil from the United States. It was at this time that the Swedish Nobel brothers introduced modern technology to the Russian oil industry, and in the mid-1870's they had begun to dominate the local market.

By 1885 it was apparent that Russian oil not only was undercutting American competition there but also was posing a challenge to U.S. firms throughout Western Europe, as contemporary reports by American diplomatic officials in Switzerland and Greece indicated. Within a year, however, Russian optimism began to turn to gloom, for the more inexpensive Baku product failed to burn as long or as brightly as its American counterpart. The U.S. consul at Beirut also reported that the Russian-made tins leaked badly. Despite these handicaps, the challenge from Russia was so great that the American consul at Batum, J. C. Chambers, accused the Russian petroleum lords of having a "quixotic ambition to drive the American oil from the markets of the world."[23] As early as 1883 the United States had begun to protest that the new Turkish petroleum duties favored Russian merchants. In 1892, however, the government of Turkey sought American technical advice in a campaign to keep Russia from Mesopotanian oil.

In 1894, the Standard Oil Company and the combined Nobel-Rothschild group attempted to divide the world's market among themselves, only to have the negotiations collapse when the Russian Minister of Finance refused to confirm the agreement. In 1897, fourteen years after the completion of the Baku-Tiflis line, which had made Russian oil competitive in Europe, petroleum output in the Caucasus reached its peak. During the next two years, the Russian production of crude oil surpassed that of the United States and in 1899 constituted more than one-half of the total world output.

After the turn of the twentieth century, Jersey Standard seriously considered participating in the Russian petroleum industry but decided not to do so for several reasons, including seemingly continuous Russian labor disputes. The oil workers began to strike in 1902, and the labor disturbances continued for the next several years, leading Dvorkovitz to proclaim in his *Petroleum Review* that "the Russian petroleum industry for the present has ceased to exist."[24] There followed a sharp decline in exports, which were not to recover their former level until 1927. As a result, in 1910 Standard Oil decided not to involve itself in Russian petroleum, even after the opening of the Grozny and Maikop fields, which were to supplement the production of the Baku field.

Four years later World War I broke out. After three years of fighting on the Allied side, two revolutions occurred in Russia during 1917: the February one brought to power the moderate to liberal Lvov-Kerensky regime, but the October one led to the inauguration of Bolshevik rule under V. I. Lenin. One of the new government's first steps was its withdrawal from World War I under the Treaty of Brest-Litovsk on March 3, 1918. Equally important, though, the Communists began to nationalize large estates, industrial concerns, and banking and communication systems, much to the displeasure of all non-Russian capitalists doing business there.

Although the U.S. reaction to the new Soviet regime was so hostile that the American government did not recognize the Soviet regime until 1933, as early as July 1920 the Department of State announced that it was removing restrictions on commerce and communication with the Moscow government. The U.S. government qualified this gesture with the *caveat emptor* that American firms would have to do business with Russia at their own risk, an indication that any American oil concern negotiating with the Soviets would not be guaranteed diplomatic support.

Because of its widespread nationalization of foreign holdings, American oil companies felt hardly any sympathy for the new Communist regime. Yet the U.S. government refrained from recognizing the counterrevolutionary regimes then operating in Russia for both diplomatic and military reasons. Standard Oil particularly opposed the recognition of the one in the Caucasus. As for Sir Henri Deterding, he not only had taken as his second wife the daughter of a White Russian general, he also had helped to finance the White counterrevolution against the Bolsheviks between 1918 and 1921.

Royal Dutch-Shell was by no means the only non-American concern interested in Russian oil. In 1920, when the Communists seized the Baku fields, they took over the petroleum properties there, including those of the Nobels, who had brought in their first producing well as early as 1874. Yet, even in 1920 Standard Oil of New Jersey officials agreed to advance $500,000 to the Nobels for the joint purchase of additional properties at Baku, in what proved to be a most reckless gamble. Jersey Standard apparently believed that anything as odious to them as the new Communist regime could not possibly survive.

Then, on July 30, Standard purchased for $11 million half of the Nobel shares in the Petroleum Production Company, a firm that at the time held investments in twenty-eight Russian subsidiaries. Since the Nobels' personal interest in the Petroleum Production Company amounted to only 20 percent, they and Jersey Standard agreed jointly to raise $5 million in order to obtain a majority interest in the firm. By December 21, 1922 Standard had paid the Nobels $6 million for 13,000 shares of the Petroleum Production Company, with the balance to be paid upon the delivery of the remaining shares, and the restoration of the Nobel holdings in Russia by Soviet officials.

The new Communist regime in Russia, of course, did survive, but the petroleum fields of the Caucasus, far from running smoothly under Soviet control, remained in a state of disarray. As a result, Lenin himself had the Council of Commissars pass a resolution early in 1921 "to approve in principle the granting of oil concessions at Grozny and Baku and at other operating fields, and to begin negotiations, pushing them forward more rapidly."[25] He then engineered the resolution through a central committee plenum and the Tenth Party Congress.

Confronted with this opportunity, the governments of those foreigners who were active as petroleum entrepreneurs met at Genoa, Italy during April 1922 in an attempt to reach an understanding with Russia over the earlier nationalizations. Although the Soviet Union refused to surrender the oil properties it had seized, it did offer part of the petroleum fields as ninety-nine-year leases or concessions to America, British, French, Italian, Belgian, and German firms. This settlement was acceptable to Henri Deterding and to Prime Minister David Lloyd George of Great Britain, who was the moving force behind the conference, but Standard Oil was opposed, as were the French and Belgian interests.

Even more objectionable to U.S. petroleum entrepreneurs was an alternate proposal advanced by the Soviets, which would have apportioned the new oil concessions in conformity with the foreign holdings there prior to nationalization. Under these circumstances Royal Dutch-Shell would have emerged the victor, and Standard Oil would have obtained virtually nothing. Again the American representatives protested in conjunction with the French and British delegations; the U.S. Ambassador to Italy stated that the United States could not accept any settlement that was out of harmony with the "Open Door" principle. Lloyd George then attempted to obtain general diplomatic recognition for the new Soviet regime, but the United States blocked this move. Acting through the French delegation, the Americans also won postponement of a discussion of property rights.

Talks at The Hague the following month were no more successful. The British government and Royal Dutch-Shell suggested giving a block concession of certain Russian fields to Royal Dutch-Shell, which in turn would settle with other foreign claimants by sharing production or through some other means. This approach, too, was unacceptable to the United States, as well as to the Franco-Belgian interests, who demanded that future petroleum discussions with Russia should be carried on jointly by the various oil groups. These three nations were among the sponsors of a resolution, adopted at this conference, stating that no single nation could accept a monopoly petroleum concession from the Soviet Union. Its passage was a defeat not only for Great Britain, Royal Dutch-Shell, and Sir Henri Deterding but also for the Communists, who would have liked this "division of interests" to continue among those nations present at the conference.

That July a group of petroleum men, including Teagle, Deterding, and Gustave Nobel, met in London. The result was the unsigned so-called London Memo of July 22, 1922. Abandoning independent negotiations with Russia in the future, the participating parties demanded complete indemnification for the seized properties but omitted any specific mention of the Soviet Union. If complete indemnification proved impossible, then the properties should be restored to the foreign oil firms, and they should also be indemnified for any damages resulting from the Russian takeover.

Although this conference marked the first step in the creation of a *front uni*—a united front—against Communist Russia, not all of the world's leading petroleum men were in attendance. The "London Memo," moreover, was somewhat vague, as it failed to list specific steps for the oil entrepreneurs to follow toward the Soviet Union. Consequently, another conference was held at Paris during September 1922. This conference, presided over by Sir Henri Deterding, was attended by thirteen petroleum firms in addition to Jersey Standard, Nobel, and Royal Dutch-Shell. But Jersey Standard again refused to sign a formal document because, as one official explained to another, "We ought to avoid at present taking any open initiative from which a blame could be fixed on us to the effect that we are at the bottom of the opposition against Russia."[26]

Despite the extensive oil fields within the Soviet Union, the new Communist government was not satisfied with its petroleum reserves, and turned its gaze southward to the equally rich oil fields of Iran. In 1907 Great Britain and Russia had come to an Iranian understanding, by which Great Britain was to have a sphere of influence in the south and Russia in the north. Although an individual named Khostaria had obtained a concession in the north during the czarist era, in 1921 Iran signed a treaty of friendship with the Soviet Union that not only voided Iranian debts to Russia but also cancelled the Khostaria concession. A year later the Russian government showed its displeasure at a proposed Anglo-Persian/Jersey Standard joint endeavor in Iran, which it feared would lead to additional British economic penetration there, by awarding an oil concession on Sakhalin to Sinclair.

It was at this time that the Sinclair interests formed an alliance with the International Barnsdall Corporation, headed by Mason Day. During July 1923 both Day and Sinclair visited Moscow, accompanied by former Interior Secretary Albert Fall and Archibald Roosevelt, the son of the former President. In November 1923 Day signed a provisional concession agreement with the Russian government. The agreement established a joint company that was to develop the Grozny and Baku fields. Not only did Sinclair pledge himself to a $115 million investment in the new concern, he also promised to float a $250 million loan in New York for the Soviet Union. As a further inducement to the Communists, Sinclair committed himself to obtaining U.S. recognition of the new Russian regime. In return, he was to obtain a virtual monopoly over the Grozny field for forty-nine years.

Unfortunately for Sinclair, however, his concession involved territory already claimed by Jersey Standard. Even more important, it was at this point in the Harding administration that the Teapot Dome oil scandal erupted, discrediting both Sinclair and Fall. As a result, the Russian scheme never became a reality.

The increasingly self-confident Soviet officials felt less and less need to rely on foreign capital to carry out their revolutionary economic program.

Nevertheless, the official Soviet publication, *Pravda*, showed considerable displeasure at the unfavorable turn of events relative to the Sinclair concession. On February 27, 1925, it editorialized: "The government of the United States made not the slightest move to facilitate the work of the company."[27] When former Secretary of State Robert Lansing approached the State Department in 1925 as a representative of Sinclair, departmental official E. L. MacMurren advised Lansing that Moscow should sue Sinclair for nonperformance because of the latter's abortive attempt to "pull their [the Russians'] political chestnuts out of the fire for them."[28] During that year, in fact, the Russian government won a law suit over the cancellation of the Sinclair concession on Sakhalin in a Soviet court with Lansing acting as the oil company's attorney.

The problems experienced by Sinclair did not discourage Jersey Standard from concluding a gentlemen's agreement with Royal Dutch-Shell in 1924 that divided the southern Russian petroleum fields between them. Although the Russians realized that the goodwill of Jersey Standard might help to bring about a resumption of diplomatic relations with the United States by 1925, they were unwilling to grant Jersey Standard a long-term monopoly over Russian oil exports.

During December of that year Joseph Stalin, who only recently had come to power, enunciated an oil policy for the Soviet Union in a report to the Fourteenth Congress of the Communist party:

> The possession of most oil will determine who will command world industry and trade. Now that the navies of advanced countries are passing over to motor drive, oil is the vital nerve in the struggle among the world states for supremacy in peace and in war. It is precisely on this issue that the struggle between the English oil companies and the American companies is a mortal one, not always bearing an open character, it is true, but always proceeding and smoldering. . .[29]

Negotiations began in Berlin in February of 1926 between agents of New York Standard and the president of the Soviet Naphtha Syndicate. The talks resumed in Paris in March, and at that time the Russians signed a contract with the Vacuum Oil Company. Under the contract Vacuum was to purchase 800,000 tons of Russian crude oil, as well as approximately 100,000 tons of kerosene. The Soviet Naphtha Syndicate pledged itself not to compete with Vacuum in Egypt. During this same period, New York Standard also bought 100,000 tons of kerosene from the Russians. These maneuvers represented a move by the American oil companies to challenge Royal Dutch-Shell in the Mediterranean-Suez Canal sector, an attempt that was successful, for Royal Dutch-Shell did fail in its maneuverings to buy the Soviet petroleum surplus.

Back in the United States, New York Standard abandoned the anti-Russian campaign it had been conducting, and Rockefeller public relations advisor Ivy Lee even wrote a book that was favorably disposed toward the Soviet Union. In this work Lee theorized that it would be possible for American capitalists to raise the Soviet economic standard through trade and investment, which in turn, he hoped, would lead to an ideological mellowing on the part of the Communists. Lee wrote former Secretary of State Elihu Root that "it would seem that the policy of drift with reference to Russia was getting us nowhere," and that what he sought was "a condition brought about under which you, and men like you, would think it wise to accord [the Soviet Union] recognition."[30]

These developments obviously displeased Sir Henri Deterding of Royal Dutch-Shell. Great Britain had officially recognized the Soviet Union in 1924; the United States was not to follow suit until 1933. Confronted with the Russian successes of the American firms, Deterding bitterly observed that "the Standard Oil Company apparently wants. . . distribution of stolen property in such a way as to destroy legitimate trade. My intention is to fight the matter to the bitter end, if necessary, over the whole world. . . ."[31] Far from being intimidated, the U.S. oil companies defended their actions on economic grounds and in 1928 extended their contracts with the Soviet Naphtha Syndicate. By this time Royal Dutch-Shell had launched a price war against the American firm, but in July of 1927 Sir Henri Deterding met with Jersey Standard President Walter Teagle in Paris. The two petroleum barons at this time agreed that their firms would not deal with the Russian government until the Soviet Union formally recognized their property rights in that country.

Despite his defiant stand, Deterding came to an understanding with officials in Moscow in February 1929. Their agreement did not mention the compensation claims, but it did provide for a 5 percent rebate on the deliveries of the Soviet Naphtha Syndicate subsidiary in Great Britain, Russian Oil Products, Ltd., with the difference to be earmarked for an indemnity fund organized for the former owners. The accord was to run for three years and enabled Russian oil to obtain a large share of the British market; on balance, it probably was more of a triumph for the Soviet Union that it was for Royal Dutch-Shell.

In 1931, as the world struggled with the Great Depression, both the Americans and the Soviets rejected a Royal Dutch-Shell plan to restrict the output of petroleum and to allocate among the major oil producers limited export quotas. There was speculation in the United States that such a plan would be against the American antitrust laws, and the goal of the Soviet government was economic expansion, not economic contraction. It was only in 1927 that the Russian oil output had again reached its 1900 level.

The leading Russian oil figures and their counterparts in the capitalist world attended an international oil conference in New York City in May 1932. Unfortunately, the participants failed to reach an agreement to regulate world oil exports. Jersey Standard proposed that it should assume all Russian petroleum exports for a period of ten years at the 1931 level and then prorate these among other oil firms for sale in their markets. Jersey Standard also recommended closing down Soviet distributing facilities abroad. The Russian delegation did not flatly reject this offer, but did balk at the dismantling of its overseas operations, as well as at the signing of a contract for more than three years.

Fearful that a great outpouring of Russian oil might glut the world market, many capitalistic representatives left the conference with plans for a "gentlemen's agreement" to boycott Soviet petroleum. In the following year, however, the Roosevelt administration recognized the Russian government, thus removing U.S. recognition as a factor in international trade as well as a consideration in granting American firms oil concessions in the Soviet Union.

The output of petroleum in Russia continued to rise until the outbreak of World War II in 1939, but exports of petroleum from the Soviet Union not only declined from 1933 onwards, they practically disappeared from the world marketplace. During the pre-World War II years, the domestic demand for oil in Russia increased disproportionately not only because of industrialization but also because the Soviet regime adopted less stringent foreign exchange requirements. Nevertheless, Socony-Vacuum signed a contract with the Soviet Union in 1935 for the purchase of Russian petroleum for resale in Egypt and the Far East. As the pressure to export goods declined, the Soviets began voluntarily to dismantle the extensive sales apparatus that had been set up to handle Russian oil abroad; in turn, the leading petroleum men of the capitalistic world, no longer threatened by Soviet oil, softened their hostility toward the regime in Moscow.

## British Lend Lease and the Abortive Oil Agreements

With the outbreak of World War II, Great Britain's access to world petroleum reserves was seriously jeopardized. By the spring of 1941 the German submarine campaign had made it quite difficult for the British to obtain oil from the Western Hemisphere, and as the war progressed, the Nazis had as many as 100 submarines operating at one time. The threat they posed to Allied shipping did not begin to subside until after March 1943.

The Chairman of the United States Maritime Commission, E. S. Land, sent a confidential memorandum to presidential advisor Harry Hopkins on April 28, 1941, in which he declared that the government of Great Britain "is becoming increasingly alarmed at the continued fall in petroleum stocks in the United Kingdom and views the situation with real concern."[32] Accordingly, FDR made fifty tankers available to Great Britain in July, even though it caused an oil shortage on the East Coast.

On the following day Deputy Petroleum Coordinator Ralph K. Davies expressed his fear to Interior Secretary Harold Ickes that the petroleum stocks of Great Britain could well reach the point of exhaustion. By September 25, however, Land was able to offer a more optimistic report to FDR in part because of the transfer to Great Britain of 500,000 deadweight tons of American flag tankers and 300,000 deadweight tons of Panamanian flag tankers. Furthermore, the average round trip of a tanker in the British trade had decreased from ninety to sixty-five days; the British themselves were building on the average three tankers a month; the Americans were helping them to repair their tankers; and Norway was in the process of transferring forty tankers to the British.

Because of these developments, Ickes reported to Roosevelt on January 12, 1942, that petroleum stocks in the United Kingdom now stood at approximately 7 million tons, a six-month supply, and that the British even were able to divert some of their ships to the Far Eastern service. Unfortunately, on December 5 the British Petroleum Representative in the United States, Wilkinson, advised Ickes that these stocks had fallen to 5.2 million tons to the alarm of the British Cabinet. In the year that followed, because of the heavy war demand for oil, the problem continued to persist. Fortunately the stocks proved sufficient to enable Great Britain to continue its counterattack.

Although American petroleum shipments to Great Britain had helped to prevent a military defeat, a bitter clash developed between the United States and Great Britain over the petroleum reserves of Saudi Arabia in the final years of World War II. When Ickes attempted to involve the American government in a Saudi oil concession in 1943, the British were considerably alarmed. Critics of Great Britain charged that the British were holding back on Middle Eastern petroleum production in Iran and elsewhere. Anglophobes pointed to the British attempts to exclude U.S. oil men from the Middle East following World War I.

Once Ickes' Saudi scheme had fallen through, Secretary of State Cordell Hull suggested holding talks on oil between the United States and Great Britain. Hull felt that the early conversations should be at the staff level because of their technical nature. Although there was some initial British reluctance, continuing pressure from Washington led Great Britain to accept the American invitation on February 7, 1944. That the U.S. Senate then failed to approve either treaty resulting from these talks does not diminish the significance of the episode, if only for the reason that it allows a comparison of the official oil policy of the United States and Great Britain at this particular moment in history.

On April 19, 1944, a group of British oil experts entered into conversations lasting for two weeks with State Department officials. At the same time Under Secretary of State Edward R. Stettinius, Jr. headed the special mission participating in negotiations in London. Among the members of the British delegation in Washington, led by Board of Trade Secretary Sir William Brown, were Royal Dutch-Shell Director Sir Frederick Godber and Anglo-Iranian Chairman Sir William Fraser. As *Newsweek* observed, the "makeup of the British delegation showed the interlocking of British Government and British oil."[33] On the American side, the State Department named a ten-man advisory committee whose members acted as individuals, despite the fact that they represented the various phases of the petroleum industry. Negotiators were confronted with the possibility that U.S. oil firms might be prosecuted under the antitrust statutes; this was a possibility that the State Department's petroleum advisor, Charles Rayner, who headed the American group of experts, tended to minimize. In the months ahead, Rayner and the State Department's economic advisor, Leo Paslovsky, were to have a continuing role in shaping an Anglo-American petroleum pact.

As early as April 29 the American and British representatives had drawn up a "Draft Memorandum of Understanding with the United Kingdom on Petroleum," which was to continue until six months after a notice of termination or until the signing of an international petroleum agreement. This document, which endorsed the principles found in the Atlantic Charter, took the position that the oil resources of the world were adequate for the years ahead. In the course of the exploratory discussions, the British delegation insisted on two points: that it be explicitly recognized that the United Kingdom was dependent upon imported oil supplies, and that the two governments should support as well as respect all valid concession contracts. But many American petroleum officials came to feel that the British were using the Allied invasion of Northern France as an excuse to delay action on an Anglo-American oil agreement. The United States also was hesitant to take joint action with Great Britain in negotiations with other governments on the validity of petroleum contracts. It believed that ample oil supplies should be made available to every country.

It was not until August 8 that Lord Beaverbrook, representing Great Britain, and Cordell Hull, representing the United States, signed a document that essentially created a petroleum cartel. Under the agreement all peaceful nations were to enjoy equal access to oil products at fair prices and on a nondiscriminatory basis. There was to be equal opportunity in the acquisition of concessions and no "raiding" once the concessions had been granted. In addition, the petroleum industry was to be freed from unnecessary restrictions by both government and private enterprise; petroleum resources were to be developed so as to facilitate the economic progress of the producing nations. Looking to the future, the agreement set up an eight-member international petroleum commission with advisory functions, four members to be appointed by each government.

The reaction of the American press to the agreement was divided. The New York *Times* observed that "provided other nations join, it will be the first time in modern history that such an essential commodity has been brought under control by either government or private industry."[34] The *Christian Century*, however, felt that other countries should have been party to the talks and to the treaty. Other commentators complained about the pact itself. Four days earlier, the *Nation* had noted that the editorial advocates of the agreement might be "a little premature in their cheers," since "who is to decide what quantity is 'adequate' and what price 'fair'?"[35]

Despite the Navy Department's support, there was immediate opposition to the ratification of the agreement from the Chairman of the Senate Foreign Relations Committee, Tom Connally of Texas, who, as the spokesman of that state's petroleum interests, proclaimed that the treaty would never win ratification. Upon studying the pact, many U.S. oil men also formed reservations, especially over those provisions representing a move toward federal regulation. Among those who attacked the agreement was Ralph T. Zook, the President of the Independent Petroleum Association of America; J. Howard Pew of Sun Oil feared that it might create a superstate cartel. Many petroleum barons did not wish to be bound in advance to accept the decisions of any international oil commission that might be set up in the future.

Because of the controversy which it engendered, FDR withdrew the treaty for revision in January 1945, stating that he wished to correct the suggestion that it was vague and to clarify the fact that the international commission was only to be advisory. After obtaining British approval of the modifications, the State Department showed it to Senator Connally, who then conferred with both government officials and oil men. On February 22 Connally announced that he believed the objectionable features of the original treaty had been removed.

On April 12 FDR died, and Harry Truman assumed the presidency. Before the Truman administration formally resubmitted the treaty to the Senate, it held additional talks with the British, which led to the signing of a revised agreement in London on September 24. By that time Germany had surrendered, and the war in the Pacific had ended.

When the terms of the new Anglo-American petroleum agreement became known in the United States in the fall of 1945, they again encountered widespread criticism. In the opinion of *Business Week*, far from clarifying matters, the revised agreement was "more innocuous, less specific, and subject to varying interpretations" than the first treaty, "a 'should' rather than a 'shall' piece of writing."[36] Although the second treaty did call for an international conference to draft a world oil agreement, it neither settled specific problems nor divided world output or markets; typical of the new treaty was the phrase, "orderly development of the international petroleum trade." There again was to be an advisory commission composed of representatives of both nations; this time there were to be six members instead of eight. This second treaty bore the signatures, not of diplomats, but rather of Secretary of the Interior Harold Ickes and Minister of Fuel and Power Emanuel Shinwell.

In assessing the new treaty retrospectively, Ickes wrote to the *New Republic* on January 1, 1946, "Our intention, if we could bring it about, was to agree with the British on certain principles affecting. . .the world that would do away with certain practices that had been of disadvantage to American oil companies generally."[37] But the prophecy made by *Time* magazine on October 8, 1945, that the Senate would ratify the pact, which already had been approved by the British, proved wrong. Despite the treaty's nominal support from the national petroleum organizations, many domestic producers were unhappy with it, fearing that it might lead to more intense worldwide competition in the future.

As a number of key Republicans also announced their opposition to it, it became increasingly apparent that the revised petroleum agreement with Great Britain would not be ratified. Acting Secretary of the Interior Oscar L. Chapman testified before the Senate Foreign Relations Committee in March of 1946 that "from the standpoint of the Nation's self-interest alone, the instrument which you gentlemen now have before you looks to me like a mighty good bargain for the United States."[38] Yet others in the Truman administration were less impressed. The Solicitor of the Interior Department, Warner Gardner, feared that Article IV

(3) (d) might lead to an allocation of international markets and international price-fixing, in effect establishing a cartel arrangement.

As a result, Gardner recommended postponing the Senate hearings. Furthermore, an Anglo-American loan agreement had been concluded on December 6, 1945, which tended to divert attention from the Anglo-American petroleum agreement. The hearings did not begin again until June 1947. Despite the powerful hostility of the Speaker of the House, Sam Rayburn, as well as Senator Tom Connally and other Senators from the Southwest on the Foreign Relations Committee, the committee eventually endorsed the Anglo-American oil agreement. The Senate as a whole, however, failed to take action on it, and thus it remained another abortive blueprint for the future. Truman himself was opposed to the creation of a UN agency to administer global petroleum resources.

## U.S. Petroleum Diplomacy with Five Neutral States

Unlike Great Britain, which was an ally of the United States during World War II, Spain had remained neutral. While Catholics in the United States had tended to sympathize with General Francisco Franco, the right-wing dictator, and victor in the Spanish Civil War of 1936-39, American liberals took an instant dislike to him, which was to continue unabated for as long as Franco was directing his nation's destiny. Spain, of course, was not the only European neutral during World War II; Portugal, Iceland, Sweden and Vichy France also did not take sides. Nevertheless, American oil shipments to Spain and Spanish wolfram shipments to Germany probably generated more diplomatic friction than did U.S. petroleum relations with any other European neutral.

On May 12, 1940, President Roosevelt informed the British Ambassador to the United States that he was willing to consider additional ways of helping the Allied governments. Eight days later the government of Great Britain suggested that its American counterpart should deny supplies, either directly or through neutral countries, to the Axis powers. When the United States expressed a willingness to accommodate the British, both Great Britain and France (which was to capitulate to the Germans that month) protested the abnormal increase during June in Spanish oil imports from the United States and the Caribbean. At this time the petroleum monopoly in Spain (Campsa) and that in the Canary Islands (Cepsa) were obtaining supplies through contracts with American companies.

Had the Spanish consumed all of this petroleum themselves, there would have been no furor, but reports had begun to circulate that some of the oil was destined for Italy. Accordingly, Great Britain asked the United States to limit the use of American tankers carrying petroleum to Spain and to restrict shipments of aviation gasoline and lubricants. Harold Ickes, who was critical of the Spanish oil monopoly, temporarily halted the departure of American flagships carrying petroleum to Spain. But since the Roosevelt administration was reluctant to move too far from its officially proclaimed neutral stance, Secretary of State Cordell Hull was hesitant to regulate openly oil exports to Spain. Instead, the Maritime Commission declared that the commerce with Spain was dangerous, thus discouraging American tankers from sailing for Spanish ports. The Treasury Department then decreed that all petroleum cargoes bound for Spain were to be inspected.

The top levels of the Roosevelt administration, however, were not united on a Spanish oil policy. Secretary of the Treasury Henry Morgenthau was far more anxious to adopt an immediate hardline stance than Secretary of State Cordell Hull. Morgenthau, in fact, advocated placing petroleum and petroleum products on the list of embargoed strategic materials. Rather than have the government openly take the lead in implementing this program, Under Secretary of State Summer Welles consulted with representatives of the oil companies involved, who agreed to keep their petroleum shipments to Spain within the usual limits. This move was directed specifically against the head of the Texaco Company, Torkild Rieber, who had close associations with Nazi Germany. According to the log of one of the Texas Company tankers, its entire cargo was transshipped to representatives of Italy, another Axis power.

In retaliation, the Spanish government seized the American-owned telephone company in Spain and used this action as a bargaining tool in its talks with the American government. When the Spanish Foreign Minister, Juan Beigbeder, told the American Ambassador, Alexander Weddell, on August 6 that the telephone controversy had been resolved, Weddell stated that his government would allow Spain to obtain as much petroleum as it could carry, and the British would exempt from seizure or search as noncontraband. Great Britain, increasingly fearful of a Spanish entry into the war on the Axis side and the subsequent loss of Gibraltar, showed a greater willingness to supply the Spanish with oil. The United States refrained from lending Spain money with which to pay for the petroleum, but it did allow American-owned tankers to sail there under foreign flags.

This Spanish-American understanding, though, did not settle matters, and the oil exporting controversy dragged on into 1944. On September 16, 1941, Ambassador Weddell suggested to Hull that "our ability to supply and withhold such petroleum products represent the trump cards in our political and economic relations with Spain" and added, "I do not consider this the psychological moment to play the latter."[39] But the Spanish Ambassador to the United States, Juan Francisco de Cárdenas, informed Ambassador Weddell in late September that Spain was in desperate straits, especially for gasoline. British investigations then revealed that large amounts of petroleum products were not making their way into Axis hands. After the British government had entered into a quota agreement with Campsa, Secretary of State Hull recommended continuing U.S. petroleum exports to Spain, subject to the quantitative limits set by Great Britain and the qualitative limits established by the United States. After additional negotiations, Spain and the United States agreed in early 1942 to a regulatory system, which would be operated jointly by the American, British, and Spanish governments.

During 1942 the diplomatic exchanges between the United States and Spain concerning petroleum continued. The new focal point of controversy was the reported accumulation of 5,000 to 15,000 tons of military gasoline in Spanish Morocco. When the American government implemented a three-month program in February to supply Spain with certain petroleum products, crude oil was not included. Nevertheless, the government agreed to furnish Spain with gasoline, kerosene, gas oil, fuel oil, and lubricating oil in varying amounts. On July 31 the Chargé at Tangier, Childs, reported to Hull that the authorities in Spanish North Africa had agreed to cooperate with the implementation of the petroleum program there and speculated that their friendly attitude was due to the possibility of an Anglo-American military landing. This, of course, happened on September 8, when General Dwight D. Eisenhower successfully led three Allied invasions of Northwest Africa.

Six months later, in March 1943, Ambassador Carlton Hayes noted from Madrid that a shortage of aviation gasoline had forced the Spanish civilian airlines (Iberia) to suspend operations, leaving only Axis airlines operating. The shortage was caused by a policy decision of the British Ministry of Economic Warfare, which had vetoed further aviation gasoline shipments to Spain. In April the State Department and the Board of Economic Warfare concluded that imports of petroleum products into Spain should not exceed 100,000 tons quarterly, the rate for the past two quarters. This relatively modest amount really did not meet Spain's normal requirements.

Such troublesome incidents aside, Ambassador Hayes concluded on May 1 that the Spanish oil policy of the American government had been successful:

> In political field it has helped to strengthen elements in the Government favorable to us and converted many others to our side. As a result we have been able to achieve many objectives which otherwise would have been extremely difficult, such as acceptance of our guarantees at time of North African landing (if they had not been accepted the soft under belly of the Axis would not have been exposed for a long time); Spanish determination to resist any Axis aggression; release of all our military internees, and of French refugees, mostly military; consent to establishment of French North African representation; and, on the economic side, smooth functioning of our broad program which has been damaging to Axis. It has created public good will which extends from lowest class to highest, excepting only minority in Falange which still clings to hatred of democracies and which would like to see our program fail.[40]

That June Secretary of State Hull agreed to supply Spain with a limited amount of 87 octane gasoline from the Netherlands West Indies for the specific use of Iberia Air Lines. A month earlier he had advised Hayes, "there is more criticism of this oil program to Spain than of any other matter of foreign policy under my direction."[41] There was, though, another alternative to supplying Iberia with aviation gasoline. Early in June the Spanish Foreign Minister, Francisco Gómez Jordana, informed Ambassador Hayes that the Spanish government agreed in principle to the operation of a U.S. airline in Spain and invited the Americans to apply for the necessary authorization to initiate such a service. On November 18

Jordana notified Hayes that the Council of Ministers had approved the American request for temporary landing rights for its commercial airlines.

A final flurry of friction occurred during the early months of 1944, when the American government suspended petroleum shipments, because Spain was not embargoing wolfram shipments to Germany. According to the Department of State *Bulletin* for January 29, "The Spanish Government has shown a certain reluctance to satisfy requests deemed both reasonable and important. . . ."[42] The British, however, were opposed to such a drastic step because they needed food imports from Spain; thus Prime Minister Winston Churchill wrote FDR at the end of March that a compromise was in order. Accordingly, Hull cabled Hayes on April 13 that the American government was abandoning its demand for a total embargo on wolfram, and on May 1 Ambassador Hayes advised Foreign Minister Jordana to this effect.

Although Spain was unwilling to embargo wolfram, it showed a willingness to accede to other U.S. demands, including the closure of the German consultate in Tangier, where the Nazis had a highly developed espionage apparatus, and the transference to the Allies of certain Italian ships in Spanish waters. In fact, after Portugal had terminated wolfram shipments to the Nazis during June of 1944, Spain followed suit. In summing up American's wartime oil diplomacy with Spain, Thomas A. Bailey observed, "As the fortunes of the Allies improved, the wily Franco gradually veered from pro-Axis nonbelligerency, to neutrality, and finally to neutrality favorable to the Allies—in short, from malevolent to benevolent neutrality."[43]

The regime which António de Oliviera Salazar had established in neighboring Portugal in many respects paralleled that of General Francisco Franco, fascist in domestic policy and neutralist in foreign policy. But Portugal, less exposed to a possible Nazi attack than Spain, had less need to adopt a cooperative attitude toward Adolf Hitler. During the first half of 1943, though, the Salazar regime became embroiled in a controversy over U.S. efforts to enforce restrictions on the sale of petroleum products to Germans and Italians in Angola. It was the position of the Angolan Governor, Alvaro de Freitas Morna, that it was contrary to humanitarian principles to deny the Germans and Italians gasoline, especially those living a long distance from medical aid. Morna declared that as a neutral country Portugal could not discriminate against German and Italian nationals residing there. But when the U.S. Consul General, Linnell, called the Governor General in early February, it was to inform him that under General Ruling Number Eleven the United States could not give petroleum to Angola for transfer to enemy nationals.

So intent was the American government on implementing this position, that it contacted the Belgian government in exile in London over limiting oil shipments from the Belgian Congo to Angola. The Belgians were sympathetic but did not want to disrupt the general Congo trade with Angola by taking such a step. When American Minister to Portugal Bert Fish showed sympathy for the Portuguese position, Secretary of State Cordell Hull informed him on May 5 that "the Department. . .intends to make allocations of petroleum products to Angola dependent upon assurances that enemy nationals will not have access to such products either in the form of pump sales or sales in larger volume except in so far as is necessary. . . ."[44]

As World War II progressed, the contribution of officially neutral Portugal to the Allied cause surpassed that of similarly neutral Spain. In addition to cutting off wolfram shipments, during August 1943 Portugal allowed the British to use Lajes Field on Terceira, as well as the port of Horta, for engaging in antisubmarine warfare. With the end of the war in Europe only six months away and after a very complex, drawn out, and frustrating series of negotiations, the United States signed an agreement with Portugal that allowed it to operate an air base on Santa Maria. No such military rights were granted to the Allies by Spain.

Further to the north, after the Germans had occupied Demark in April 1940, Iceland declared its independence from the latter. As Great Britain's own survival was at stake at this time, Winston Churchill and Franklin Roosevelt reached an understanding under which American troops, beginning in July 1941, gradually replaced their British counterparts in Iceland. Although it was undeniable that the Nazis would have liked to bring Iceland under their control, the fiercely nationalistic Icelanders viewed the British and American military forces with considerable misgivings.

During the decade or so before World War II British firms had supplied Iceland with gasoline, oil, petrol, and kerosene; beginning in September 1942, however, the U.S. Navy took their place under an Anglo-American agreement reached independently of the Icelandic government. Friction then ensued when the U.S. Navy raised the price of these petroleum products in the wake of the Icelandic battle against inflation. Yet, far from attempting to gouge the Icelanders, the Americans were charging a price equivalent to the cost of delivery as determined by the War Shipping Administration. In fact, they were operating without the subsidy that the British firms had enjoyed.

Nevertheless, there was an overriding need to maintain the goodwill of the Icelanders, who grudgingly were tolerating the American military occupation. Thus, in February 1943 the naval commandant in Iceland was told to reduce petroleum prices to approximately the level charged earlier by Great Britain. When the Icelandic Minister to the United States requested that the American government also refund the monies paid since September in excess of this rate, the United States agreed to credit the Icelandic oil companies with the differential amount.

A somewhat different situation existed in Sweden, which was sending both iron ore and steel ball bearings to Germany. There was a bitter battle in Washington over allowing the Swedes to have any petroleum at all. While Secretary of State Henry Stimson and Secretary of the Navy Frank Knox were fearful that American oil might end up in German hands, Secretary of State Cordell Hull doubted that this would occur. He suggested instead that U.S. trade with Sweden would probably reduce Swedish economic assistance to Germany in the future. In any event, he argued, the oil would come from Caribbean and Gulf areas, where there was no shortage, and would be transported in Swedish ships.

Hull found a supporter for his position in the Board of Economic Warfare, which presented its report to President Roosevelt on November 16. A majority of the members adopted a resolution that favored continuation of trade with Sweden, including the shipment of petroleum products. Both the Secretary of the Treasury, Henry Morgenthau, and the Under Secretary of War, Robert Patterson, who considered Sweden to be too closely tied to Germany, dissented. The Joint Chiefs of Staff also favored commercial relations, but only if Sweden consented to a trade agreement with both the United States and Great Britain, which would remedy a number of unsatisfactory aspects in the Swedish commercial policy. As the Board of Economic Warfare admitted in a three-page report, entitled "Sweden's Contribution to the Enemy's War Effort," "about 90% of Sweden's current exports go to the enemy. This country receives none of Sweden's exports."[45]

In mid-November Roosevelt decided to permit the oil shipments because Sweden already had granted substantial concessions to the United States, including the Swedish termination of arsenic shipments to Germany in the summer of 1942. The American petroleum quota for Sweden was set at 30,000 tons quarterly, strictly for Swedish military use. The U.S. attitude toward Sweden was much more liberal than that of the British, who were demanding a reduction in the transit of German troops and material back and forth between Norway and Germany.

The short-lived Vichy regime, headed by the World War I hero, Marshall Henri Pétain, was set up following the German military defeat of France during the summer of 1940 and the subsequent German occupation of northern and western France. In January 1941 the Vichy regime requested the unblocking of French funds to allow the purchase of a tanker load of petroleum products in Mexico, which it would then ship to Morocco, where they were urgently needed. Hull granted this request on February 5 with the provision that the French should obtain the oil from some Western Hemisphere source other than Mexico.

Diplomatic friction ensued during the following month, however, when the American government investigated reports that petroleum shipments from French North Africa were reaching the Axis powers. On March 30, Admiral Jean Darlan, the Vichy Vice Premier, told the U.S. Ambassador to France, Admiral William D. Leahy, that not only had Italy requested 5,000 tons of gasoline from North Africa but also that Germany had asked for 10,000 tons of aviation gasoline, 4,000 tons of motor oil, and 15,000 tons of fuel oil. Darlan stated that he was obligated to ship 5,000 tons of gasoline to Italy under a prior agreement but said that he would try to get this from continental France. Leahy advised Darlan that continuing North African oil shipments to the Axis powers would make it difficult for the United States to aid continental France and its African colonies. At the end of the year Darlan admitted that the French were shipping oil to Nazi forces in Libya but claimed that this was in exchange for material France had received from Germany and that Germany had threatened to occupy Morocco unless it received this petroleum.

Throughout 1942 the controversy over oil shipments from French North Africa to the Axis powers raged unabated, generating a thick stack of diplomatic correspondence. On March 27 the British Government sent an *aide-mémoire* to the Department of State protesting the shipment of oil products to North Africa, unless the French in turn exported strategic raw materials from there to the United States and the United Kingdom. The French Embassy in Washington replied in a letter to the State Department, dated April 8, that the Axis powers would intercept shipments of cobalt and other North African minerals to the United States and that North Africa could not endure without outside help such as the American government was providing. The American government then asked France to stop cobalt shipments to Germany. The Vichy regime responded that it would be unable to do this officially without making the Germans suspicious but that it would halt shipments of cobalt and molybdenum to Germany in French bottoms. Despite this running dispute, the State Department agreed on October 21 to furnish French North Africa with 40,000 tons of petroleum quarterly. By this time, though, the Allied invasion of North Africa, which was to bring about the fall of the Vichy regime and the Nazi occupation of France, was imminent.

## Russia, Eastern Europe, and the World

The Soviet Union was an ally of Nazi Germany under a nonaggression pact from August 23, 1939, to June 22, 1941, when Adolf Hitler unleashed a military attack against Russia. During this two-year period the export of oil and aviation gasoline from the United States to the USSR came to a halt, although there were shipments of regular gasoline to Siberia. When Michigan Democratic Representative Frank Hook wrote Cordell Hull about the extent of American petroleum exports to the Soviet Union, Hull replied on February 24, 1940, that there was no existing law prohibiting oil exports from the United States to any other country. Hull did add the qualifying observation that the State Department "has taken steps to discourage the further delivery of plans, plants, manufacturing rights, or technical information required for the production of high quality aviation gasoline to countries the armed forces of which are engaged in the unprovoked bombing or machine-gunning of civilian populations from the air."[46] This was in obvious reference to earlier Russian attacks.

American attitudes towards the Soviet Union began to shift in a more positive direction after the German invasion of the latter country. The liberal columnist I.F. Stone complained in an article entitled: "Russian Lives and Oil Patents," published in the *Nation* on September 26, 1942, that U.S. petroleum companies were reluctant to share American processes and patents with the Soviet Union. Stone added that several key officials in the State Department also were holding back Russian access to these, despite three presidential directives ordering all possible aid for the USSR. He identified the key governmental obstructionists as Adolf A. Berle, Jr., Max W. Thornburg, and Loy W. Henderson. Russia had been seeking this sort of technical information since January 1941, when it was still an ally of Germany.

According to Stone, the oil men were afraid that the Soviet Union might use these processes and patents to obtain a competitive edge in marketing oil in the postwar world. In addition, various State Department officials questioned Russian competence in producing petroleum and doubted that America could spare the materials. Nevertheless, by the end of September in response to the Stone article, Ickes had written to Secretary of the Treasury Henry Morgenthau, outlining the need for an aviation gas plant for the Soviet Union. This led Oscar Cox to advise Harry Hopkins on September 29 that "Morgenthau's announcement on the Russian oil plants may start a long war between him and Ickes."[47] Cox at that time was the General Counsel for the Lend Lease Administration.

Not everyone in the United States was as concerned as I. F. Stone that Russia was not producing enough petroleum products. On December 30, 1942, Democratic Senator Francis Maloney of Connecticut wrote to Lend Lease Administrator Edward Stettinius and to Secretary of State Cordell Hull that he had been informed by the Socony-Vacuum Company that the Soviet Union had been supplying oil and gas to Japan as an act of appeasement. Although Hull denied the charge, the Japanese had been producing petroleum from a concession on northern Sakhalin that had been granted originally in 1925. On May 24, 1943, Ambassador to Russia William Standley informed Hull that the USSR had asked the U.S. to accelerate the shipment of equipment for four oil refineries and to increase the components for aviation gasoline. There was not total unanimity within the government over extending such assistance to Russia, however. As Kermit Roosevelt of the State Department wrote in a March memo following a meeting at the Office of Lend Lease Administration:

> With regard to the oil drilling and other equipment needed by the Soviets to restore production in the Caucasian fields, it was pointed out that it is very difficult to handle this program piecemeal. The War Production Board and other agencies concerned are reluctant to grant high priorities on individual requests when it is possible that the Russians will later present further demands of even greater urgency.[48]

Although the Soviet dictator, Joseph Stalin, was far from satisfied with the amount of lend lease that the United States had sent the USSR, frequently over circuitous and hazardous routes, the highly suspicious Russians did not always reveal their most pressing needs. Furthermore, the American government in the long run was so generous in supplying lend lease aid to Russia that Stalin questioned U.S. motives. Thomas A. Bailey suggested in his American diplomatic history textbook that the United States should have considered discontinuing lend lease to Russia after 1943, when the tide had begun to turn in favor of the Allies. Such a hard-line approach, he believed, might have induced the Kremlin to make certain concessions that it otherwise would not have made.

Even before the war had ended, the Russians had begun to seize the petroleum well equipment of Rumania and to ship it to the Soviet Union. Not only did the Soviets wish to establish friendly Marxist political regimes in Eastern Europe, they also hoped to set up an economic sphere of influence there. For this reason, the developments pertaining to oil in Austria, Hungary, and Rumania are quite important. (In Austria, like Germany, the Russians were able only to exert a dominant influence over a single zone.)

If they were thwarted in their efforts to obtain as much petroleum equipment as they had hoped from America, their occupation of Rumania allowed them to confiscate the oil equipment there and transport it to the USSR. When the American and British governments learned of this action late in 1944, they protested strongly to the Russian government, as the property of both American and British companies was involved. Vice Commissar Andrei Y. Vishinsky took the position that the property in question (23,000 tons of tubes, parts, and other equipment) belonged to the Germans and was in excess of what the Rumanian oil industry needed for rehabilitation. Vishinsky went on to say that his government looked upon the confiscated items as war booty, and he suggested that they should be written off as a minor lend lease shipment.

Unfortunately, Vishinsky's ideas did not coincide with those of the U.S. government, which suggested that a tripartite commission of oil experts should be created to survey the situation in Rumania. The Potsdam Conference set up such a commission, and its first meeting took place on August 18, 1945. Compared to other issues facing the victorious Allied leaders, Stalin was of the opinion that the Rumanian oil question was a "trifling matter."

By the beginning of the summer, the State Department had begun to display a more conciliatory attitude toward the Soviet Union over Rumanian petroleum. As Acting Secretary of State Joseph Grew observed to the American representative in Rumania, Burton Berry, on June 2, the "Department realizes that American-owned companies, like other oil companies in Rumania, will be required to contribute part of current output for payment of Rumania's reparation obligation to U.S.S.R."[49] on a nondiscriminatory basis. One key factor that hindered the three-nation Petroleum Commission from reaching a mutually satisfactory settlement was the actual value of the casing and other equipment that the Russians had seized from Romano-Americana. While Standard Oil placed this amount at $800,000, Romano-Americana estimated its book value at $1,971,896 and its replacement value at $2,542,159.

When the Rumanian peace treaty was before the United Nations in 1946, British Foreign Minister Ernest Bevin, a member of the Socialist Labour Party, stated that Great Britain was not opposed to nationalization *per se* but that it expected all nationals to receive equal treatment in Rumania. Despite their differences of opinion, both the United States and the Soviet Union were critical of the British attempt to include a special provision relating to petroleum in the annex to the peace treaty. Both countries felt that Article 24, which dealt with UN property in Rumania, made adequate provision for compensation.

The controversy continued to drag on unresolved. On July 21, 1947, the American Embassy in Moscow protested to the Russian Foreign Office the inability of the joint U.S.-Soviet petroleum commission in Rumania to reach a settlement. Unfortunately, the failure of this body was only one of the numerous bitter fruits of the post-1945 "Cold War."

In September 1945 U.S. diplomats in Austria had informed the State Department that the Russians were trying to persuade the new coalition government, headed by Socialist Karl Renner, to nationalize crude oil deposits. The USSR hoped to obtain a large share of these by the creation of a joint Austro-Soviet corporation. In defending this scheme, Soviet Diplomat Andrei Y. Vishinsky informed the U.S. Chargé in the Soviet Union, George Kennan, that the petroleum enterprises in question were located in the Russian zone of Austria. Using the recent Potsdam decrees as a justification for the Soviet plan, Vishinsky added that these oil properties had once been the property of the Germans. When the Renner government, as a result of pressure from the United States and its allies, suspended talks with the Russians, the USSR cut off deliveries of gasoline from the region of Zistersdorf, thus causing a gasoline shortage, and started to regulate the American and British oil companies operating under Soviet rule in Austria more strictly.

The U.S. interest in Austrian petroleum operations was far more than academic. By the end of World War II Anglo-American oil investments there exceeded $100 million in value. Socony-Vacuum and Royal Dutch-Shell once held the RAG, an Austrian producing company, but the Germans had seized its holdings without compensation in July 1940 under the terms of the Bitumen Law. It was these rights that the Russians took over after World War II.

Soviet authorities confiscated eighteen American tankers en route between the Vienna and Lobau refineries in late September of 1946, and it took the intervention of the highest Allied military authorities to force their return. Even more serious was the Russian failure to pay the Americans and the British for their crude and their finished products; by September 1 the Soviet indebtedness had reached $2 million. During this time, the Americans, British, and French continued to oppose the Russian scheme to participate in a new state petroleum firm.

At the outbreak of World War II, Hungary, unlike Austria, was not a part of Germany. In late 1940 I. G. Farben had approached Jersey Standard about the purchase of its Hungarian subsidiary, Magyar Amerikai Olajipari Reszvenytarsasag, abbreviated as Maort. Maort had been organized in 1938 as a wholly owned subsidiary of the European Gas and Electric Company, which in turn was almost totally owned by Jersey Standard. Negotiations had progressed to the point where the German firm and its American counterpart met in Rio de Janerio in July 1941 and agreed to a purchase price of $24 million in gold. Secretary of State Cordell Hull, however, was opposed to the proposed transaction, and the Treasury Department was also hesitant to approve this deal, believing that "to have approved the proposed sale. . . would have been tantamount to an approval of Germany's new economic order in Europe."[50]

Rather than siphon off its profits for the benefit of its investors, Maort had reinvested most of its earnings. By 1941 it had become one of the most important industrial firms in Hungary. In this same year, though, the Hungarian government seized Maort and diverted the output from its fields to Germany. After the war Maort was returned to its original American owners. By 1948 it was employing approximately 3,800 Hungarian nationals and had accumulated properties with a book value of $20 million.

Unfortunately for the U.S. petroleum firm, Hungary then became a satellite of the Soviet Union. Between September 18 and September 25 of that year the Hungarian police held Raul Ruedemann and George Bannantine, both American citizens, in custody on the grounds that these two individuals had been engaging in economic sabotage against Hungary. They later were released and expelled from the country. Moreover, while they were in prison on September 24 the Hungarian Prime Minister, Lajos Dinnyes, issued a decree under which the government seized the holdings of Maort. This step was taken, according to a contemporary Jersey Standard document, "in order to prevent willful sabotage of the production of crude oil, which is of first rate importance from the viewpoint of national economy, and in order to secure undisturbed production."[51]

On November 30 the American legation in Budapest protested the seizure to the Ministry of Foreign Affairs. In its complaint the U.S. government denied the allegations made against the Jersey Standard subsidiary and complained that their captors had forced Reudemann and Bannantine to copy in longhand certain documents condemning Maort. If there had been a reduction in Maort petroleum output, the United States noted, it was because during the summer of 1947 the Secretariat of the Hungarian Supreme Economic Council and experts at the Hungarian Ministry for Industry had felt that this step was necessary.

Despite these protests, the United States was unable to reverse the nationalization of oil and other properties in Hungary and other Eastern European countries following World War II. In a significant *volte face* a quarter of a century later, the Export-Import Bank extended a direct loan in September 1972 to Impexmin, an agency of the Rumanian government, for the purchase of a jack-up rig. The $1,191,809 loan was to be for five years, at 6 percent interest. Although the government of Rumania had sought such a loan for some time, the Export-Import Bank had refrained from making loans to Eastern European nations for a period of five years, as the result of a 1971 legislative ban that withheld government credits from those countries supplying North Vietnam.

According to at least one report the Soviet Union believed that the oil discoveries in southeastern Asia played a major role in shaping U.S. policy in Indochina around 1970, a theory that the American government challenged. Certainly the impact of petroleum on Cuban-American relations in 1960 is less a matter of conjecture. When three U.S. oil companies operating in Cuba were reluctant to refine the crude oil the Russians were shipping there under a new trade agreement, the Castro regime took over the American petroleum holdings and operated them itself.

It has been Middle Eastern oil, however, that the Soviet Union has most ardently wished to divert to its own use in the twentieth century. Following the close of World War II, the Russians balked at withdrawing their military forces from northern Iran until Iran signed an oil agreement with them. The Iranian government entered into such an understanding with the Russians in April 1946 but repudiated the agreement in October 1947. In the early 1950s the Communist Tudeh party hoped to gain control over Iran after the expropriation of Anglo-Iranian's petroleum holdings in 1951, but their hopes collapsed following the firing of Prime Minister Mohammed Mossadegh in 1954. Seventeen years later, in 1971, Russia advanced a $224 million loan to Iraq to build two oil pipelines and a petroleum refinery. This Soviet exercise in foreign aid ultimately proved unsatisfactory to Iraq, which declared war against Iran in 1980. Given the continuing instability of the region, it is not surprising that there has been talk in the United States in recent years of a preventive military strike in the Middle East to forestall a possible Soviet attempt to take over its petroleum fields.

Despite its interest in Middle Eastern oil, the fact remains that the Soviet Union itself has been and is a major producer of petroleum, which has been and is a threat to free world oil markets. One president who went to great lengths to circumvent the challenge, after such U.S. oil company executives as Gordon W. Reed of Texas Gulf and George F. Getty II of Tidewater had complained to the American government, was John F. Kennedy, who ordered the State Department to take steps to halt this Russian penetration. Among other things, the State Department sent communiqués to several U.S. embassies, instructing them to discourage Russian oil imports into these countries. The effectiveness of the campaign is to be seen in the embargo that the North Atlantic Treaty Organization nations imposed on shipments of large diameter pipe to the Soviet bloc on the grounds that this item would enhance the Russian war potential.

By this time, however, the Soviet Union had surpassed Venezuela to become the world's second-ranking producer of petroleum. In November 1961 Secretary of the Interior Stewart Udall requested an investigation of the long-range consequences of Russian oil exports by the National Petroleum Council. The following year, the twenty-three member committee released a two-volume report. The report concluded that these exports would proliferate, unless the free world took action. The report appeared at a time when there were an increasing number of advocates of "East-West" trade in the United States. But in 1983, as in 1962, the most disturbing aspect of the Russian oil riches is not that the Soviet Union will undermine free world markets but that, unlike the Axis powers during World War II, it has the petroleum potential with which to fight a major war against the United States and its allies.

## Western European and British Developments: 1945-1973

Between 1945 and 1973 U.S. interest in the marketing of petroleum in Great Britain and Western Europe peaked on three occasions. These were at the time of the inauguration of the Marshall Plan in 1948; at the time of the oil embargo that followed the Suez Canal crisis of 1956; and finally, at the time of the 1967 Arab-Israeli war, which also involved the closing of the Suez Canal. On all three occasions the American government helped to alleviate what otherwise would have been a severe oil shortage.

The Marshall Plan, also known as the European Recovery Program (ERP), went into operation at a time of a global petroleum scarcity. After a number of members of Congress had become concerned that there might not be enough oil to supply American domestic needs, the U.S. representative in Europe to the ERP, W. Averell Harriman, pointed out that under the ERP the European producers could expand petroleum production. These producers, moreover, did use ERP funds to build additional refineries to process the less expensive imported crude oil.

American oil companies then began to pressure the American government to stop ERP funds. By the end of 1949 the Economic Cooperation Administration (ECA) not only was withholding its approval from European crude oil production schemes but also reducing its financial assistance to refineries there. One-third of these refineries were built and owned by U.S. firms. Dr. Oscar Bransky, the acting chief of the Economic Cooperation Administration's petroleum branch, stated that the European controlled refineries "would seriously reduce market outlets for American-owned oil and thus jeopardize American concessions in foreign lands."[52]

As a result, by the middle of 1950 highly priced oil constituted 11 percent of the ECA shipments to Europe. Walter Levy, who had resigned from Socony-Vacuum in July 1948 to direct ECA's petroleum division, observed to the National Petroleum Council in 1949 that "without ECA aid Europe would not have been able to afford during the last year, and could not afford during the next three years, to import large quantities of American oil—from either domestic or foreign sources controlled by American companies."[53] Not surprisingly, U.S. petroleum firms expanded their oil refining and marketing facilities in Europe in only a few years.

According to the thirtieth ECA report, which covered the period from April 1948 to November 1950, procurement authorizations for petroleum and petroleum products totalled just over $1 billion. No less than $724 million were for Great Britain, France, and Italy, where Standard Oil, Royal Dutch-Shell, and Anglo-Iranian were in an ascendant position. During the forty-five months the European Recovery Program was in operation, the ECA financed $11.4 billion in imports for the participating nations, and petroleum made up over 10 percent of the total. Most of the petroleum imported consisted of offshore oil, which could be delivered to Europe more cheaply than petroleum from the United States. Since control over this oil rested most entirely in the hands of multinational petroleum companies, the entire ECA oil operation merely continued the dominant international role of the great petroleum firms.

Unfortunately, oil imports into Great Britain during 1949 and 1950 became entangled in a complex sterling-dollar rivalry. During July 1949 the British, who were facing a dollar deficit, decided to restrict the dollar outlays of the sterling bloc by 25 percent; for that year alone the gold and silver deficit of the sterling bloc was to reach $1.5 billion, and petroleum was responsible for approximately 45 percent of the deficit. The fact that the British firms had to purchase much of their oil equipment from the United States contributed to the problem. Yet, the advantage was by no means totally with the Americans. The British exchange control system made it almost impossible for U.S. oil firms to sell petroleum to nations outside the sterling area, even to such countries as Argentina and Norway.

When the British announced to Jersey Standard, Socony-Vacuum, California Standard, and Texaco that they were going to reduce the dollar share of their oil imports, the American government and the oil companies pressured Great Britain into postponing this planned cutback from January 1 to February 15, 1950. Devaluation of the pound, which made dollar petroleum even more expensive, further complicated matters. Again compromising in reaction to heavy U.S. pressure, Great Britain agreed to reduce her total dollar imports by a quarter and to withhold the implementation of any further restrictions as long as the European Recovery Program was in progress.

The American oil companies were far from satisfied. In early 1950 the National Petroleum Council issued a report, charging that Great Britain, instead of safeguarding its reserves, was attempting to obtain long-term commercial advantages for British petroleum firms. On the floor of Congress, enraged members introduced amendments to the European Recovery Program legislation that were designed to protect American petroleum firms against such discrimination. The Truman administration, though, was well aware that a total abandonment by the British of discrimination would probably deal a fatal blow to British dollar reserves. While the administration did persuade Congress that the implementation of such restrictions was the responsibility of the Economic Cooperation Administration, it also presented the British with a note in April, affirming that American firms must maintain their prerogative of trading anywhere in the sterling area. On February 1, Secretary of State Dean Acheson had issued a press release in which he stated that Great Britain had acted without adequately consulting American petroleum companies; he added that the British should have implemented any restriction on dollar oil imports in a gradual manner.

Fortunately for both parties, by the summer of 1950 a truce had last been reached in the Anglo-American oil imports feud. Rather than launch a price war against their British petroleum rivals, Standard-Vacuum and Caltex won back 1.5 million of the 4 million tons lost under the British substitute plan by agreeing to a British incentive scheme. Under the plan, they could market their oil for sterling provided that they expended the proceeds on British-built tankers, drilling machinery, and refinery equipment, none of which could be used in Great Britain itself. Jersey Standard and Socony made agreements of their own that differed somewhat from the Standard Vacuum-Caltex deal. The outbreak of the Korean War during the summer of 1950 helped to resolve the Anglo-American oil import feud as it seemed certain to eliminate the threatened global petroleum surplus underlying it.

In March 1951, however, the Iranian government expropriated the holdings of

Anglo-Iranian, following a long dispute. The United States officially adopted a neutral position toward this seizure, which would have global consequences, thus irritating Great Britain. But the United States did set up a Foreign Petroleum Supply Committee that summer, consisting of nineteen American oil companies operating abroad. The group attempted to combat the impending shortage of petroleum in Western Europe and elsewhere, by accelerating the production of crude oil and the manufacture of refined petroleum in a number of countries. Finally during the summer of 1954 an oil settlement with Iran was reached with the assistance of the American government. Under the settlement, the group of American, British, and Dutch multinational firms known as the "Seven Sisters" were to operate the oil fields of Iran under a consortium agreement in conjunction with a French petroleum company. This arrangement soon was modified to permit the participation of a number of independent American oil companies, which profited greatly therefrom.

Europe again faced an oil shortage in the summer of 1956, when Egypt, led by President Gamal Abdel Nasser, blocked the Suez Canal, thus jeopardizing the delivery of petroleum. In the aftermath of this event the U.S. Office of Defense Mobilization drew up a plan of action that it then turned over for elaboration to a Middle East Emergency Committee, which included representatives from fourteen oil companies. The result was probably the most gigantic operation in petroleum logistics ever attempted. Not only were tankers routed from the Persian Gulf around the Cape of Good Hope to Great Britain—an 11,000 mile journey—but the United States, Canada, and several Caribbean nations sharply increased their oil exports to Western Europe. The Middle East was to supply Western Europe with approximately 800,000 barrels daily and the Western Hemisphere (especially the United States) with around 400,000 barrels. Since neither Great Britain nor Western Europe had the dollars to pay for the petroleum they had been purchasing for sterling, it was necessary for them to obtain hundreds of millions of dollars in credit from the United States.

Although the American government helped to maintain the Western alliance by assuring oil supplies to Great Britain and France, this was hardly reassuring to the Arab world, which already was irritated by the United States support for a Jewish national state in Palestine. Saudi Arabia would not allow any of its petroleum to be shipped over the Tapline through Jordan and Syria to the coast of Lebanon, if the oil was to go to either Great Britain or France. Nor was the Arab world reassured when Great Britain and France, taking the side of Israel, invaded Egypt just prior to the American presidential election, in which Eisenhower won a decisive victory. If the United States proceeded with its European oil supply program too vigorously, there was a real danger that the Arab states might cut off their petroleum shipments to the free world.

The closing of the Suez Canal was not the only obstacle confronting the oil-consuming nations of Western Europe. On November 3 the Syrian army blew up several pumping stations that belonged to the Iraq Petroleum Company, thus further interfering with the flow of petroleum to the West. The Syrians, however, only blocked the British pipeline from Kirkuk to Tripoli; they did not interfere with the American pipeline from Dhahran to Sidon.

Given these circumstances, President Eisenhower was well aware that the international situation with which he was dealing was a highly treacherous one. Thus at a bipartisan legislative meeting held on November 9, Eisenhower brought up the matter of the sabotage of the Middle Eastern oil pipelines. At this same meeting, however, in obvious reference to the recent Russian invasion of Hungary, Eisenhower also observed, "We must never forget that Moscow remains the big enemy."[54]

One of the few groups to express its approval of the developments culminating in the oil shortage were the independent U.S. oil producers. Long forced to compete with the large quantities of crude petroleum imported from the Middle East, independent American oil producers now could improve their position, and they exploited this opportunity to the fullest.

Seventeen nations met in Paris that November to form an oil pooling agreement that was to control petroleum imports by means of a collective, cooperative effort at the governmental level. But rather than have Europeans buy their oil directly from the U.S. government, American officials preferred for them to purchase it on the open market from private U.S. oil firms. To help facilitate the process, the Justice Department approved a scheme that would enable fifteen American petroleum companies to form a single marketing combine, which then could supply Europe with oil without facing possible U.S. antitrust prosecution.

When the French and the British announced that they were withdrawing their troops from Egypt—they were to complete the process by Christmas—it was easier for the American government to give its full support to the European oil supply program. By this time the industries of Great Britain and Western Europe were beginning to feel the lack of petroleum, and unemployment stood at several millions. The Middle East Emergency Committee, which had ceased to hold meetings when Great Britain and France invaded Egypt, now began to meet again, especially after the Soviet Union hinted that it might ship oil to Europe. There were by then petroleum shortages in France, Spain, Denmark, and West Germany, as well as oil rationing in Great Britain, Sweden, and Turkey.

Many individuals in Great Britain and France became highly embittered by the American failure to support those nations at the time of the Egyptian invasion. While some Frenchmen felt that the United States was trying to force France out of North Africa, British M.P. Julian Amery focused his observations on an area further to the east, specifically tying in petroleum competition:

> Ever since the war U.S. policy has been whittling away at British prestige and power. They pushed to get us out of Palestine. They pushed us out of Abadan. They pressured us out of the Suez Canal, and advised us to leave the Sudan, and let a war start over the Buraimi Oasis between their oil satellites and our oil satellites.[55]

These expressions of discontent aside, by March 1957 Assistant Secretary of the Interior Felix Wormser was able to report to the House Committee on

Interstate and Foreign Commerce that the oil lift was now supplying 90 percent of Europe's overall normal requirements. Even the more conservative European estimates of motor gasoline, gas oil distillates, and heavy fuels set the amount at about 80 percent. The growing effectiveness of the oil lift had become manifest in February, during the course of an unusually mild winter. It seemed that the peak of the crisis had passed by this time, although supplying petroleum to Europe remained a serious problem. It was at this time that Egypt agreed the salvage operations should be completed, in order to reopen the Suez Canal in April. Syria also consented to the repair of the sabotaged Iraq Petroleum Company pipeline. By the middle of the year the canal was in full operation, and the pipeline was in a state of reasonable repair. It therefore became possible to terminate gasoline rationing in Great Britain and France.

In view of their threatened access to Middle Eastern oil, why didn't the nations of Western Europe begin to explore alternate sources of energy at this time? In 1952 France, the Netherlands, West Germany, Italy, Belgium, and Luxembourg had joined forces to set up the European Coal and Steel Community (ECSC), which had become operative in the following year, but between 1956 and 1967 the Western European dependence on coal, instead of lessening, increased to keep pace with its growing industrialization. Multinational oil refineries in Western Europe, moreover, produced large quantities of fuel for sale at low prices, which led to the near collapse of the coal industry.

As the oil-producing nations moved toward an alliance that would culminate in the establishment of OPEC in 1960, the countries of Western Europe were themselves forging a closer economic union with the creation of the European Economic Community (EEC) in 1957. The organization, which became operative in 1958, consisted of the same six nations that formed the European Coal and Steel Community; at the same time that they set up the EEC, the six nations also formed the European Atomic Energy Community (EAEC). In 1959 Great Britain, Denmark, Norway, Sweden, Portugal, Switzerland, and Austria joined forces to inaugurate the European Free Trade Association (EFTA). Finally in 1961 twenty nations attempted to promote economic integration between North America and Europe through the establishment of the Organization for Economic Cooperation and Development (OECD). By this time friction had developed between the EEC and the EFTA, which had been set up partly because of the breakdown in negotiations to align Great Britain with the EEC.

Within several years the Middle East was on the verge of another eruption. In December 1966 following a dispute, Syria closed the Iraq Petroleum Company pipelines. Nearly six months later, on May 23, 1967, President Gamal Nasser of Egypt decided to block the Strait of Tiran to prevent Israeli ships from passing through. As a result, the State, Defense, and Interior departments again were concerned about the availability of petroleum products for Europe, if conditions should worsen.

The six-day Arab-Israeli war broke out in early June 1967, lasting from June 5 to June 10. As in 1956 global petroleum commerce was disrupted. After Egypt and Jordan had charged that Great Britain and the United States had supplied war planes to Israel, the oil-producing Arab countries, led by Libya and Iraq, embargoed petroleum shipments to both countries. Egypt again closed the Suez Canal, but Europe fortunately had enough oil. The growing use of giant supertankers produced a shipping surplus rather than a shortage this time.

Although the 40 million barrels of Arab petroleum the United States was using to fight the Vietnam War could not be replaced easily, the closing of the Suez Canal in conjunction with the Arab-Israeli war of 1967 caused fewer economic dislocations than its shutdown at the time of Arab-Israeli war of 1956. It is true, though, that the higher cost of oil from the Western Hemisphere not only caused economic problems for the Europeans but also aggravated Great Britain's balance-of-payments crisis. Yet, after the oil-producing nations of the Caribbean and the United States increased their output to meet the needs of Europeans, a petroleum glut, which depressed prices, occurred when the canal was reopened.

It was during the fall of 1969, in an attempt to counter any similar future emergencies, that the State Department broached the possibility of developing a common approach to future energy problems at a meeting of the Organization for Economic Cooperation and Development. At the OECD meeting, held in Paris in May 1970, Assistant Secretary of State Philip Trezise proposed that these problems should be considered in a multilateral context, but he received only a lukewarm response. Most delegates felt that Europe and Japan were not in danger of losing Arab oil, as they adopted a pro-Israel stance. Perhaps the two strongest defenders of the American position were Great Britain and the Netherlands, countries closely identified with two of the multinational petroleum giants, Anglo-Iranian (now British Petroleum) and Royal Dutch-Shell.

As late as 1967 one European minister observed, "American companies brutally conquered our market; if they do not keep us supplied at all times, they will be expelled."[56] Yet, by 1971 relations between the U.S. oil companies and the European oil-consuming nations had become more friendly, in part because both the petroleum firms and the State Department now kept the various European governments informed of OPEC's activities. As the challenge from OPEC mounted, such cooperation became not only desirable but a matter of survival, especially during the great Arab oil embargo in 1973.

Apart from such established oil-producing areas as the Middle East, a new petroleum area in the North Sea appeared in the late 1960s. In 1969 Phillips struck oil in the Norwegian sector, and British Petroleum made its first significant discovery in the British zone a year later. The British government then attempted to favor British firms and independent companies rather than major American firms. Nevertheless, by 1973 the five U.S. "sisters" controlled a majority of the North Sea oil; British Petroleum, with 20 percent, and Royal Dutch-Shell, with 15 percent, enjoyed a total share of only slightly more than one-third of these potentially enormous reserves. The economic efficiency and long-standing technological superiority of the great American petroleum firms, so frequently displayed elsewhere, at this time manifested itself in the backyards of Great Britain and Western Europe.

# 3.

# Canada and Mexico

## Canada to 1945: The Canol Project

Petroleum did not begin to play an important role in Canadian-American relations until World War II. Such standard works as Edgar W. McInnes' *The Unguarded Frontier: A History of American-Canadian Relations* (1942) and Hugh L. Keenleyside and Gerald S. Brown's *Canada and the United States: Some Aspects of Their Historical Relations* (1952) do not even mention oil. It was not until the late 1940s that the discovery of large-scale petroleum and natural gas fields in Alberta and the prairie provinces ushered in a new era in Canadian history. In the decade that followed, the longest oil and natural gas pipelines in the world were constructed from these fields to Ontario in the east and to Vancouver in the west, with additional links southward to the American market. Even with these fields, Canada still imports a considerable amount of petroleum from both Venezuela and the Middle East; it is the relatively isolated Atlantic Provinces that are the most dependent upon foreign oil.

At the turn of the twentieth century, Standard Oil obtained its most significant acquisition in Canada, the Imperial Oil Company, Ltd., which had been in operation since 1880. After its takeover of Imperial Oil, Standard Oil controlled approximately 75 percent of the petroleum refining in Canada, and Canada came to rely on the United States as the major source of its oil imports.

In 1919 Imperial Oil successfully drilled for oil around Fort Norman, a minor outpost 100 miles south of the Arctic Circle in the Northwest Territories. Unfortunately only three of the six wells drilled actually produced petroleum with the result that after a few years Imperial Oil shut down its operations there, including a small refinery. When a minor boom occurred near Fort Norman in 1932, Imperial Oil reopened its oil wells and refinery, bringing in a seventh well. Total production, however, was limited to the range of 400 barrels a day, which hardly constituted an economically viable field.

Six months after Pearl Harbor, wartime pressure for petroleum led American and Canadian officials and oil men to sign a contract under which the U.S. Army would participate in the development of the Fort Norman field. The key component of this project, known as "Canol," was a 550-mile-long pipeline to transport the crude oil to Whitehorse in the Yukon for refining. From Whitehorse there was to be a 110 mile long pipeline to Skagway on the Alaskan coast and still another to the Watson Lake air base. In addition, there were plans to truck the refined crude oil from Whitehorse to Fairbanks in central Alaska along the Alaskan highway. The realization of Canol was slow, and consequently the resulting oil production was minimal.

As early as May 29, 1942, Interior Secretary Harold Ickes had begun to complain to President Roosevelt that "from any operating standpoint, this development would seem to be decidedly impractical."[1] Ickes advised the President that experts viewed the Canol project most critically. In his reply, dated June 13, FDR admitted that Canol was not commercially feasible but noted that the recent Japanese attack on Dutch Harbor, Alaska, had made clear the need for more petroleum, and therefore the Canol project had his full approval. Having been rebuffed, the Interior Secretary turned to Secretary of War Henry Stimson, writing to him on June 22 that storage bases should be erected at Fort Norman as a safeguard; Ickes was concerned about the operating hazards of the pipeline itself. Then on June 29 Ickes again wrote the President, stating that it would be desirable to supplement the Fort Norman output with another means of supply.

During the fall of 1943, though, Ickes' call for the immediate abandonment of the "impractical and wasteful" $34 million project led the Truman Committee to hold hearings on Canol. This was a special U. S. Senate committee to investigate the national defense program, headed by Democrat Harry Truman of Missouri. Among other things, the committee learned that the U.S. Army had approved the project without adequate consultation; that experts at Imperial Oil were pessimistic about production and transportation; that the projected cost was now far in excess of the original $25 million estimate; and that in the long run the Canadian government and Imperial Oil would be the real beneficiaries of the project. It was this final point—that the U.S. government would not have a postwar share in the Fort Norman field that it had both financed and developed—that was the major argument against Canol, along with the exorbitant cost of the project.

Among the witnesses testifying before the Truman Committee other than Ickes—who told that body that he had learned of the project by accident—was James H. Graham, a seventy-three-year-old former dean of engineering at the University of Kentucky, who was now a dollar-a-year advisor to the chief of the Army Service Forces, Lieutenant General Brehon B. Somervell. Graham, it was learned, had been the author of the one-page memorandum to Somervell that had been the basis of the Canol project. Graham, who had not consulted with petroleum experts before recommending the project, took the position that he never considered costs in wartime. Still another defender of the scheme was Under Secretary of War Robert Patterson, who boldly asserted that his department was proud of Canol. On November 23 he told the Truman Committee:

> The Canol project, as far as I know, is the first major venture of the United States government in oil production on foreign soil. After our experiences with gasoline rationing and the serious depletion of our own crude-oil reserves, I believe that the American

people may favor similar ventures in the future. . . . Our participation in oil reserves on foreign soil is necessary to our national welfare and our future defense.[2]

The actual production of the Northwest Territories oil fields had increased from 131 barrels for the first four months of 1942 to 57,729 barrels for the first four months of 1943. By the end of 1943 two dozen new wells had been sunk and were capped and waiting. The petroleum, moreover, was of high quality; according to *Time* magazine, "It is good oil, of high gasoline content and such low pour-point that the pipeline can be laid on the surface."[3] Still, by the end of 1943 the pipeline from there to Whitehorse remained uncompleted, although the line from Whitehorse to Skagway was in operation.

In all fairness to the Canadian government which allegedly was to benefit disproportionately from Canol, it should be noted that it had made available the essential lands and necessary rights of way, as well as having waived, during the course of the war, import duties, territorial and sales taxes, and equipment and supplies charges. In addition, the Canadian government had promised to remit, during the course of the war, all royalties on the Fort Norman oil production. It was this aspect of the operation that did not receive proper emphasis in the American press, which tended to dwell on the project's negative aspects.

Early in January 1944 the Truman Committee rendered its judgment on Canol: an inexcusable waste of the taxpayer's money. Although the committee conceded that it might be possible to forgive the original mistake, there was no excuse, they felt, for the continuance of the operation, since the U.S. Army had been warned by its own engineers, the American oil companies, and Harold Ickes. "The Committee," the report lamented, "has never seen a similar situation in either business or in Government."[4] Rather than meeting the onslaught head on, Lieutenant General Somervell contented himself with referring to the biblical parable of the wise and foolish virgins in the Book of Matthew.

At the beginning of May, when the construction of the refinery was finished, the Canol project was proclaimed officially complete. The final link of the 550-mile-long pipeline had been laid on February 18. While optimists predicted that the Norman wells might one day produce as much as 50 to 100 million barrels, the short-range prospects were far less favorable. By the first part of 1945, yearly production was only 1 million barrels of oil and gasoline. With the end of the war in Europe in sight, Lieutenant General Somervell decided to abandon the operation effective June 30; in giving Canada first option to purchase the facilities, he observed that he would charge the whole business to the "waste of war." In fact, the War Department had only enough money to operate Canol until July 1, and such key Senators as Republican Homer Ferguson of Michigan had stated that they would challenge any further expenditures on the project. Even though Somervell had estimated the total cost of Canol at only $101 million rather than the $134 million suggested by the Truman Committee, one assessment of the final year of Canol found that it cost $3.70 to produce a single barrel of petroleum, a rather astronomical figure.

Summing up the episode, Edward A. Harris observed in an article entitled "Canol, the War's Epic Blunder," that "the sorriest chapter of the American war effort on the home front has just been completed."[5] By this time the tanker situation had improved greatly, and there was substantial progress in the Pacific war, which was being fought far closer to Japan than to Alaska. Although the Canol project did turn out to be a "white elephant," the far more economically justifiable Alaskan pipeline of the present day also has generated a storm of criticism.

**Canada Since 1945: the Alberta Boom**

Between 1946 and 1959 American petroleum investments in Canada increased no less than 1,400 percent. Aside from Imperial Oil, other active U.S. petroleum firms included McColl-Frontenac (Texaco), Socony-Vacuum, California Standard, and Union Oil, along with various independent oilmen from Texas and Oklahoma. As a result, the Canadian share of total U.S. foreign investment during this period rose from 10 to 24 percent; in 1945 American investors held 95 percent of the total foreign-owned petroleum and natural gas properties in Canada. Although this figure did fall to 85.5 percent in 1961, the total value of these properties increased 2,500 percent in the span of sixteen years.

The increase in U.S. investment in Canadian petroleum occurred despite the widespread negative image left by the Canol project, which future Democratic Senator Richard Neuberger of Oregon once described as "a Rube Goldberg gadget on a colossal scale."[6] The highly negative tone of his article was typical of the opinion held by most American observers. Far more sympathetic was corporate historian Richard Finnie who declared that Canol not only "was one of the most stupendous construction feats of its kind ever undertaken. It was also one of the least understood and most maligned."[7] Observing that as a job of pioneering Canol surpassed the Alaska highway, Finnie stated that it was justified as both a wartime measure and a peacetime project; there were now more overland roads, airfields, and telephone lines in that part of Canada because of Canol. Furthermore, the project had furnished valuable experience in transportation and construction under sub-Arctic conditions.

In August 1947 Imperial Oil, Ltd.—which was operating the Norman Wells field—made a successful bid of $1 million for the Whitehorse, Yukon refinery. Built in connection with Canol, it had stood idle for over two years. Although the materials used to construct the refinery had cost $6 million, in April 1947 one bidder had offered as little as $150,000 for the refinery, a sum that the American government had rejected as an insult.

Further to the south, in Alberta, entrepreneurs had discovered petroleum in such enormous quantities in February 1947 that the town of Redwater, thirty-five miles northeast of Edmonton, quickly became the headquarters for the biggest oil boom in Canadian history. It was hoped—and expected—that the petroleum discovery would help to reduce the dollar deficit that Canada was experiencing. In 1948 the country had spent $200 million on crude oil imports alone; only coal and industrial machinery ranked ahead of petroleum. For the United States, however, the key consideration was the role this oil might play in the future defense plans of the two nations.

U.S. petroleum activity in Canada continued to flourish during the next decade, and there were no new major developments in America's oil policy. Although President Eisenhower's special cabinet committee technically did not recommend that Canadian oil should be subjected to discrimination, Canadian petroleum shipments did begin to decline for a number of reasons, including increased supplies of crude petroleum from the Middle East and Far East, the availability of cheap ocean tankers, and a general levelling off of the oil market due to an economic recession.

On Christmas Eve in 1957 Eisenhower extended the earlier voluntary restrictions on petroleum imports to five western states, including Washington and Oregon, which were not then producing oil. The government claimed that it was reducing these imports to help stimulate the development of adequate petroleum reserves within the country in case of an emergency. The official justification ignored the fact that while Venezuela and other nations were shipping oil to America by sea and thus were vulnerable to an enemy attack, Canada transported its petroleum over land.

To the surprise of no one, Ottawa protested bitterly. Finance Minister Donald Fleming declared angrily, "The Canadian government cannot accept the view that there is any justification for U.S. limitations on oil coming from Canada on either economic or defense grounds."[8] The Canadian Embassy in Washington pointed out that these restrictions on oil imports were in violation of the General Agreement on Tariffs and Trade, the American-Canadian principles of economic cooperation on defense matters, and the recent North Atlantic Treaty Organization understandings. In terms of the immediate future, the embassy's statement continued, Canadian petroleum exports to the Pacific Coast probably would not decline drastically, but the restrictions would prevent that section of the United States from absorbing Canada's rising petroleum output in years to come, which in turn would discourage the search for additional oil reserves in Alaska and western Canada.

Canadian petroleum suffered still another blow on March 10, 1959, when a system of mandatory controls went into effect in the United States. Under the system, there was to be a reduction in Canadian oil exports to District V (the American West), which then was importing petroleum from the new oil fields of western Canada. At the end of April, however, the American government decided to exempt Canadian petroleum from this system of mandatory controls. The only other country thus favored was Mexico, which, like Canada, could ship its oil by land under a series of modifications that went into effect on June 1, 1959. This time, though, the mandatory program did not exempt ocean shipments from British Columbia to refineries in California, as it had in the aftermath of the Suez crisis.

Canada now began to develop its own program for petroleum. In February

1961 the Canadian government made public a national oil policy that called for the production of 640,000 barrels per day of crude oil and natural gas during that year, with 160,000 additional barrels to be produced daily by 1963. Since Canada itself could not absorb this output, it had to increase exports to the United States. To facilitate the exportation, the Canadian government established the National Energy Board, which was to work in close cooperation with the U.S. Department of the Interior.

The Canadians were confident that its petroleum would enjoy relatively easy entry into the American market, and in fact it did expand into District V (the Pacific Coast) and District II (the Middle West) during the first two years that the national oil policy was in effect. This trend persisted even though the Interior Department had attempted to limit Canadian petroleum shipments into the United States as early as 1961. After the government of Canada had balked at such restrictions, Interior Secretary Stewart Udall asked President John F. Kennedy to reduce the Canadian quota, only to have JFK reject his suggestion.

Leading the opposition to increasing Canadian oil imports was the Independent Petroleum Association of America, which was backed by such organizations as the American Coal Association. Critics apparently were more numerous in the Middle West than in the Pacific Northwest, and opposition was apparently the strongest in North Dakota, Kansas, and Montana. During May 1962 domestic lobbyists in the United States began to place heavy pressure on Congress to terminate the Canadian exemption from the mandatory oil quota program.

In November President Kennedy did limit imports east of the Rockies to 12.2 percent of domestic production for the first six months of that year, but he excused Canadian petroleum from these restrictions. In December the Interior Department decided to restrict Canadian oil imports for the first half of 1963 to a 4 percent increase over the average for the second half of 1962, which was the percentage increase from the first to the second half of that year. A year later, in December 1963, the Interior Department sanctioned Canadian crude imports of 274,500 barrels daily for the first half of 1964 and 250,000 barrels daily for the last half of 1964.

Crude oil imports from Canada into both the United States as a whole and District V in particular continued down to 1967, when the outbreak of the third Arab-Israeli war caused the American government to take a more positive attitude toward Canadian petroleum. It was at this time that the United States and Canada reached a secret agreement over the level of Canadian oil imports. The understanding was not made public until February 1969, when the Clark Oil and Refining Company attempted to import Alberta crude for refining in Chicago. The 1967 agreement stated that no Canadian crude oil was to be allowed in the Chicago market before 1970 and that exports to Districts I, II, III, and IV were not to exceed 28,000 barrels daily in 1968 or increase after that date more than 26,000 barrels annually. There were no formal import controls, as there was no formal machinery for implementing the agreement; not surprisingly the understanding did not work well, as imports from Canada were not held within the recommended levels. A further source of irritation to the United States was that third country crude was entering the country from Canada.

By 1969 Canadian petroleum exports to the United States had reached 465,000 barrels daily. Canadian production had doubled since 1960, and by the end of the decade Canada was exporting 40 percent of its oil to the United States. But the discovery of petroleum at Prudhoe Bay in Alaska had disturbed the Canadians, who then expressed an interest in developing a continental oil policy with the United States that would eliminate both tariff and import quotas. At this time Jersey Standard, Gulf, Texaco, Sun, and other American petroleum firms controlled approximately three-fourths of the Canadian oil industry.

On May 5, 1969, *The Oil and Gas Journal* headlined an article, "Canada, under pressure, curbs oil flow to U.S." The Canadian government was informed by its American counterpart that it must "agree to a voluntary cut-back or get used to the idea, quickly, of rigid and permanent controls."[9] Yet on October 27 this same publication headlined a story: "Overland import policy falling apart." American refiners, who could save money by processing cheaper oil from Canada, were doing so, and 50,000 barrels in excess of the suggested limit were entering the United States daily.

A new era in Canadian-American petroleum relations began on March 10, 1970, when President Richard Nixon imposed temporary official limits on crude imports from Canada into Districts I, II, III, and IV. The formal daily limitation to 395,000 barrels was far in excess of the previously agreed on 332,000 barrels, which the current flow of oil from Canada already was exceeding. Not surprisingly, these import quotas irritated a number of American refiners. Newcomers resented the quotas that older importers had received, while the older importers complained that some of the refiners thus favored didn't need Canadian oil.

The Liberal government of Canada set up licensing procedures for imports of gasoline in 1971, and in early 1973 it imposed export controls on shipments of crude oil and gasoline to the United States. These steps were followed in October with a charge on petroleum exports to the United States. A year later, in November 1974, the Canadian government announced a program to limit progressively crude exports to the United States. With the establishment of a state corporation, known as Petro-Canada, it became even more obvious that the Trudeau government was placing domestic needs first and that the United States could no longer count upon Canada to balance the American oil deficits.

## The Díaz Era and the Mexican Revolution

Unlike Canada, petroleum has played so important a role in Mexican-American relations that entire books have been written on the topic. One outstanding example is Lorenzo Meyer's *Mexico and the United States in the Oil Controversy 1917-1942*. Although there had been other petroleum expropriations elsewhere in the world before 1938, the Mexican seizure unquestionably attracted the most publicity in the United States. This was the case, not only because of its geographical proximity but also because of the generous compensation that the

American and British oil firms unsuccessfully demanded from the government of Lázaro Cárdenas.

Before the growth of Mexican nationalism after 1910, the Mexican government had greeted wealthy foreigners with open arms. For a generation prior to that date, Porfirio Díaz had guided the destiny of Mexico; Díaz had made available to foreign investors lucrative concessions and contracts for the development of his nation's petroleum, mines, land, and railways. Standard Oil operated in Mexico through its affiliate, Waters-Pierce, and one of America's greatest petroleum entrepreneurs, Edward L. Doheny, arrived on the scene in 1900. By 1910 American entrepreneurs had invested approximately $1 billion in Mexico, and other foreign interests had followed suit. But a revolution that was to transform Mexican society broke out in November of that year, led by Francisco Madero. This culminated in the resignation of Díaz, who left for Europe, and forced a number of Americans to abandon their holdings.

It was during the Madero years that the Texas Company made its first investments in Mexico. The Gulf Oil Company soon followed suit with its first foreign investments anywhere. At the same time, Standard Oil of New Jersey and Standard Oil of New York obtained 400 acres of land around Tampico through the Magnolia Oil Company, which these two firms controlled. Jersey Standard also entered into purchasing contracts with Edward L. Doheny in 1911 and Lord Cowdray in 1912, although Cowdray rejected a 1913 offer made by Jersey Standard to purchase his petroleum properties.

Writing retrospectively in 1935, Samuel Guy Inman observed that "from the beginning of the revolution in 1910 until Dwight Morrow went to Mexico in 1927, the oil kings—always closely connected with other great financial interests—more often than anyone else directed the State Department's policy towards our southern neighbor."[10] While there is much truth in this generalization, the exact degree of their influence is hard to pinpoint, just as it is difficult to ascertain the exact role the American oil companies played in the revolution that drove Díaz from power. Their real or alleged political manipulations have caused widespread criticisms, but some important governmental officials have also noted their contributions. One might cite a favorable estimate of one of the leading petroleum barons (Edward L. Doheny) by a leading American diplomat (Henry Lane Wilson): ". . . under the benevolent and generous administration of [his] directing spirit . . . the treatment accorded the native labourers and their families had not only been fair and just, but would compare favourably in this respect with any industrial enterprise in the United States."[11]

If the American government favored American petroleum interests in Mexico, it did so for a number of reasons, including its resentment toward the presence of the British oil firms there. Both Bryan and President Woodrow Wilson accused Mexican President Victoriano Huerta (1913-14) of favoring the British petroleum companies. Bryan once observed to Sir William Tyrrell, the secretary to the British Foreign Secretary, Sir Edward Grey, that the only reason that Great Britain was interested in Mexico was its petroleum and that the British Foreign

Office had turned over its Mexican policy to the British "oil barons"; Tyrrell, in turn, accused Bryan of acting unknowingly as a spokesman for Standard Oil. Given the violent animosity that Bryan frequently displayed toward Wall Street, this charge may seem rather surprising. Bryan, however, was by no means the only influential figure in the Wilson administration to voice suspicions of Great Britain. Presidential advisor Colonel Edward House wrote of Lord Cowdray, after a talk with Wilson, "We do not love him, for we think that between Cowdray and Carden a large part of our troubles in Mexico has been made."[12] Sir Lionel Carden was at this time the British Minister to Mexico.

Given this attitude on the part of American government officials, one might have expected the Department of State to have looked the other way when Standard Oil attempted to meddle in Mexican politics to the possible disadvantage of Great Britain. This, though, did not prove to be the case. As there was not much liquid capital available in northern Mexico, revolutionary leaders there could not have acquired sufficient capital for their planned overthrow of the Díaz regime without expropriating property. Since they did not resort to expropriation, the money for the large quantities of arms and ammunition they purchased in the United States must have come from some source. The accusing finger was quite naturally pointed at Standard Oil, which apparently had backed Francisco Madero after Díaz had favored Lord Cowdray. Not to be outdone, Cowdray seemingly supported the Huerta coup that overthrew Madero, only to have Standard Oil in retaliation allegedly finance the revolution Venustiano Carranza then launched against Huerta.

Contacts between Standard Oil and Madero probably commenced as early as April 1911, when, according to Justice Department agents, Standard Oil made a loan to Madero in return for 6 percent gold bonds. The U.S. government was most concerned about the loan because the Supreme Court had ruled that loans to revolutionaries were in violation of the American neutrality laws and thus could not be enforced in the courts. Standard Oil officials denied that any deal had been made. It is possible the U.S. government scared the Rockefeller interests into ending negotiations or that the rebels' need for capital ended with the fall of the border city of Juarez. Existing fragmentary evidence indicates that the agent for Standard Oil may have been Henry Pierce of Waters-Pierce, the autonomous marketing subsidiary that conducted Standard Oil's operations in Mexico. It is noteworthy that the revolutionary leaders apparently were unwillingly to grant perpetual oil concessions to their American financiers and instead insisted on the right of cancellation after a number of years.

If lending money to Mexicans was one way to influence government policy, withholding the payment of taxes to the Mexican government was another. After the Wilson administration refused to recognize the Huerta regime, Edward L. Doheny refused to pay taxes to it. Shortly thereafter, moreover, Doheny agreed to the demand for $10,000 made by General Higinio Aguilar of the Carranza Constitutionalists; the transaction was made by Doheny under protest but with the knowledge and consent of the American presidential representative in Mexico, John Lind. On other occasions, however, the transactions occurred on a more voluntary basis. Thus, Doheny gave a Constitutionalist representative $100,000 in cash at the Hotel Belmont in New York City and advanced $685,000 in fuel oil to Carranza on credit. In justifying his actions, Doheny took the position that every American corporation active in Mexico at this time gave some form of assistance to Carranza.

While American oil men were courting Carranza, their British counterparts were approaching Huerta. At this time Royal Dutch-Shell was interested in acquiring some of the oil properties in the Tampico area. The Huerta regime gave these European entrepreneurs the right to operate there; they in turn organized the La Corona Oil Company at The Hague. By the beginning of 1914 one of the British wells near Panuco began to gush petroleum in large quantities. Despite the sharp competition between British and American oil men, both U.S. and British ships cruised the waters off Tampico in December of 1913 in an attempt to protect the foreign oil interests near that city. As the New York *Commercial* observed, Lord Cowdray "put the Monroe Doctrine to the test," by personally requesting that the American government safeguard his properties there.[13] In April 1914 U.S. naval forces seized Vera Cruz.

With World War I nearing, Great Britain decided that friendly relations with the United States were more important than Anglo-American rivalry for Mexican oil. On June 2, 1914, the British and U.S. governments accordingly agreed that existing oil concessions in Mexico should be protected, provided that the parties involved had complied with contractual obligations and legal requirements; an exception was made for default or noncompliance due to political unrest or military activities. On the same day the American and Dutch governments signed a similar understanding.

By that summer Europe was at war. The production of the Tampico field was to prove vital to the Allies, and some American oil men apparently urged the U.S. government to seize that part of Mexico to safeguard its output. But President Woodrow Wilson turned down the suggestion, observing "Germany raised the same point when she invaded Belgian territory. We cannot do the same thing."[14] That the petroleum of this area was in jeopardy is attested to by the destruction of 100,000 barrels of Huasteca oil when the forces of Pancho Villa clashed with those of Venustiano Carranza in 1915.

As the war progressed, President Wilson became increasingly disillusioned with the motives of the American businessmen then operating in Mexico. By 1916 he had come to the conclusion that "certain owners of Mexican properties" and "unscrupulous influences" had been disseminating false statements about his Mexican policy.[15] It is interesting to note that Arthur S. Link, Wilson's most exhaustive biographer, has concluded that the President badly misjudged the intentions of these business men, who were convenient targets for critics.

### The Constitution of 1917 and Subsoil Rights

The year 1917 marked a watershed in Mexican oil policy, if only because of the promulgation of the Constitution of 1917, which had a far-reaching impact upon the petroleum industry. The first Mexican law dealing exclusively with oil was formulated in 1901, shortly after significant commercial production had begun. The measure provided for the granting of concessions on government lands. Eight years later, in 1909, another Mexican law gave the surface owner exclusive title to the pools or deposits of mineral fuels on his or her property. The law was a direct repudiation of colonial Mexican mining law under which all rights to subsoil metals and minerals belonged to the royal family. It was against this background that the Constitution of 1917 was written.

As early as the beginning of 1916 the State Department had begun to fear that Venustiano Carranza was about to seize foreign oil holdings in Mexico. But when a special U.S. agent approached the Mexican leader in this connection, Carranza asserted that his government was not considering such a step. He repeated this assertion in private to the American Ambassador, Henry P. Fletcher, in August 1917.

Earlier that year, in February, delegates had assembled at Queretaro to write a new constitution to replace that of 1857. The proposed constitution contained a number of progressive features, but the article that caused the most controversy was number 27. This article declared that the state was the inalienable owner of all minerals, phosphates, oil, and hydrocarbons. Foreigners could become property owners in Mexico only if "they agree before the department of foreign affairs to be considered Mexicans in respect to such property, and accordingly not to invoke the protection of their governments in respect to the same, under penalty in case of breech, or forfeiture to the nation of property so acquired."[16]

Distasteful as such provisions were to the oil barons, the further provision that no foreigner was to obtain direct ownership within fifty kilometers of the sea coast was even less acceptable, since most of the oil wells were located in that area. While the petroleum interests did find some solace in article 14, which declared that "no law shall be given retroactive effect to the injury of any person whatsoever,"[17] still another provision in article 27 gave the executive the right to determine whether or not it was absolutely necessary for commercial stock companies to hold certain lands. It was quite apparent that this document was open to various interpretations not only from president to president but also from day to day.

Following the promulgation of the Constitution of 1917, American oil men began to make regular payments to General Manuel Peláez, a local landowner who held title to certain properties that had been leased to the Mexican Eagle Oil Company and who opposed the new constitution. With his own private force, Peláez essentially operated a state within a state. The Wilson administration was disturbed by the possibility that a force of Carranzistas attacking from the interior might attempt to dislodge the Peláez forces from the Tampico area and in the process either shut down or destroy the petroleum wells. Since the channels surrounding Tampico were shallow, it was not possible to defend the port with naval vessels. Fortunately for the United States, Carranza did not disturb the delicate balance existing in the Tampico region.

On February 18, 1918, however, Carranza issued a decree "for the imposition of certain taxes on the surface of oil lands, as well as on the rents, royalties, and production derived from the exploitation thereof."[18] The State Department declared that this amounted to confiscation, and Edward L. Doheny accordingly refused to file certain documents that showed the basis of his concession titles with the Mexican government. Carranza then issued supplementary decrees on July 31 and August 12 that gave additional weight to his original decree, which had imposed a 10 to 50 percent tax on foreign-owned oil leases. He also had the Mexican police close down several firms that attempted to drill for petroleum without permits.

By the summer of 1918 Ambassador Henry P. Fletcher had arranged with Carranza for James R. Garfield and Nelson O. Rhodes to meet with government officials in Mexico City. While it was hoped that they could resolve the controversy that had erupted over the new Mexican oil policy, the meetings did not bear much fruit. During the course of the year, U.S. investors organized two new organizations, the Association of Producers of Oil in Mexico, headed by Edward L. Doheny, and the National Association for the Protection of American Rights in Mexico. Both associations engaged in propaganda on behalf of U.S. interests.

The State Department, in a protest to the Carranza regime in June 1919, noted the possibility that the regime might grant new concessions covering those areas operated by U.S. individuals and firms who already had received concessions from the Mexican government. Great Britain earlier had lodged similar protests on behalf of its nationals. Even these gloomy prospects did not prevent several American petroleum companies from sinking new wells in Mexico that spring without authorization. Following a series of diplomatic exchanges, Carranza did agree to issue some provisional drilling permits in July, but by October the Mexican government had again put a stop to the sinking of unauthorized wells.

Back in the United States, President Wilson found in the *Nation* a journalistic backer of his Mexican policy. In April this periodical charged that oil companies and Morgan bankers were attempting to stir up public opinion in favor of American intervention in Mexico. The periodical also claimed that American Catholics were being brainwashed to support a vigorous anti-Mexican policy by stories of anticlericalism and religious persecution under the Carranza regime. Still another supporter of the cautious Wilsonian approach was the League of Free Nations; the membership of this organization included Dr. Samuel Guy Inman, a prominent Latin Americanist, and Bishop James Cannon, a leading Methodist.

But in Congress the anti-Mexican mood continued to prevail. With Wilson's archfoe Henry Cabot Lodge of Massachusetts chairman of the Senate Foreign Relations Committee, it was only a matter of time before the Senate took further steps to show its displeasure with the President. On August 8, 1919, Lodge obtained senatorial approval for his committee to investigate "in general any and all acts of the Governments of Mexico and its citizens in derogation of the rights of the United States or of its citizens."[19] He then appointed a subcommittee, headed by fellow Republican Senator Albert Fall of New Mexico, to make the pertinent inquiries.

The Fall subcommittee hearings lasted from September 8 to May 28, 1920. In the opinion of Howard Cline, an authority on Mexican-American relations, Fall was "fishing for materials useful in the forthcoming presidential campaign" and "played directly to the gallery." Cline also complained that most witnesses "spewed hearsay and slander."[20] One of the key testifiers against Wilson was Edward L. Doheny, already the head of the Association of Producers of Oil in Mexico; pro-Wilson spokesmen, such as Dr. Inman, were treated rather harshly. In its final report the subcommittee concluded that it might be necessary for the American government to send a police force to Mexico.

In November 1919, while the Fall subcommittee hearings were being held, General Álvaro Obregón and two other revolutionaries formed an alliance to drive Carranza from power. By the spring of 1920, a full-fledged revolution had developed, and on May 21 Obregonistas killed the fleeing Carranza. Obregón, who up to that time had appeared to be a moderate on the question of foreign oil properties, assumed the presidency later that year. The U.S. government, though, refrained from officially recognizing the Obregón regime and continued to withhold its official recognition in an attempt to force the new president to sign a treaty protecting those American property rights in Mexico threatened by article 27 of the 1917 constitution. American manufacturers and exporters of consumer goods tended to favor recognition in spite of the general opposition of U.S. investors in Mexican oil and agrarian properties. Since the eventual extending of recognition was more the result of concessions on the part of Mexico than of business pressures in the United States, historian N. Stephen Kane has concluded that events during the early 1920s serve to confirm the theory set forth by such scholars as Bernard C. Cohen, Raymond Bauer, and Lester Milbrath, who maintain that businessmen have had little real influence over American foreign policy.

There were approximately 150 American petroleum companies active in Mexico in 1920, but they by no means were united on a strong antirecognition policy. Smaller independent companies were just as fearful and critical of Jersey Standard and Doheny's Mexican Petroleum Company as they were of the Mexican government. It was their larger competitors, after all, who controlled the oil pipelines in that country and the marketing facilities outside of it. Among the smaller firms that not only supported recognition of the Obregón regime but also questioned the legality of the titles held by the oil giants were Mid-Continental, AGWI, Pierce Oil, and the Normal Oil Corporation. Still another critic of the nonrecognition policy was the Independent American Oil Producers Association, which was displeased at the privileged status of Mexican crude oil imports and thus favored a tariff on crude oil from Mexico.

Given this division of opinion within the ranks of the petroleum men themselves, it is not surprising that former U.S. Ambassador to Mexico Henry P. Fletcher was quoted in 1921 as saying, "It is absolutely untrue that oil interests are determining the action of the American Government."[21] Nor did other economic interest groups in the United States blindly support the petroleum companies. Samuel Gompers of the American Federation of Labor asserted in the same year that his organization "has repeatedly expressed the opinion—and backed it with proof—that oil interests were the prime cause of difficulties between the American and Mexican nations."[22]

On March 4, 1921, a new Republican administration, headed by Warren G. Harding, took office in Washington. Harding's choice for Secretary of State was Charles Evans Hughes, who immediately came out in favor of a new treaty with Mexico. Accordingly, on May 27, 1921, the American Chargé, Summerlin, presented a draft treaty to Obregón that declared, among other things, "that neither the Mexican Constitution which went into effect on May 1, 1917, nor the Decree of January 6, 1915, to which the said Constitution refers, is retroactive in its operation."[23]

Another member of the Harding cabinet who showed great interest in the Mexican situation was former Senator Albert Fall of New Mexico, who became the Secretary of the Interior. On March 21 he sent a letter to Henry Cabot Lodge in which he complained that British petroleum companies were attempting to ingratiate themselves with Obregón at the expense of American oil firms. This was going on, Fall maintained, at the same time that the British government allegedly was supporting its U.S. counterpart in its protests over the confiscatory decrees of the Mexican government.

Obregón himself had a dignified statement published in the New York *World* on June 27 in which he declared that the rights of the Mexican people—90 percent of whom lived in ignorance and poverty—must be respected. He went on to say that although he welcomed foreign capital, he would not extend to it excessive privileges. Nevertheless, he claimed that private holdings in Mexico would not be confiscated:

> This falsehood is the work of those who resent our policy of nationalization because it blocks further campaigns of exploitation and monopoly. Every private right acquired prior to May 1, 1917, when the new constitution was adopted, will be respected. Article 27. . . will never be given retroactive effect.[24]

Having extended the olive branch, Obregón signed a decree, effective July 1, that increased the petroleum export tax, the proceeds of which had been applied to paying off the external debt. In Obregón's view, the tax was hardly confiscatory, but American oil interests nevertheless complained about the decree. Although the Mexican government was in dire need of funds at this time, American and other foreign petroleum firms were reluctant to pay taxes to a country torn by revolution, for they theorized they might be forced to pay again, should there be a change in government.

Obregón retorted that without the export tax increase, he would not have the funds to pay his soldiers and maintain power. The oil firms, however, were unmoved by Obregón's plight, and rather than limit themselves to a protest against the tax, they organized a shutdown of petroleum production in Mexico. To complicate matters further, there was a possible threat to Obregón's presidency in the Tampico area in the person of Manuel Peláez. Peláez not only had received large sums from the American petroleum firms there, but he also had become a shareholder in these and had acquired arms and ammunition in the United States.

That July 4 five units of the American Navy arrived off Tampico on the pretext that the U.S. Shipping Board needed fuel oil. According to Secretary of the Navy Edwin Denby, it was necessary to protect petroleum tankers bound from Tampico and Tuxpan. Denby also claimed that a shutdown of the oil fields might lead to labor troubles, which in turn would jeopardize American property. U.S. labor leader Samuel Gompers, among others, found this argument offensive and thus cabled Secretary of State Hughes his opposition to the American Navy being "exploited by the employing interests for the avowed purpose of overawing the workers who are now engaged in a lockout imposed upon them."[25] Still another critic of the naval excursion was Progressive Republican Senator Robert LaFollette of Wisconsin, who was disturbed by this use of the U.S. fleet without the approval of Congress. Before opposition could proliferate, though, the last American warship departed from Tampico, only nine days after the first one had arrived.

Obregón fortunately had made General Peláez the commander of the Mexican forces at Tampico and therefore faced no challenge from him at this time. Peláez's presence there was allegedly a deterrent to the landing of American marines. With the U.S. Navy gone, Obregón announced that Mexican oil exports would cease on September 1, unless he had reached an agreement with U.S. petroleum firms in the interim. He claimed that the British companies that were exporting oil and paying the export tax were making money, but the American petroleum barons challenged this claim.

As for Congress, Republican Representative Thomas Chandler of Oklahoma was unsuccessful in his attempt to rewrite the Fordney-McCumber tariff bill to discriminate against cheap Mexican petroleum imports. Chandler's maneuverings incurred the displeasure of Edward Hurley, Chairman of the U.S. Shipping Board, southwestern railroads, and New England manufacturers and shippers. Hurley's agency used a large amount of crude oil each year, and there was a regional concern about petroleum availability and prices. The most prominent foe of the Chandler proposal, however, was President Harding himself, who wrote Chairman Joseph Fordney of the House Ways and Means Committee, a Republican from Michigan, that future military and domestic needs necessitated American access to overseas petroleum resources.

Convinced that Obregón was not bluffing, the leading U.S. oil men involved in Mexican petroleum operations made a joint trip to Mexico to confer with him. Among the members of this party were such prominent figures as Teagle of Standard, Doheny of Mexican Petroleum, Sinclair of Sinclair, and the heads of

Texaco and Atlantic Refining. By September 3 these oil executives had reached an agreement with Obregón by which the production tax was to be 14 cents per barrel but the export tax was to vary. There also was to be a sliding specific tax of 4.93 to 7.95 cents per barrel. The agreement was complicated by the provision that the petroleum companies could pay their taxes in Mexican bonds, which were selling for approximately $40 apiece, although the par value plus the accumulated interest was three and one-half times as much, $135. The bankers, who held $150 million in these bonds, balked at this deal, since it would jeopardize their profits. As a result, U.S. oil firms halted the practice of paying their taxes in bonds. Prior to the signing of the final agreement, there was talk that the oil men might float a Mexican loan, using the petroleum export tax as security, but both J. P. Morgan's senior partner, Thomas W. Lamont, and Secretary of State Hughes urged them not to do this, since the Mexican government already had pledged export tax revenues toward the payment of the external debt.

It is perhaps not coincidental that in the same month that the petroleum entrepreneurs came to an agreement with Obregón, the Mexican Supreme Court made public its decision in the *amparo* case of the Texas Company. Ruling that the fourth paragraph of article 27 was not intended to be retroactive either in letter or in spirit, the court upheld the company, although it also held that ownership of subsoil petroleum prior to 1917 was not privately vested unless some "positive act" relative to the granted oil rights had taken place before the adoption of the new constitution. Eight months later, in May 1922, the Mexican Supreme Court in four additional *amparo* hearings handed down decisions paralleling the Texaco decision.

In April of 1922 another contingent of American oil leaders conferred with Mexican officials in Mexico City about the petroleum situation. Walter Teagle of Jersey Standard summed up the position of the American oil men for Secretary of Finance Adolfo de la Huerta:

> With regard to the further development of Mexican petroleum resources, we endeavored to make it plain that the oil industry will have no future in Mexico unless an intensive effort to find new fields of production be undertaken and successfully prosecuted; that such effort cannot be undertaken until and unless the government shall have completely removed the unusual hazards created by domestic legislation, oppressive taxation and unreasonable and unnecessary departmental regulation and supervision, and shall extend to the oil industry its cordial co-operation and encouragement.[26]

Unfortunately, because of Mexico's foreign debt, foreign investors had the country in a compromising position. On June 16 Thomas W. Lamont reached an agreement with de la Huerta that set the foreign debt at slightly over $500 million, including railroad indebtedness. Repayment of the adjusted external debt was to commence in 1928 and extend over a forty-year period. In reaching this settlement the Mexican government agreed to turn over to the International Committee of Bankers on Mexico all petroleum export taxes for five years, as well as the net earnings of the national railways.

Once the agreement had been signed, de la Huerta approached Lamont privately about extending a loan to Mexico. After Lamont informed him that it might be difficult to obtain this from the bankers even if the American government officially recognized the Obregón regime, de la Huerta approached the oil executives.He proposed a $25 million loan as an advance against the petroleum taxes and promised that his government would adhere to the September 1921 agreement. When a representative of the oil companies, however, approached Secretary of State Hughes about the scheme, Hughes was somewhat less than enthusiastic. His coolness was unquestionably a factor in the petroleum firms' hesitation. Moreover, due to the strong opposition of some of its members to the agreement, the Mexican cabinet had yet to ratify it, and Hughes' negative stance toward the loan left Obregón with no choice but to pressure his administration into accepting the financial settlement. Without the latter, there would be no official recognition by the United States and in all likelihood no foreign loan.

On May 2, 1923, Harding appointed two American commissioners, Charles Beecher Warren and John Barton Payne, to discuss the question of property rights with the Mexican government. Warren and Payne began conferring with Mexican authorities in Mexico City later that month about the seizure of agricultural lands and the nationalizing of subsoil mineral deposits. The main point of controversy was property owned prior to 1917 by American nationals who had not performed some "positive act" to utilize the subsoil deposits. Under the final settlement, they were to receive preferential consideration over the claims of third parties. Those U.S. investors who had executed "positive acts" in Mexico received protection against expropriation. Lands acquired after 1917, though, were to be subject to the agrarian legislation that followed. A system of special, general, and mixed claims was set up to deal with claims dating both from the prerevolutionary and revolutionary periods that had been challenged.

When the Mexican government gave its pledge on August 2, 1923, that it would not enforce article 27 retroactively, it cleared the way for official U.S. recognition of the Obregón regime. This action took place on August 31 with Great Britain and France following suit shortly thereafter. Early in September the American and Mexican governments signed two conventions; the first dealt with general claims since 1868, the second with special claims dating from the revolutionary period. It appeared that a new era had dawned in Mexican-American relations.

## Restoring Tranquility with Dwight Morrow

As a result of his increasingly cooperative attitude toward the United States, Obregón received the support of the American government when the Mexican Secretary of Finance, Adolfo de la Huerta, launched a revolution in December 1923. Not only did the U.S. government send the Obregón regime a large quantity of surplus war equipment, it also placed an embargo on arms shipments to de la Huerta and even allowed Mexican federal troops to cross Texas at El Paso. Some American oil men, though, did back de la Huerta.

Unfortunately for the United States, in 1924 Plutarcho Calles assumed the presidency and began to move in the direction of applying article 27 retroactively. The new chief executive began to differ with the American Ambassador to Mexico, James R. Sheffield, who had succeeded Charles Beecher Warren; Sheffield sent back to the United States rather negative assessments of the Calles regime. Relations between Calles and the new Secretary of State, Frank B. Kellogg, were even more acrimonious. In June 1925 Kellogg indiscreetly announced to the American press that "the government in Mexico is now on trial before the world."[27] Not only did Kellogg expect the Mexican government to restore U.S. properties and to indemnify their owners for losses, but he warned Calles that the American government would support his regime only for as long as it protected U.S. lives and properties in Mexico. To the liberal publication the *Nation*, this statement was "a naked club, publicly brandished in the face of a friendly government."[28]

Calles, filled with resentment at Kellogg's comments, declared that his government would "reject with energy any imputation that Mexico was on trial in the guise of a defendant. . . which in essence would only mean an insult."[29] At the end of that year the Mexican Congress passed two new acts, the Petroleum Law and the Alien Land Law, which Calles then signed; these measures placed new restrictions on American oil men operating in Mexico. Owners of petroleum property were required to exchange their original concessions for fifty year concessions by January 1, 1927, or face the confiscation of their holdings. In addition, there was to be no foreign ownership of property closer than thirty miles from the sea coast or sixty-two miles from the frontier, an obvious attack on the holders of oil concessions. When the larger American petroleum firms resisted this new legislation, the Mexican government cancelled their drilling permits, resulting in a decline in oil production.

Back in the United States, Democratic Representative John Boylan of New York demanded that the American government "withdraw its recognition from this Bolshevik and robber republic to the South until it amends its Communistic Constitution."[30] But when the American critics of the Calles regime suggested possible U.S. military intervention, they encountered strong opposition. Among the Senators opposed to intervention were Robert La Follette of Wisconsin and William Borah of Idaho, both Republicans; Democrat Burton Wheeler of Montana charged that the Coolidge administration was "bullying" Mexico, and Democratic Representative George Huddleston of Alabama declared, "I am not willing that a single American boy shall be sent to Mexico to lose his life in order that the oil interests may pay dividends."[31] Newspapers in every section of the country favored peace, and numerous Protestant church groups raised their voices against the possiblity of war between the two countries.

Unable to mobilize American support for an attack on Mexico, Coolidge and Kellogg chose a conciliatory approach. By extending the olive branch, they mirrored the sentiments of a later ambassador to Mexico, Josephus Daniels; in 1927 Daniels pointed out that "the purchase of an oil well does not carry the right to a warship from Uncle Sam to control the government of that country."[32] Yet he was highly skeptical of the administration's choice of an ambassador, Dwight W. Morrow, a Morgan partner. Upon departing from Mexico, Morrow observed, "I know what I can do for the Mexicans. I can *like* them."[33] In the months ahead his frank and friendly personal conferences with President Calles restored harmony between the two nations.

Just prior to the arrival of Morrow in 1927, General Arnulfo Gómez launched a counterrevolution that Calles was able to thwart, even though the State Department applied an arms embargo against the Mexican government. Although Gómez had stated that he would take a more sympathetic attitude toward U.S. oil interests, only a minority of the American petroleum leaders supported his revolt. Calles' victory over Gómez, fortunately, did not harden his position toward the United States; Mexico was suffering from a decline in oil production, as well as a fall in the price of silver, its second most important export. It also was torn with religious strife, since the Calles regime had unleashed an anticlerical campaign.

Morrow did not formally present his credentials to President Calles until October 29, but as early as January 9 Calles had given signs of adopting a conciliatory attitude by revealing his willingness to submit to The Hague Tribunal the unresolved controversy surrounding the U.S. oil concessions. Coolidge allegedly had said to Morrow: "My only instructions are to keep us out of war with Mexico."[34] Although some of the Mexican leaders, as well as some of the U.S. oil men, had misgivings about Morrow at first, President Gerardo Machado of Cuba swayed Calles with a letter dated October 1 praising Morrow highly. Morrow himself also undermined, if not refuted, the charge that he was a puppet of J. P. Morgan by terminating his long association with that firm at the end of September.

Unlike past Mexican-American diplomatic negotiations, which more often than not were quite prolonged, Morrow quickly confronted Calles with a solution to the U.S. petroleum concessions dispute. The basis of his solution was to be the Texas Company case of 1921. Since there were a number of similar cases currently pending in the Mexican courts, Morrow suggested that a decision upholding the Texaco opinion would pave the way for a settlement. Calles then promised a Supreme Court ruling within two months and probably exerted pressure on that body behind the scenes to hand down a ruling parallel to the 1921 Texas Company decision. Calles also informed Morrow that "he wished all substantial rights of the oil companies to be observed. . .[35]

On November 17 the Mexican Supreme Court handed down a ruling favorable to the American petroleum firms in the case of the Mexican Petroleum Company of California. Although the court held that requiring the confirmation of acquired rights without alteration did not violate constitutional guarantees, it declared unconstitutional and void those portions of the petroleum legislation that demanded foreign oil companies apply within a year for fifty-year concessions. The decision received a generally enthusiastic reception in the American press, although the Louisville *Courier Journal* skeptically viewed the ruling "not as a diplomatic victory except upon the cynical theory that the Mexican Supreme Court functioned to save the administration's face."[36] Nevertheless, some of the American petroleum entrepreneurs complained because the opinion did not protect those properties acquired prior to 1917 on which they had performed no "positive act."

So as to strengthen the impact of this ruling, President Calles submitted a bill to the Mexican Congress the day after Christmas, promising "confirmatory concessions" for those private oil holdings where "exploitation works" had begun prior to May 1, 1917. Applications for these were to be made by January 11, 1929. Early in 1928 the Mexican government altered the petroleum regulations and in effect abandoned the "Calvo clause," declaring instead that any attempt to transfer a concession to an alien or to a foreign nation would be invalid. It seemed, then, that with the passage of the Calles bill and the change in the rules governing concessions the oil controversy between the United States and Mexico was nearing a settlement.

But within Mexico itself there were those who disagreed with the government's oil policy. Among these was *El Universal* of Mexico City. In an editorial on January 10, 1931, it declared:

> Our petroleum policy, as demonstrated by realities which are as sad as they are eloquent, has succeeded in producing but two lamentable concordant results: It has depressed the petroleum industry in Mexico and helped it in other countries. Not only do we not produce oil to the limit of our capability, which is abundant, but we import it, which is the same as importing diamonds into the Transvaal or lumber into Sweden.[37]

During the depression years the dispute over Mexican petroleum simmered rather than boiled. By 1934, however, there were indications that the Calles-Morrow settlement was beginning to come apart. During the spring of that year oil company representatives began to complain *en masse* about the problems they were experiencing with the Calles regime over the confirmatory concessions. Josephus Daniels was now the American Ambassador to Mexico; he had replaced J. Reuben Clark (1930-33), who in turn had replaced Dwight Morrow.

The American petroleum firms attempted to enlist Clark as their spokesman in talks with President Calles, whose term in office expired in that year. Their attempt irritated Ambassador Daniels, who wrote Secretary of State Cordell Hull, "I do not agree with the public man who said 'all oil stinks,' but we have seen so many evil practices growing out of the greed for its possession and the power it gives that we are warned to be cautious when we are asked to further the desires of the oil interests."[38] Jersey Standard, though, attempted to pressure the State Department into supporting a Clark mission.

Talks between Clark and Calles did take place eventually, but by that time Lázaro Cárdenas was President. Unfortunately for the U.S. oil firms, a break between Calles and Cárdenas sharply reduced whatever leverage Calles may have hoped to exert over Cárdenas. While the State Department did not become involved with the Calles-Clark negotiations, it did inform Josephus Daniels in March of 1935 of its concern over the confirmatory concessions and recent Mexican tax increases.

Labor problems then began to plague U.S. oil firms. In May the leading Jersey Standard subsidiary in Mexico, the Huasteca Petroleum Company, complained about the "excessive" demands of striking workers employed by the Tampico refinery. Unless conditions improved, Jersey Standard threatened to pull out of Mexico, a gesture that was probably part bluff. Organized labor in the United States, meanwhile, tended to be rather sympathetic toward the cause of the Mexican workers during the Cárdenas years.

For the remainder of 1935 and most of 1936 events in Mexico built inexorably toward a climax. On November 13, 1936, a federal expropriation law was passed. The law allowed the seizure of privately held property of public utility "to satisfy collective necessities in case of war or interior upheaval." To safeguard against the possibility that Mexican jurists might not interpret this measure as planned, President Cárdenas had Congress pass another law stripping judges of their life tenure and restricting the duration of their terms to that of the chief executive who had appointed them. When Daniels queried Cárdenas over the new expropriation law, the Mexican president affirmed that he "would not engage in any suicidal policy," and stated that the Mexican government "would not, for instance, endeavor to take over the oil fields or the mines, since that would be impractical and would place the Government in the situation with regard to foreign investment which it intended to avoid."[39] Cárdenas suggested that he wished to have legislation on the statute book in advance of a shutdown.

Unfortunately for both the American government and U.S. investors, the Cárdenas regime then began to expropriate agricultural and grazing properties held by American nationals; apparently it was confident that it could do so with a minimum of U.S. interference. On June 23, 1937, the Mexican government seized the Mexican National Railways with only a minimum of protest from the American side of the border. This subdued U.S. reaction was not surprising for several reasons. First, the Mexican government held a controlling share in the Mexican National Railways; second, other, more profitable, foreign-owned railways were not affected by the decree; and finally, the Mexican National Railways was facing bankruptcy so that the minority American stockholders and bondholders had little to lose and possibly something to gain from the takeover. Nevertheless, it is possible that if the U.S. government and American investors had made more of a protest at this seizure, the Mexican government might not have expropriated foreign oil holdings in 1938.

In the interim, the Mexican government, realizing that an understanding with British oil investors might fracture the Anglo-American solidarity over oil rights that had been in evidence on more than one occasion, attempted to reach such an agreement. In November 1937 the government did enter into an understanding with El Aguila, or the Mexican Eagle Oil Company, the subsidiary of Royal Dutch-Shell in Mexico. Under the terms of the agreement Mexico was to grant confirmatory concessions to El Aguila in the Poza Rica fields in return for a pooling arrangement that would give to the Mexican government a share of the petroleum production.

Even with this settlement Mexican finances were in a rather shaky state at the end of 1937. The worldwide depression had ravaged Mexico, Cárdenas' reform program was expensive, and foreign capital had been fleeing the country. The American oil companies had been partly responsible for the latter phenomenon, for they had been withdrawing their cash reserves from Mexico, as well as refusing to sell oil domestically on credit. Ambassador Josephus Daniels therefore suggested to Secretary of State Hull in November that he confer with Secretary of the Treasury Henry Morgenthau over the possibility of extending financial aid to Mexico. During the next month Finance Minister Eduardo Suárez headed a delegation that went to Washington in search of a direct loan and a buyer for a large amount of silver.

The Cárdenas regime found a sympathetic respondent in Secretary Morgenthau, who believed that it was better for Mexico to deal with the United States than with the Fascist powers and so advised President Roosevelt. Having convinced FDR, Morgenthau agreed on December 29, 1937, to the monthly purchase of 35 million ounces of Mexican silver. The U.S. petroleum companies, eager to keep Mexico in a precarious financial position, were highly displeased with Morgenthau's efforts.

That same day Jersey Standard had issued a press statement indicating that it would not accept the December 18 ruling by the Board of Conciliation and Arbitration, raising the salaries of the petroleum workers. Jersey Standard declared that the increased labor costs would force it into bankruptcy, and appealed the board's decision to the Mexican Supreme Court. That body, however, upheld the 26.3 million peso wage increase on the last day of February 1938 and also denied the petroleum companies' request for an injunction against the Board of Conciliation and Arbitration award. Despite this ruling, Jersey Standard announced on March 15 that it would be unable to implement the decision. It was not until after the British Minister to Mexico had spoken with President Cárdenas and El Aguila officials had conferred with their Jersey Standard counterparts that Jersey Standard agreed to the wage increase, and even then it was unwilling to accept other features of the award. Jersey Standard's counteroffer was not acceptable to the Mexicans; for one thing, union leaders doubted their workers would received the full salary increase. Although at one time the British oil interests apparently were willing to accept the terms proposed by the Cárdenas regime, the resistance the American petroleum leaders displayed may have led them to adopt a more inflexible position. The stage was now set for the great confrontation.

## 1938: Lázaro Cárdenas' Petroleum Expropriation

On the evening of March 18, 1938, President Lázaro Cárdenas went on national radio to announce that the Mexican government was expropriating the property of a number of foreign petroleum companies. Rather than restricting his attack to the labor question, Cárdenas raised the old charge that the U.S. oil firms had given aid and comfort to the rebel factions in northern Vera Cruz and on the Isthmus of Tehuantepac between 1917 and 1920. Cárdenas went on to state:

> This is a clear and evident case obliging the Government to apply the existing expropriation Act, not merely for the purpose of bringing the oil companies to obedience and submission, but because, in view of the rupture of the contracts between the companies and their workers pursuant to a decision of the labor authorities, an immediate paralysis of the oil industry is imminent, implying incalculable damage to all other industry and to the general economy of the country.[40]

The action triggered an international controversy, the repercussions of which were felt for years. As Raymond Vernon later observed, "The conflict transcended questions of rights and wrongs; the struggles simply represented a clash between two different orders having different norms and different systems of values."[41] In the years ahead the Mexican seizure of foreign oil properties would serve as a model for similar actions by other governments.

Generally overlooked today is the fact that the holdings of Gulf Oil in Mexico were not expropriated. At an earlier date this firm had signed liberal work agreements with the Mexican nationals in its employ. Thus when its employees had an opportunity to join a union, they did not, and consequently Gulf Oil was not a party to the labor dispute that led directly to Cárdenas' decree. Had Jersey Standard and the other U.S. companies been equally generous in dealing with their workers, expropriation might not have occurred.

Why did the Mexican government choose expropriation rather than the imposition of additional restrictions? According to J. Richard Powell:

> . . . three factors stood in the way of this approach to Mexico's oil problems. First, there was not sufficient agreement among Mexicans as to what the specific economic problems were, nor was there sufficient understanding of the economic principles involved. Second, some of the companies were already losing interest in Mexico for investment purposes, which suggests that these companies were not receiving "excessive" returns. Third, the oil problem was not simply economic. There were political and social aims as well. Mexicans were willing to make economic sacrifices, if necessary, to achieve these other ends.[42]

Rather than wait for a Mexican Supreme Court ruling on the legality of Cárdenas' action, Under Secretary of State Sumner Welles called in the Mexican Ambassador to the United States, Francisco Castillo Nájera, on March 21 and informed him that the policy of his government toward the oil companies was "absolutely suicidal." Having pointed out that the United States was purchasing silver from the Cárdenas regime, Welles observed that the Mexican government could not make a profit from the petroleum industry, since the oil companies owned most of the world's petroleum tankers. If it were to find a market abroad, it would be in such Fascist nations as Germany, Italy, and Japan whose dominant political ideologies were far removed from that of the reformer Cárdenas. Castillo Nájera's response was that he had not favored the step the President had taken but that the Mexican government could hardly repudiate the expropriation decree now.

At the Treasury Department, Henry Morgenthau opposed using the possible termination of the silver purchase program as a weapon against Mexico, but Secretary of State Hull persuaded him to change his mind. Hull pointed out that Colombia and Venezuela might follow Mexico's example, if the American government did not take a hard line toward the Cárdenas regime. On March 25 Morgenthau announced that the monthly silver purchases would cease until further notice.

The United States was the only purchaser of Mexican silver at this time, and the proceeds from these sales supplied Mexico with needed dollar exchange. During the three-year period of the American boycott up to Pearl Harbor, the U.S. Treasury purchases of Mexican silver remained at a minimal level, totalling only 500 thousand ounces in 1941. In 1938 the American government had bought 44 million ounces. Although William Scroggs observed in that year "silver rather than oil is the basis of Mexican prosperity," the Mexican economy survived.[43] Furthermore, the U.S. petroleum men acquired a new enemy in the silver-mine owners, most of whom were American citizens; they blamed the oil concerns for the anti-American atmosphere that had developed in Mexico.

Until April 1, 1938, FDR remained somewhat aloof from this controversy, but on that day at a press conference he announced that he did not fully endorse the support the State Department had extended to U.S. petroleum firms with Mexican holdings. He offered his opinion that some of the damage claims had been inflated out of proportion, and that the oil companies should not expect compensation for prospective profits. In taking this position FDR in effect undercut Sumner Welles, who earlier had informed Ambassador Castillo Nájera that the petroleum holdings in Mexico were worth hundreds of millions of dollars.

As for the U.S. oil companies, they did not wait for governmental action. They seized the initiative on their own. At a conference of tanker owners assembled in Oslo, Norway, in April, Jersey Standard and Royal Dutch-Shell pressured the other participants into not selling or renting petroleum tankers to Mexico, which, of course, made it more difficult for the Cárdenas regime to transport its oil. The two companies then persuaded foreign nations not to buy any petroleum from the Mexicans. Even this did not satisfy them, however, for they further hampered petroleum production in Mexico by organizing a boycott of equipment needed by the oil industry there such as pipes, drilling machinery, and other items.

The Mexican government was even harassed by the petroleum firms in American courts. Through the American courts, the Eagle Oil and Shipping Company, a Canadian firm, was able to force an El Aguila tanker docked at Mobile, Alabama, into temporary receivership in April. It was not until seven and one-half months later that the United States Fifth District Court of Appeals freed the vessel. During the same year, El Aguila in a suit brought before the United States Southern District Court in New York charged that the Eastern States Petroleum Company had taken 1.7 million barrels of crude oil from Mexico that was the property of El Aguila. Although El Aguila had challenged the act of expropriation, the court ruled on April 3, 1939, that it had no jurisdiction over foreign confiscations and that as a Mexican company El Aguila should resort to the Mexican courts.

In the months ahead Roosevelt held to his restrained course of action. In June FDR made it clear to a Standard Oil attorney that the expropriation would not be reversed and that the petroleum firms should turn their attention to a financial settlement. At the end of the year the President repeated this advice and again stated that compensation should be geared to actual investments rather than potential profits. When a friend of Roosevelt's criticized Ambassador to Mexico Josephus Daniels, the President observed, "It is largely because of the ownership of Mexico by 'successful business men' for so many years that the somewhat unhappy transition period of the last twenty-five years became inevitable."[44] Daniels himself tended to be sympathetic to the Mexicans, most of whom supported Cárdenas' expropriation of the petroleum firms. "Doheny and Pearson [Lord Cowdray]," he wrote Roosevelt on March 22, "obtained oil leases for a song from the corrupt Díaz government" and added, "most revolutions are bred in hunger and privation."[45]

During 1938 General Saturnino Cedillo, the political boss of San Luis Potosí, provided additional excitement by launching an abortive revolt against the Cárdenas regime. Despite the expropriation of foreign holdings, Franklin Roosevelt made it clear that the American government would in no way endorse Cedillo, who was also rebuffed by the petroleum firms. Jersey Standard reasoned—and doubtless correctly—that should the Cedillo revolt fail, its position in Mexico would be even more precarious.

Although the existence of ties between Cedillo and Adolf Hitler remains mere speculation, it is known that Germany had been taking increasingly large quantities of Mexican oil for several years. In return Germany had shipped steel, heavy machinery, and oil equipment to Mexico. A lesser amount of Mexican petroleum went to Italy, which sent in return rayon and tankers to Mexico. Under an agreement concluded with Italy in October 1938, Mexico was to supply the Azienda Generale Italiana Petrole (Agip) with $5 million worth of oil.

Elsewhere in Europe, the British government was in a less conciliatory mood toward the Cárdenas regime than its American counterpart. The Manchester *Guardian Weekly* observed,: "The Mexican Government now knows that whatever else the British government will suffer tamely, it will not suffer the expropriation of British property—by a small state."[46] When Owen St. Clair O'Mallery, the British Minister to Mexico, pushed the Cárdenas regime for the payment of a $361,737 debt that was four months overdue, the debt was paid but nothing was paid on the $500 million the British oil companies were seeking. Foreign Minister Eduardo Hay pointed out to O'Mallery that there were nations more wealthy than Mexico that had defaulted on large debts—an obvious slap at Great Britain's failure to pay off its World War I debt.

Conservative members of both major political parties in the United States now began to champion the oil interests with increased vigor. Republican Senator Styles Bridges of New Hampshire was one of the most vocal critics of American diplomacy; he charged that the Roosevelt administration had "encouraged and even connived at the establishment of communism in Mexico."[47] Even more extreme in his claims was isolationist Democratic Senator Robert R. Reynolds of North Carolina, who claimed that Mexico was moving toward Communism by giving Leon Trotsky asylum, and Fascism by trading with Germany and Italy. Isolationist Republican Representative Hamilton Fish of New York examined the implications of FDR's Mexican policy for all of Central and South America, and concluded: "If we do not take a firm position and stop this communism and this confiscation of property, including the oil properties, it means the end of our investments not only in Mexico but throughout Latin America."[48]

Even though this remark was uttered by an individual whose relationship with the President was one of mutual animosity, Secretary of State Cordell Hull did suspect that Mexico was encouraging other Latin American states to seize American investments. Unlike Sumner Welles, who felt that the U.S. petroleum firms should do their own negotiating with the Mexican government, Hull thought that it was necessary for the United States to exert strong pressure on the Cárdenas regime. At the same time Hull was opposed to congressional investigations of the oil negotiations. Early in 1939 several resolutions to this effect were introduced into both houses, but the Secretary of State told Senate Foreign Relations Committee Chairman Key Pittman of Nevada that such an inquiry would have "unfortunate repercussions" on the settlement of the petroleum controversy. As a result of Hull's opposition, there were no public hearings.

## Mexican-American Negotiations 1939-1942

Before the outbreak of war in Europe on September 1 negotiations had begun in Mexico between spokesmen for the U.S. oil companies and the Cárdenas regime. The petroleum firms had chosen as their agent Donald R. Richberg, the former head of the National Recovery Administration. Although Richberg had many New Deal ties, which was one reason why Walter C. Teagle had hired him, he was highly supportive of the oil companies. Thus Richberg observed of Teagle's scheme for settling the expropriation question: "It is hard to imagine any more fair or generous proposal."[49] While he accepted the seizure of the petroleum firms as irreversible, he by no means approved of Cárdenas' action.

Richberg arrived in Mexico in March, in time for the first anniversary celebration of the foreign oil firms expropriation. He held three weeks of talks with Cárdenas, with Ambassador Castillo Nájera acting as interpreter. When he returned in April for additional discussions, Cárdenas invited him to review a revolutionary May Day parade at Saltillo. Following this event, Richberg presented to the president a four-point program that provided for the creation of several new firms with Mexican majorities on the boards of directors. These companies were to receive long-term contracts from the Mexican government, which was to guarantee reasonable labor demands and fixed tax obligations. This, it was hoped, would allow the newly established "Big Four" petroleum companies to operate the oil industry with assurance. The U.S. petroleum firms had been willing to settle for fifty-year contracts, after which the oil properties would become the exclusive properties of the Mexican government.

Cárdenas thought over the four-point program until July, when he accepted it as the basis for a settlement but with the qualification that the Mexican government would not only appoint a majority of the boards of directors but also the presidents of the four companies. This obviously was unacceptable to the American petroleum barons, who concluded that further talks with the Mexican government would be useless. Sumner Welles then made a compromise proposal under which the two parties would each appoint one-third of the boards of directors, while these directors would in turn select the remainder of the boards from neutral nominees who were neither Mexican nor American. Both parties rejected the plan. Roosevelt next suggested that impartial arbiters might settle the controversy, and Ambassador Daniels presented this solution to Cárdenas on September 8, one week after the start of the war in Europe. The Mexican President, though, was hesitant to submit to arbitration, since public opinion in Mexico still felt that a negotiated settlement was possible. Under a measure signed by Cárdenas on December 30, 1939, with minor modifications later approved by President Ávila Camacho on May 2, 1941, the oil industry of Mexico was declared to be a public utility. The legislation reaffirmed the principle found in article 27 of the 1917 Constitution, which stated that the national government enjoyed dominion over all hydrocarbons. It also provided that the government could take over surface lands, subject to the payment of compensation.

As for developments in the United States, the growing willingness of Sinclair by the winter of 1939-40 to make a separate settlement with the Mexican government dealt a severe blow to the U.S. oil companies who wished to present a united front. It also was becoming apparent that Donald R. Richberg would not be able to reach a comprehensive settlement on the basis of the various proposals then being considered. Accordingly, Patrick J. Hurley proposed on behalf of Sinclair that the Mexican government offer the U.S. firm 30 million barrels of oil as compensation for its holdings in Mexico; the American company then would purchase another 15 million barrels of petroleum from the Mexicans. Although Mexico regarded this proposal as too demanding, Hurley's scheme served as a basis for additional serious negotiations. The fact that Hurley not only was direct and candid in his conversations with President Cárdenas but also appeared to be sympathetic toward the Mexicans was a positive factor, as was the fact that both he and Cárdenas were military men.

Hurley reported to Secretary of State Cordell Hull on March 1 that Sinclair and Mexico were on the verge of signing a contract, but Secretary Hull still formally proposed arbitration to the Mexican government in a note dated April 3. Some

Washington analysts concluded that he was trying to influence the Mexican presidential election of 1940, but the German legation in Mexico City felt that his objective was to sway the American presidential contest of that year. Not surprisingly, Jersey Standard praised Hull for proposing arbitration, which only intensified anti-American sentiment in Mexico. As the *Ultimas Noticias* of Mexico City observed, "The international Pan-American New Deal of Roosevelt No. 2 is nothing but a branch grafted in the trunk of the big stick of Roosevelt No. 1."[50] Twenty thousand laborers paraded before President Cárdenas in support of expropriation, while the Mexican army followed suit on a number of occasions, during which it also commemorated Mexican resistance to the American invasion of Vera Cruz in 1914.

It took the Mexican government one month to reject the U.S. arbitration proposal. On May 1 the Cárdenas regime declared that it believed in arbitration but that the petroleum controversy was a purely domestic matter. To prove that a negotiated settlement was possible, it announced that it had reached an understanding with Sinclair; under this agreement the American firm was to receive $8.5 million over a three-year period and was to purchase 20 million barrels of petroleum from Mexico over the next four years at bargain prices. This placed the total compensation that Sinclair would receive between $13 and $14 million.The biggest winner, however, was probably Hurley, who received a large fee from the oil company for his services and the Aztec Eagle decoration from the Mexican government.

Even though arbitration was now a dead issue, during May and June a number of labor unions on the Pacific Coast protested that the State Department's "demand" for arbitration violated international law. Most of these protests took the form of mimeographed resolutions. Among those unions that threw their support to the antiarbitration cause were those of the fishermen, ship scalers, and agricultural workers.

The reaction of the U.S. Congress was quite different. Among the more extremist indictments of the Cárdenas regime was the observation of Republican Representative Dewey Short of Missouri, who claimed that "a horse-high, hog-tight, bull-strong case of robbery is standing against Mexico. . . .No one arbitrates with a thief."[51] Members of both houses suggested various reprisals, including restrictions on Mexican immigration, letters of marque and reprisal against Mexican shipping, and an end to Mexican silver purchases. A bill encompassing the latter objective even passed the Senate, as did an obscure amendment to the National Stolen Property Act, which would have in effect invalidated the Sinclair agreement with Mexico. The House of Representatives, however, failed to enact either, even though Secretary Hull went so far as to assert that he had no objection to the latter.

Standard Oil of New Jersey, on the other hand, vented its wrath against Sinclair, claiming its Mexican purchases of petroleum were contrary to the American antidumping statutes. When Sinclair made a bid to provide the U.S. Navy with 560,000 barrels of oil, Jersey Standard complained to the State

Department that petroleum which Sinclair was importing from Mexico fell into the "stolen property" category. After conferring with Secretary Hull, the Navy decided to postpone a decision on the Sinclair bid. Although Hull informed Secretary of the Navy Frank Knox on October 14 that the State Department could find no complaint with the oil, the Navy decided to avoid a possible controversy by awarding Jersey Standard the contract.

While Jersey Standard was criticizing Sinclair's role in Mexico, it also may have been meddling in the Mexican presidential election of 1940 by offering clandestine support for the right-wing general Juan Andreu Almazán. Almazán had the support of not only the large landowners and several Fascist organizations, but also the Germans. Some Americans hoped that, if elected, he would return the expropriated oil properties. After Almazán lost badly to General Manuel Ávila Camacho, he claimed that the petroleum companies had promised him $200,000 for purchasing arms and that both Franklin Roosevelt and his son, Elliot, were sympathetic to his cause. In 1975 Harry Stegmaier, Jr., an American historian, charged that not only did Jersey Standard help finance Almazán but also that Ambassador Josephus Daniels was aware of its involvement.

The interests of the American government and Jersey Standard, though, were not always the same. When the Mexican government seized all German and Italian ships in Mexican ports at the beginning of April 1941, it not only revealed to the U.S. government that it was favorably inclined toward the Allied cause, but it also augmented its tanker fleet, much to the displeasure of Standard Oil. The Jersey Standard lobbyist in Washington, John Bohanon, complained that Mexico might now ship its petroleum to other Western Hemisphere republics at a reduced price, which of course would undercut the Jersey Standard market.

By this time the American government was coming around to the position that the oil companies were asking too much for their nationalized holdings. The Interior Department already had advised the State Department that Sinclair probably had received more from Mexico than its properties were actually worth. Nevertheless, there still were important officials in the State Department who were sympathetic to the U.S. oil companies. Among these were the economic advisor, Herbert Feis, and the petroleum advisor, Max W. Thornburg. In the opinion of Thornburg, the Interior Department appraisal was worthless, a hoax that was "either incompetent or insincere"; as for Feis, he felt that a settlement should be postponed until Mexico recognized its need for foreign investments. Even Interior Secretary Ickes feared that a Mexican agreement might trigger additional Latin American expropriations.

Fortunately, the settlement of the expropriation controversy was now at hand. On November 19, 1941, Secretary of State Hull and Mexican Ambassador to the United States Castillo Nájera signed documents containing the following provisions: Mexico would pay $40 million over a period of fourteen years to settle all general and agrarian claims; both nations would appoint experts to determine in conjunction the amount due to the U.S. petroleum firms; the United States would help to stabilize the Mexican peso by purchasing Mexican silver and negotiating a Mexican trade agreement; and finally, the Export-Import Bank would lend Mexico $30 million for road construction to be applied primarily to that part of the Pan American Highway running between Mexico City and Guatemala.

In the process of nominating candidates to serve as the American referee, the Roosevelt administration consulted a number of individuals, including Chief Justice of the U.S. Supreme Court Harlan F. Stone and Associate Justice William O. Douglas. The six names presented to President Roosevelt were: Morris Llewellyn Cooke, Lloyd K. Garrison, Isaiah Leo Sharfman, Isaiah Bowman, G. O. Mulford, and J. R. Lotz. FDR selected the sixty-nine-year-old Cooke, a Pennsylvania Democrat who had served as an advisor on power to Roosevelt when he was governor of New York. In designating Cooke, however, FDR merely sent a brief memorandum to Under Secretary of State Sumner Welles on December 18 saying, "I think Morris Llewellyn Cooke would be my first choice as the United States expert to evaluate oil properties in Mexico."[52] He offered no further explanation.

The two referees, Cooke and Manuel J. Zevada, had little difficulty in agreeing on approximately three-fourths of the award, but Cooke was unable to prove convincingly that Standard Oil had invested $400 million in its properties. As Cooke and Zevada worked toward the final settlement, Wallace Pratt (a Jersey Standard board member) approached the State Department with the proposal that his company "would be willing to see that Mexico received 50% of the profits through royalties, taxes, et cetera, and that a working arrangement would be better than a fair or unfair evaluation."[53] While it obviously was too late for this type of arrangement to be accepted by the Mexicans, it showed unusual generosity on Jersey Standard's part, considering that American petroleum firms had yet to endorse any such fifty-fifty profit sharing agreements. Despite this offer, as late as September 27, 1941, W. S. Farish, the President of Standard Oil of New Jersey, advised Secretary of State Hull during a conference in Washington that a settlement that renounced American property rights was not advisable. Hull's angry response was that the national security of the United States came first, but Farish remained adamant in a meeting on October 28.

The British government joined Jersey Standard in resisting a compromise. After breaking off diplomatic relations with Mexico, following the seizure of British petroleum holdings in 1938, Great Britain took the position that it would not resume diplomatic relations with the Ávila Camacho regime until there was a settlement of the oil dispute. Farish, moreover, had emphasized in his talks with Hull that Great Britain was leading the defense of property rights on the international scene. Even so, the British and the Mexicans reestablished diplomatic relations on October 21, 1941, seven years before the oil settlement.

While these discussions were going on, Cooke and Zevada pursued their inquiries. It took them only five months to determine the amount of compensation the petroleum firms would receive. In their final report, dated April 17, 1942, they set the total compensation for American petroleum firms at $23,995,991. Of this amount, Jersey Standard was to receive over three quarters: $18,391,641.

The amounts were far less than the inflated figures the U.S. oil companies had proposed, but they were far closer to the actual value of the Mexican holdings. In a radiogram to FDR endorsing the settlement, President Manuel Ávila Camacho of Mexico referred to ". . . my Government's attitude in its eagerness to grant full guarantees to the participation of private capital, whether Mexican or foreign, in the extraction and development of the material resources of this republic."[54] With the dispute over expropriation now resolved, it appeared that Mexico was preparing to adopt a conciliatory stance toward U.S. and British petroleum firms.

### World War II: The Oil Refinery and Other Issues

On September 1, 1939, when the Germans had attacked Poland, they threatened the European markets for Mexican petroleum. As a result, Mexican officials began to make inquiries about possible sales of Mexican oil in the United States. Officials in Washington stated that they would not interfere with such sales, and by 1940 American imports of Mexican crude had reached 12.3 million barrels. In addition, U.S. firms resumed exporting petroleum well machinery to Mexico. Under an agreement reached in July of 1941, the United States would buy all the strategic war materials produced by Mexico at the market price.

With the announcement of the Cooke-Zevada settlement nine months later, the American and Mexican governments could now direct their full attention to the war effort. At this time Secretary of the Interior Harold Ickes pointed out to President Roosevelt that despite the great concern the White House had shown over the delay in U.S. financing for a proposed 100 octane Mexican petroleum refinery, progress was still minimal. Oil man Edwin Pauley offered to build the plant at a cost of $15 million—and then to operate it—in return for a 10 percent share of the gross sales, only to have Secretary of Commerce Jesse Jones complain to FDR that Pauley's management fee was "entirely too high." Jones went on to say, "It does not occur to me that it is in line with your policy that the Secretary of the Democratic National Committee [Pauley] should use his position to make money out of the Mexican government."[55]

When a group of oil experts, headed by Everette De Golyer, visited Mexico later that year, they recommended expanding the refinery in Mexico City. The State Department, however, appeared to be adhering to a policy of procrastination at this time. Secretary Hull took the position that a plan had not yet been formulated that would meet the needs of Mexico, while Under Secretary Welles also favored a "go slow" approach. Furthermore, because of the war, there was a shortage of petroleum equipment in America.

For these reasons, it was not until May 1943 that the United States finally made a $10 million loan for the construction of a new, high octane oil refinery near Mexico City, as well as a series of related projects. In March of that year the head of the Mexican state oil monopoly (Pemex), Efraín Buenrostro, had come to Washington with three assistants for talks with the State Department. The loan was a part of the package the negotiators drew up. Unfortunately, the amount of the loan failed to meet Mexican expectations and Interior Secretary Ickes was dissatisfied because he wanted the American government itself to build and operate the plant.

There was not widespread confidence in the United States in the Mexican state oil monopoly. Pemex, in fact, had suffered a series of financial reverses due to its highly irregular output, the loss of various export markets, rising production costs, and the unavailability of oil drilling materials and equipment. Among the American government officials who felt that Pemex would not be able to develop Mexican petroleum resources adequately over the next half century was the Ambassador to Mexico, George Messersmith, who played a key role in talks between the two countries.

Four months earlier, on January 30, the United States had signed an agreement with Mexico that permitted unlimited oil exports to America and reduced duties to 10.5 cents per barrel. This understanding paralleled the Venezuelan agreement of 1939, but it also abolished the petroleum quota established at that time. Due at least in part to the new measure, U.S. oil imports rose from 64 to 114 million barrels between 1943 and 1945; the Mexican agreement remained in effect until the closing years of the Truman administration. The placement of Mexican tankers carrying oil to American ports under lease to the War Shipping Administration was still another factor stimulating petroleum exports to the United States during the war.

In 1944 Mexico sought a loan from the United States to exploit its petroleum resources, but the American government was quite hesistant because of the lingering Mexican hostility toward foreign capital. Thus, on July 1, 1944, George Messersmith complained to Under Secretary of State Edward Stettinius about the "dilatory and negative practices of General Cárdenas [the former Mexican president], who is the one principal element now standing in the way of the development of a reasonable and constructive oil policy by Mexico."[56] The American government, though, was willing for private petroleum firms to lend money to Mexico to help create a special military oil reserve for the United States. In this connection FDR informed Hull on July 19 that he was not concerned about the petroleum the public consumed but rather about that destined for the armed forces. On December 19 Roosevelt again affirmed his opposition to a governmental loan for the commercial development of the Mexican oil industry, and his successor in the White House, Harry Truman, confirmed his stance on October 13, 1945.

## Pemex and the United States since 1945

During the early postwar period, the Mexican state oil monopoly continued to face a series of problems that prevented it from attaining its true potential. "Until 1945," concludes J. Richard Powell, "Pemex would have been judged a failure against any test."[57] As of 1947 Pemex had failed to develop a single petroleum field, and it was not drilling enough wells to replace those in which the oil had been, or was being, exhausted. The total number of wells since expropriation was 355, and most of them were in established fields. Pemex was losing money at the rate of $3 million a month largely because an increase in the payroll had resulted only in a decrease in production. In 1947 Pemex drilled 20 wells, while oil men in Texas drilled 9,000.

In all fairness to Pemex, it should be recognized—as Leonard Engel pointed out in the *Nation* on December 3, 1949—that the Mexican state oil monopoly had inherited a "sick industry." In the last decade Pemex had suffered from both a shortage of technical personnel and a boycott on the part of foreign marketers and equipment suppliers. After World War II Pemex oil production did increase 10 to 15 percent each year and had reached 70 million barrels annually by 1949. Part of this success resulted from the placing of Senator Antonio Bermúdez, an able administrator, in charge of Pemex late in 1946. The presidential support of Miguel Alemán during a confrontation with workers who were illegally on strike unquestionably enhanced Bermúdez' position.

But even with increased production, Mexico was consuming most of its domestic oil production rather than exporting it. Just prior to expropriation only two-fifths of Mexican petroleum was consumed at home; by 1943 the percentage had more than doubled to 94 percent, while in the postwar era it remained in the four-fifths range. Certainly the highly nationalistic and mercantilistic inscription on the façade of the Pemex building was most appropriate: "Consume that which the country produces; produce that which the country consumes."

In 1947 Pemex signed the first Mexican-American petroleum contract since 1938. Under this agreement J. Edward Jones, a veteran oil royalties dealer, was to drill 100 wells, supply the machinery and laborers, and finance the operations. In return, Jones was to receive Mexican bonds to cover all of his expenses and an additional ten percent. These bonds were to be convertible into oil at a preferential price.

Then in 1948 Cities Service signed a contract with Pemex through which the American firm would finance the exploration and development of approximately 1 million acres of oil lands in the vicinity of Tampico. In return Cities Service received the right to buy half of the petroleum it discovered at one agreed price. One of Cities Service's subsidiaries, however, had held leases on this land prior to nationalization, so it was familiar territory to the company. While the American firm did take a gamble in that it agreed to assume all the drilling risks, it was known that there were productive oil fields in that area, although there never had been a petroleum strike on its acreage. In addition, Cities Service obtained the prerogative of withdrawing from the scheme, should it become dissatisfied.

Texaco, too, offered Pemex an exploration and production deal, but it was rejected. Suspicious Texaco officials concluded that Bermúdez had drawn up terms that were bound to be unacceptable to foreign oil firms so that he could approach the U.S. government for a loan, stating that he had tried to reach a deal but had failed to do so. Texaco also believed that Bermúdez had presidential ambitions, an opinion with which the Economic Counselor of the Mexican Embassy, Bohan, agreed. Bohan told a group of foreign service officers and State Department officials that it was widely believed former President Lázaro

Cárdenas would support Bermúdez for president in 1952, if he remained firm on nationalization. But if Bermúdez did hope to become president, he never succeeded.

During the years that followed, Mexican-American petroleum relations continued to follow a pattern of conflict alternating with cooperation. In March 1949 a new firm, the Mexican American Independent Oil Co., won a twelve-year concession to drill for oil along the Yucatan tidelands and elsewhere. The leading figures behind the Mexican American Independent Oil Co. were California oil man Edwin W. Pauley, Ralph K. Davies of the American Independent Oil Company, and Samuel B. Mosher of the Signal Oil and Gas Company. According to the terms of the contract, the new firm would receive half of the petroleum production until it had paid for its drilling costs, after which its share would fall to 15 percent. Unfortunately, by 1951 it had drilled only one truly productive well on its Mexican properties.

On the other hand, in the spring of 1949 the United States refused to advance a petroleum development loan that was requested by Pemex chief Antonio Bermúdez in conjunction with a Mexican program costing $470 million over a five-year period, even though in 1943 Pemex had obtained $10 million from the Export-Import Bank to build a refinery. Unfortunately, in November of 1947 an interdepartmental committee of the State, Interior, Commerce, Army, and Navy departments had written a report, worldwide in scope but focusing on Mexico and Bolivia, which stated that under no circumstances should America loan public funds to nationalized oil industries abroad. Leading the attack on the proposed 1949 deal with Bermúdez was Max W. Ball, a former director of the Interior Department's Oil and Gas Division; Ball declared that Mexican petroleum was not vital for U.S. defense, and added that Mexico would need perhaps $2 to $4 billion to exploit her oil resources. Ball's lack of enthusiasm was shared by such powerful figures as Senator Robert Kerr of Oklahoma (himself a wealthy oil man) and former Governor Lester Hunt of Wyoming. These two Democrats were moving heaven and earth in an attempt to repeal those Mexican petroleum statutes that discriminated against foreign firms.

Nevertheless, the House Interstate and Foreign Commerce Committee of the Eightieth Congress favored the loan, since it would provide an additional emergency oil source for America. The chairman of this committee, Republican Representative Charles Wolverton of New Jersey, had visited Mexico with the entire committee and staff in the summer of 1948 as guests of Pemex. Rather than cooperate with U.S. diplomatic officials already there, the Wolverton Committee had aroused the ire of the Ambassador to Mexico Walter Thurston by ignoring the American Embassy. The Wolverton committee pointed out that such a loan to Mexico might have consequences for Latin America as a whole. According to its report, the United States could use a petroleum exploration loan to Pemex, "to evince our good neighbor policy, to assist in the maintenance and stabilization of the Mexican economy and the creation of a Mexican exportable surplus commodity. . . ."[58] Wolverton's successor as chairman was Democrat Robert Crosser of Ohio, who also favored such a loan.

With prospects for the loan apparently quite favorable, Antonio Bermúdez paid a visit to the United States only to have the State Department balk at considering half of his project. During the ensuing discussions the amount of the projected loan was scaled down to $203 million and then to $100 million. By July negotiations had broken down, and the governments of Mexico and the United States announced that they had elected to suspend the talks without prejudice to their resumption. Even though President Harry Truman had recently endorsed the loan, thus reversing his original opposition to it, the State Department demanded as a condition of any loan that Mexico admit American oil firms to the petroleum fields, and at the same time provide an acceptable legal basis for their participation. The reaction in Mexico to this request was the expected one, and well expressed by *Newsweek*: "The return of expropriated American companies alone would have meant repudiation of the Mexican revolution and political suicide for President Miguel Alemán."[59]

One of the factors that contributed to the undermining of the negotiations was the somewhat dubious claim of the Sabalo Transportation Company, which was demanding $200 to $400 million for its expropriated properties, even though the Mexican-American joint commission theoretically had disposed of all U.S. oil claims in 1942. Prior to 1938 Sabalo had held the creeks and the watersheds of the Poza Rica field, a Royal-Dutch Shell project; having obtained its valuable concession at little cost, it had done little to develop its holdings. As a result, the claims commission awarded Sabalo a mere $896,000. Sabalo had refused to accept this amount and had then appealed its award to the Mexican courts.

On June 6, 1949, Ambassador Walter Thurston sent a letter to President Miguel Alemán in which he requested the establishment of a special commission to negotiate the settlement of the Sabalo claim. Earlier that year, on February 24, the Third Supreme Court of Justice had ruled that the Mexican government had not expropriated the rights of Sabalo. The letter from Thurston and a State Department *aide-mémoire*, dated July 6, had irritated the Mexican government, but Secretary of State Dean Acheson continued to press for the settlement of the Sabalo claim.

When Sabalo lawyer Allen Dulles met with Antonio Bermúdez on August 11, Dulles told Bermúdez that the Sabalo claim was worth $40 million, based on its possible profits rather than its investments. Dulles maintained that the Cooke-Zevada settlement had greatly undervalued Sabalo's equipment, but Bermúdez countered with the opinion that the $40 million figure was outrageous and exaggerated. Dulles returned on the next day, armed with an estimate that the Sabalo investment at Poza Rica was worth $6.1 million. By December 10 Sabalo expressed a willingness to settle for as little as $13.5 million plus accumulated interest, but Bermúdez found this proposal totally unacceptable. Sabalo's offer expired on December 31, 1949.

With the breakdown of the talks and the impasse over the Sabalo claims, those in the United States who favored a loan to Mexico for petroleum development began to express their dissatisfaction. One of the most vocal of these was maverick Republican Senator Wayne Morse of Oregon, who maintained that an antiloan policy was inconsistent with past ECA grants. Truman himself advised Morse in March that "I have made it perfectly plain to all concerned that the loan to the Mexican government for pipelines and refineries should be promptly made."[60] Prior to his letter to Morse, Truman had written, on a letter dated January 23, 1950, from Secretary of State Dean Acheson: "Watch the successors of Teapot Dome and see if we can't help Mexico and the Mexican people."[61] There was, however, a rising sentiment in the State Department that to lend Mexico money for the construction of either pipelines or refineries would encourage other Latin American governments to pass expropriation measures or to set up state monopolies.

When Allen Dulles again conferred with Antonio Bermúdez on June 27, 1950, he stated that the Sabalo claim had nothing whatsoever to do with the oil loan but added that the settlement of the claim would help Mexico to obtain the loan. By this time Sabalo had reduced its demands to $5 million plus interest. That August 28 Wayne Morse complained to Truman that the State Department was using the Sabalo claims to intimidate Mexico, but shortly thereafter the Export-Import Bank agreed to extend $150 million to Mexico for economic development.

As the controversy continued to rage over the petroleum loan and the Sabalo claim, the long and barren search for oil in Mexico was ending. In January of 1950 Tortuguero well number one had gushed oil south of Vera Cruz in the Gulf of Mexico. This was the first producing well in a dozen years to be explored and drilled with American capital. During the year before, moreover, Pemex had brought in 180 producing wells. As a result of this marked expansion in oil production, Pemex was able to turn over to the Mexican treasury $50 million in profits for 1950.

By 1951 Pemex was able to export oil surpluses for the first time since 1938. To handle the increasing petroleum output, Mexico built three new refineries, a lubricants plant, and a network of pipelines. Because of this progress, Interior Secretary and Petroleum Defense Administrator Oscar Chapman took a personal tour of the Mexican oil industry during the summer of 1951. Chapman, who apparently had in mind a steel for oil trade, was favorably impressed with the Pemex installations that he visited.

But there was a dark side to the Mexican petroleum story as well, for Antonio Bermúdez as the head of Pemex had to contend with a tradition of graft, which, because of political pressures, he was only able to reduce rather than eliminate. Furthermore, the Mexican petroleum industry employed too many laborers, and any attempt to reduce the work force encountered the strong opposition of the unions. These two factors alone were enough to hold down Pemex's profits, should the government oil monopoly emerge from red ink into the black. In addition, the Mexican government—unlike American petroleum entrepreneurs—was reluctant to drill in untested areas, even though there might be oil there.

The termination of the Mexican-American petroleum agreement in 1952 in effect reinstituted the tariff quota provisions of the agreement the United States had signed with Venezuela in 1939. Nevertheless, when a modified system of mandatory oil imports went into effect on June 1, 1959, the American government exempted Mexico, since like Canada it could ship petroleum to the United States by land. Mexico eventually agreed to maintain a ceiling on oil shipments voluntarily.

During this decade Pemex had continued to make progress. In 1948 Mexico had reached an expropriation settlement with British oil interests that totalled $81 million. In turn the settlement facilitated discussions with foreign petroleum firms throughout the 1950s. In 1953 the World Bank concluded that Pemex had achieved three goals: the meeting of domestic demand; the improvement of internal petroleum transportation facilities; and the avoidance of net imports of refined products. On March 18, 1959, shortly before his resignation as the head of Pemex, Antonio Bermúdez spoke eloquently of the achievements of the state oil monopoly:

> . . . with this spirit, within our economic possibilities, we have not lessened our efforts to place within reach of every worker those means which raise the level of man's life, namely, just wages, clean and comfortable living quarters, medical services, education, and a human, respectful treatment of his individuality—benefits which all men of all latitudes and under all conditions have a right to receive.[62]

Because of its enhanced status, Pemex had been able to obtain a $50 million loan from U.S. investors in June 1962. The loan was divided between American investment houses and the Chase Manhattan Bank; Mexico offered as collateral in this transaction the gas export contract between Pemex and the Texas Eastern Transmission Corporation. Still, as late as 1961 Pemex only had been able to borrow for three years at a time and at rates of interest that were by no means minimal, despite the favorable assessments of Bermúdez and the World Bank. Not only were there numerous cancellations of supplier contracts, but American sources of credit hesitated to make new commitments. Because in part of the funding of short-term liabilities, however, Pemex by 1962 was able to present an image of improved financial health, which led to the $50 million loan.

Assessing the status of Pemex in August 1963 from the American side of the border, *Fortune* concluded that despite the costly benefits the trade unions had won there and other heavy expenses,

> [Pemex] is well along on the hard road from shaky beginnings to solid respectability. It has become the vertebrae of Mexico. The largest integrated oil and gas producer in Latin America, it had revenues last year of $540 million. At the same time it is the largest dispenser of social services outside the federal government itself.[63]

Throughout the 1960s the rather devious means by which Mexican oil entered the United States—the so-called "Brownsville Loop" arrangement—was a focal point of increased attention and criticism on both sides of the border. By 1962 the State Department felt obliged to defend this unexpected byproduct of the oil import control program, which had been designed to discourage petroleum shipments from Mexico to Cuba. Under this program, after the Mexican crude oil and unfinished oils had arrived by ocean tanker at Brownsville, Texas, and landed in bond, they were placed on a truck, carried over the international bridge to Mexico, brought back, reloaded on the ocean tanker, and shipped to the Atlantic coast of the United States. This second entry—unlike the first—qualified for the overland exemption.

Prior to the oil import program, Pemex had sold petroleum in amounts comparable to those now trucked at Brownsville to New York City importers. These shipments, however, were taken over by a major international oil company, which preferred to import its petroleum from Trinidad. Thus, the "Brownsville Loop" became in effect a substitute for the lost 30,000 barrels-a-day New York City market, although direct tanker shipments from Tampico to the Atlantic Coast had been more economical and involved less handling.

In the summer of 1970, President Richard Nixon decided to terminate the "Brownsville Loop" at the end of the year and to allow oil from Mexico to enter the United States under some other arrangement. The arrangement took the form of a country-of-origin quota. In the summer of 1972 the American government took another key step in developing its Mexican oil policy, when the Export-Import Bank agreed to participate in the financing of a loan to cover 90 percent of the cost of exporting $46 million worth of equipment and services to Pemex. The Export-Import Bank previously had made five loans to Pemex totalling $41 million.

At this time the Arab oil embargo was only a little more than a year away. That catastrophic event in world oil relations was followed by the discovery of new petroleum fields in the Chapas-Tabasco region of Mexico in 1974. Additional discoveries at a later date enabled Mexico to claim that it possessed one of the largest petroleum reserves in the world. By the end of 1974 Mexico assumed the role of a net exporter rather than a net importer of oil.

Yet the petroleum bonanza brought with it problems as well as blessings. Mexico realized that any drastic alteration in its economy as a result of an abnormal increase in oil pumping and refining might have severe political and social repercussions. Therefore, much to the displeasure of many Americans, the Mexican government took the position that the future well-being of Mexico should take precedence over supplying their neighbors with huge amounts of petroleum. Therefore, if there is to be an oil deal between the Mexican and American governments, it must be on the basis of negotiation with equals or partners, not on a patron-client basis. The likelihood that such an understanding may well encounter obstacles is shown by the Carter administration's controversial rejection in 1977 of a natural gas importing agreement with Mexico. President Jimmy Carter was of the opinion that the latter cost too much. This action, which greatly offended the Mexican government and people, placed the United States in a position where it would have to pay more for both Mexican natural gas and petroleum in the future.

# 4.

# The Noncontiguous Western Hemisphere

## Venezuela under the Gómez Dictatorship

Like Mexico, the petroleum of Venezuela long has attracted the attention of the United States. During the twentieth century the country has been torn politically in two directions: dictatorship, exemplified by such totalitarian leaders as Juan Vicente Gómez (1908-35) and Marcos Pérez Jiménez (1953-58), and more recently democracy, typified by reformist President Rómulo Betancourt (1945-48 and 1959-64).

Unlike many other Third World countries, Venezuela theoretically possesses the economic resources needed to build a stable, democratic society. Its income from oil has so increased during the twentieth century that it now dominates the export market and is the chief source of government revenues. In 1965 Venezuela had the highest per capita income of any Latin American nation, but in this decade before the 1973 OPEC embargo it also had the fastest growing population, much of which lived in poverty and illiteracy.

During the years following World War I Juan Vicente Gómez continued to maintain the autocratic control over Venezuela he first had established in 1908. The pro-German sympathies which Gómez had manifested before the war, however, now gave way to a more friendly attitude toward the United States in general and American business firms in particular. Apparently worried about growing British power in Venezuela, Gómez looked upon the United States as a potential check on Great Britain. The first beneficiary of his new policy was Sun Oil, but between 1920 and 1922 Jersey Standard also obtained extensive land holdings in Venezuela.

Despite this change in attitude on the part of Gómez, in 1920, at the urging of Development Minister Gumersindo Torres, the Venezuelan Congress passed a new petroleum law that American oil men found objectionable. Among its key provisions was one granting to concession holders only the temporary right to exploit, rather than outright proprietorship. Other provisions included an increase in the surface taxes, a raise in the minimum royalty, and a limitation on the area to be exploited. Torres was attempting to play upon the Anglo-American oil rivalry to strengthen his country's position in petroleum negotiations.

In the eyes of American oil men, perhaps the most objectionable provision was article 50, which required them to select and exploit parcels of land within a three-year period, a rather short time. The Minister to Venezuela, Preston McGoodwin, argued their case before Gómez, and the petroleum men themselves refused to buy leases from Venezuelan landholders who held exploration permits.

Back in the United States, Republican Senator Henry Cabot Lodge of Massachusetts complained that Royal Dutch-Shell was encroaching upon the American position in both Venezuela and Colombia and suggested that U.S. firms in those countries might soon pass into foreign hands unless they received government protection. In the spring of 1921, when the Venezuelan government was contemplating annulling certain concessions held by the London-based Colón Development Company, Ltd. on the basis that the British firm had failed to fulfill its contractual obligations, the State Department took the position that McGoodwin should not openly support Colón. The State Department's anti-British frame of mind at this time is apparent in its instructions to McGoodwin:

> To certain companies operating in Venezuela, such as the Caribbean Petroleum Company, the Colón Development Company Limited, the Venezuelan Oil Concessions Limited, the Carib Syndicate and others, which are known to be either British controlled or closely affiliated with British controlled companies, no diplomatic assistance should be given, generally speaking, without clear and specific instructions from the Department.[1]

Due to the various pressures which both the American oil men and the U.S. diplomatic corps placed upon the Gómez regime, the Venezuelan Congress passed a new petroleum law in 1921 that in many ways was more favorable to American companies. Not only was the maximum exploitation area doubled, but the taxes were reduced as well. Yet some U.S. petroleum leaders still were not satisfied because it remained unclear what parts of the mining code were applicable to the petroleum industry.

Rather than refusing to accede further to the demands of U.S. oil men, the Venezuelan government dismissed its Development Minister, Gumersindo Torres, and asked the advice of three American companies on drawing up another petroleum law. This 1922 measure followed both U.S. and Mexican practices and defined the petroleum industry as a public utility, totally independent of the mining code. The biggest loser under this new law was the Venezuelan Congress, since only the President or his ministers now had the authority to grant concessions. The Venezuelan courts also obtained complete and final judicial powers over all oil disputes, and foreign petroleum firms could only seek diplomatic assistance at the risk of having their concessions nullified. Although the new oil law in many respects was the most favorable mining law in Latin

America at this time from the standpoint of the foreign companies, its provisions were still rather strict, and it more than adequately protected the interests of Venezuela.

Under the new measure, concessions were limited to forty years, rather than thirty, and they were to be larger. Employers were held accountable for the health and well-being of their employees, and required to build hospitals, supply medicine, and provide for accident compensation. While all American concessionaires holding grants under the former laws took advantage of the opportunity to comply with the new measure, British firms operating under the 1907 and 1912 contracts and the 1910 mining code did not follow suit.

By this time, though, the new Republican administration in Washington had begun to adopt a less hostile attitude toward the activities of British oil firms in Latin America, and had become in general more solicitous of the Open Door rights of foreign nations. Accordingly, during 1922 the State Department instructed the Caracas legation to support British and Dutch representations in the disputed Caribbean Petroleum Company concession; this was a multinational enterprise financed by American, British, and Dutch capital. In 1923 the State Department again intervened when the Venezuelan government threatened a Royal Dutch-Shell subsidiary, British Controlled Oilfields, Ltd., with the cancellation of its Buchivacoa concession. Jersey Standard was directly involved in the event, since it had obtained an option to develop the eastern third of this area. Standard's agents had the U.S. Minister to Venezuela arrange a personal meeting with Gómez. At the meeting Gómez indicated there would be an amicable settlement of this dispute, and joined by British diplomats, the Americans were successful in obtaining a complete reaffirmation of the concession.

While the United States was becoming less hostile to British oil operations in Venezuela, it was casting an increasingly jaundiced eye on the activities of a German group, the Stinnes interests. Up to World War I, German-Venezuelan relations had been rather friendly, and Germany now was unwilling to adopt a do-nothing policy while American, British, Dutch, and French oil firms gobbled up the petroleum reserves and potentially petroliferous areas of Venezuela and other Latin American nations. Accordingly, in February of 1924 a group headed by Hugo Stinnes obtained an option to purchase a 25 percent interest in the Compañía Venezolana de Petróleo. Since certain lands belonging to the Compañía were adjacent to the exploitation parcels held by American companies, the Germans could take advantage of the development work done by the U.S. firms and draw on common underground reservoirs. Although the State Department had its Caracas legation complain to the Venezuelan government, Jersey Standard decided to act decisively by purchasing the national reserve hectarage still up for sale. This action undercut the position of Stinnes, but left Jersey Standard wondering whether Gómez had not encouraged the Germans in order to force the Americans to buy additional concessions at maximum prices.

Having made these purchases, Jersey Standard rejected an opportunity in 1926 to obtain a large share in the Colón Development Company, Ltd., a subsidiary of

Royal Dutch-Shell. At this time Walter Teagle of Jersey Standard observed, "We have invested so much in wildcatting in Venezuela without result, that the acquisition of producing property would. . . place us in a position of actual producers."[2] A key factor in the company's decision was Teagle's belief that Henri Deterding would not sell anything valuable, although Deterding had asserted that the offer was made because his company held too much land in Venezuela. It should be noted that the concession later produced 20,000 barrels of oil daily.

By the time of the Great Depression, other petroleum firms aside from Royal Dutch-Shell and Standard Oil of New Jersey had become active in Venezuela. Standard of Indiana was present, as was Gulf Oil. In 1927 Atlantic Refining Company obtained a 50 percent share in the four million acres of the Andes Petroleum tract, while California Petroleum and Union Oil of California signed a contract to spend $7 million in developing the Pantepec Oil tract. Anglo-Persian also held a large concession in the state of Falcon. Thus it is not surprising that Venezuelan oil production, which was minimal at the end of World War I, increased sharply during the following decade, with its output often doubling from year to year. By 1927 Venezuela was the world's second-ranking producer of petroleum.

Only a small part of this oil was refined in Venezuela, however. Fearing a change in policy would occur when Gómez died, the U.S. and British oil barons channelled their oil outside of the country. Royal Dutch-Shell sent most of its Venezuelan output for refining to Willemstadt, Curaçao, just off the coast of South America. In 1928 the Deterding firm completed another refinery at Oranjestadt, Aruba, another Dutch island near to Curaçao, and in the following year Standard of Indiana completed its own refinery on Aruba. Rather than admitting that political considerations motivated the location of the refineries in the Netherlands Antilles, the petroleum companies pointed out justifiably that the bar across the Gulf of Maracaibo prevented ocean-going tankers from crossing it.

Venezuelan oil imports into the United States had now reached such a level that certain American producers, especially the midcontinent operators from the Great Plains states, Texas, Oklahoma, and Kansas, became disturbed by the threat they felt the imports posed to them. On the eve of the Great Depression they marched on Washington, lobbying for a tariff on the Venezuelan product. The subsequent economic collapse, if anything, intensified their agitation. In 1930 the Republican-controlled Congress enacted the highest tariff in American history, the Hawley-Smoot Tariff, but it was not until June 6, 1932, that Congress specifically imposed a tariff on imported petroleum.

The results of the tariff were not as favorable as had been hoped from the American standpoint, although imports of Maracaibo crude did decrease and Venezuelan gasoline practically disappeared from the American market. By 1932 adverse circumstances forced Indiana Standard to sell both Lago Petroleum and its Aruba refinery to Jersey Standard for $140 million; Jersey Standard then marketed the Venezuelan oil in Europe and South America rather than in the United States. Gulf Oil sold its surplus to Royal Dutch-Shell, which in turn sold it in Europe. World petroleum markets accordingly began to manifest a "new look," with Venezuela replacing the United States as the leading supplier of oil to Europe.

Friction between the American oil companies and the Venezuelan government then developed during 1931, as a result of certain discrepancies in reports made by the U.S. petroleum firms to the Venezuelan government and to the United States Tariff Commission. In 1927, for example, Indiana Standard had reported to the Venezuelan government that it had cost 68 cents a barrel to transport petroleum from the oil fields to the Atlantic Seaboard market, but it had informed the United States Tariff Commission that the cost was only 33 cents. Development Minister Torres, who had been reinstated earlier, calculated that Indiana Standard had defrauded the Venezuelan government out of £26 million British sterling and Gulf had defrauded the government out of £30 million British sterling. Torres accordingly presented the two American firms with bills for these amounts. The U.S. petroleum companies, though, maintained that their royalty payments had been in conformity with the agreements they had signed in 1927 and accused Torres of launching an antioil industry offensive. Although Torres did enjoy the support of the Venezuelan press, he lost his position as a result of this uproar, and the 56 million British sterling claims against the petroleum firms faded as an issue.

In 1935 the Gómez era ended. After more than a quarter of a century of dictatorial rule Gómez, who like Porfirio Díaz in Mexico had encouraged foreign investment, died. On June 30, 1950, a top secret Department of State policy statement on Venezuela summarized the economic developments under Gómez:

> While the dictatorship of Juan Vicente Gómez (1908-35) gave no opportunity for the development of democratic experience and anticipated in many respects some of the most odious features of the Nazi-Soviet police state, his economic policies laid the foundation for the growth of the country into the financially strongest nation of Latin America. Gómez paid off substantially all of Venezuela's foreign debt and sedulously avoided disputes or entanglements with foreign powers. He had legislation enacted which made it possible for foreign companies to develop Venezuela's vast petroleum resources to such an extent that income from activities of the oil companies currently supply Venezuela with 95% of its foreign exchange and 72% of its total governmental revenue.[3]

## The Prewar Years and World War II

Gomez's successor was Eleazar López Contreras, who had been Secretary of War. During his five years as chief executive López Contreras popularized the slogan "Sow the Petroleum," that is, use the earnings from oil to invest in other sections of the economy. In 1936 and 1938 new petroleum laws went into effect that were less favorable to the oil firms; the second of these measures permitted the government of Venezuela to enter the petroleum industry, and raised the exploitation tax. A Creole Petroleum internal memorandum dating from this period noted that with the rise in the exploitation tax it did not "appear probable that any serious operator would undertake the risk of [new] operations."[4] Equally disturbing to the oil companies was a new constitutional provision that allowed the Venezuelan government to impose taxes on exports.

Venezuela also adopted new labor legislation in 1936 legalizing labor unions and providing for an eight-hour day. These measures inspired Henry E. Linam of Jersey Standard to observe: "It's against my principles to let anybody else run my business."[5] Under the new law companies had to make housing, schools, and hospitals available to their workers; although the oil firms were already providing these voluntarily, they did not wish to be forced to act under duress. Oil companies feared that the new rulings on wages, hours, and working conditions would bankrupt them, but a rise in the price of crude oil calmed their apprehension somewhat.

Despite Linam's criticisms of the López Contreras regime, his company was more than cooperative with the new government, thus blunting an attack on foreign firms by the extreme nationalists. Jersey Standard arranged to build roads, bridges, and hospitals in exchange for the payment of customs duties; in the years that followed both American and British petroleum companies managed to keep their Venezuelan concessions intact. In fact, during September 1936 the López Contreras regime granted new concessions to both Jersey Standard and Socony-Vacuum.

On December 15, 1937, Gulf Oil sold a 50 percent interest in its Mene Grande subsidiary to the International Petroleum Company, a Jersey Standard subsidiary, for $100 million. International Petroleum in turn sold half of its share to Nederlandsche Olie Maatschappij, a Royal Dutch-Shell subsidiary. The net result of this transaction was to remove Gulf from the Venezuelan oil scene as a significant factor. Henceforth Mene Grande produced only 100 barrels for every 345 barrels that Jersey Standard produced, and the disparity between Mene Grande and Royal Dutch-Shell was almost as great.

Even though the Venezuelan government encouraged the foreign petroleum companies to continue operating there, it did suggest that a shake-up in their personnel would be most desirable. The American Minister, Meredith Nicholson, agreed: "Just as the country itself is undergoing a new deal, so would the affairs of the oil companies benefit by a new deal in personnel; in the opinion of the Legation such a change should also afford an opportunity of improving the standing of Americans and the reputation of the United States in Venezuela."[6] Under Secretary of State Sumner Welles, however, felt that the State Department should not make a formal suggestion along these lines to the petroleum firms at this time. In the following months the oil companies did raise wages and improve facilities, but as of June 1938 only one of them changed its personnel in Caracas.

That March the Cárdenas regime in Mexico expropriated the holdings of various American and British petroleum firms, but the Venezuelan government did not follow suit. Instead, the Supreme Court of Venezuela handed down a ruling in April that required Gulf Oil and Mene Grande to pay approximately $5 million as a penalty for benefitting from an illegal tax reduction, dating back to

1925. But the Venezuelan Supreme Court was not entering into a program of harrassment, for in April it continued the Lago Petroleum Company's exemption from customs duties on various types of equipment and materials.

As the months passed and the petroleum controversy with Mexico remained unresolved, it became increasingly apparent that a guaranteed flow of oil from some other nation to the United States was desirable. Therefore in November 1939, just after the outbreak of war in Europe, the American and Venezuelan governments concluded an executive agreement, to become effective provisionally on December 16, that halved the 1932 import tax from 21 cents per barrel to 10.5 cents on petroleum in amounts up to 5 percent of the processed crude oil production in the United States in the previous year. Amounts in excess of that percentage were to be taxed at the older rate.

Under the accompanying quota system, Venezuela received a share equal to 70 percent of American petroleum imports; Curaçao and Aruba combined received 21.3 percent, Colombia 3.2 percent, and all of the remaining countries 5.1 percent. Since most of the oil shipped from Curaçao and Aruba originated in Venezuela, the quota system, in effect, gave Venezuela an almost complete monopoly over American petroleum imports. In contrast, Mexico was not even mentioned specifically among the oil exporting countries that received quotas, and thus emerged as a big loser from this arrangement.

Venezuela's privileged status encountered some opposition in the United States, especially from the petroleum firms of the midcontinent field and various political leaders. On November 17, 1939, for example, the Governor of Kansas, Payne Ratner, wrote a letter to Secretary of State Cordell Hull, in which he complained that the Venezuelan oil agreement was "detrimental to the domestic petroleum industry and in particular to that industry in Kansas."[7] Hull's reply emphasized that Venezuela in turn had agreed to reduce the duties on hog lard and wheat flour.

Despite its advantageous position under the American tariff, in the first part of 1941 the Venezuelan government challenged Mene Grande, charging that it owed $15 million in back taxes for the years between 1927 and 1931. Mene Grande admitted that it owed some taxes but not that much. When President Eleazar López Contreras informed Frank P. Corrigan, the American Ambassador, that he would present the claim to the courts, if necessary, Under Secretary of State Sumner Welles placed pressure on Corrigan to convince López Contreras not to take such a drastic step at the end of his presidential term. Welles also held talks with the president of Gulf, the parent company of Mene Grande, in Washington and arranged for a $10 million settlement that the Venezuelan cabinet then approved.

Hemispheric defense was also a concern of the American government at this time. The General Staff, fearing that Axis agents might attempt to destroy the oil tanks or refineries of nearby Aruba or Curaçao, desired to store bombs and station skeleton companies at two places in nearby Venezuela. The Venezuelan government, though, opposed the quartering of American troops there because it would have a negative affect on public opinion. It did agree to two U.S. Army noncommissioned officers serving as the custodians of the bombs, which it would allow the American government to store on Venezuelan soil.

During August of 1942 the friction between the government of Venezuela and the American oil companies recurred, when the Venezuelan government requested that the oil firms pay a larger share of their profits. Isaías Medina Angarita, the president of Venezuela since 1941, told Roosevelt that he was planning to reform the petroleum industry. At this point the State Department intervened in the controversy, although it did not make its role public at that time; petroleum advisor Max W. Thornburg held talks with oil company executives and the Venezuelan Attorney General, Gustavo Manrique Pacanins. Before the negotiations had ended, officers of American and British petroleum firms were participating, as well as Arthur Curtice and Herbert Hoover, Jr., who were acting as oil consultants to the Venezuelan government. Royal Dutch-Shell proved more willing to enter into a compromise with Venezuela than Jersey Standard, but Jersey Standard eventually did agree to a settlement.

As these talks progressed, a special Venezuelan commission was drafting a new petroleum bill that became law on March 13, 1943. Among its major features were an increase to 16 2/3 percent in the royalties due the Venezuelan government and the establishment of a new base more favorable to Venezuela. The oil companies were pleased because the measure settled the question of alleged title defects, granted new concessions for a forty-year period, and simplified government regulations. Looking to the future, there was an agreement that American and British firms would construct a series of oil refineries in Venezuela within five years after the end of the war.

Although technically it was not part of the law, another key feature of the Venezuelan petroleum settlement was the 50-50 profit splitting formula. According to Leonard M. Fanning, the major oil companies agreed to this idea in principle in 1943, although Standard Oil later denied it had done so. It was not until 1948, however, that the 50-50 concept was placed on the Venezuelan statute books, and it then was incorporated into the income tax law.

With the new oil law now in effect, Jersey Standard brought all of its Venezuelan holdings together into the Creole Petroleum Corporation on August 23, 1943. By 1945 Creole was producing half of Venezuela's oil. California Standard also reentered Venezuela during the final years of the war, while Phillips Petroleum made its initial investments in the country at that time.

Despite the fact that petroleum from Venezuela made a major contribution to the Allied war effort, Venezuelan-American relations were not always harmonious. In January 1944 Roosevelt and President Medina Angarita attended a dinner in Washington at which FDR stated that he would like to see a large refinery constructed on the Venezuelan mainland, as soon as the necessary materials were available. Medina Angarita's reply was that he had no preference among the foreign oil companies that might be interested in building these facilities. In April, however, the Venezuelan government unveiled a new plan under which 10 percent of the nation's crude oil would be refined within the country, while the remaining 90 percent would be shipped out of the Caribbean area, thus bypassing the Dutch islands of Curaçao and Aruba. Even though Royal Dutch-Shell would have suffered more from this plan than the American petroleum firms, the U.S. government still protested.

On June 8, 1944, the Acting Secretary of State, Edward Stettinius, advised Ambassador Corrigan that "it [the State Department] cannot regard the prohibition against the refining in the Caribbean area of Venezuelan oil from new concessions as other than a restriction of trade and a discrimination against a specified geographic area, which area includes certain United States territory."[8] Angered by the diplomatic protests, the Venezuelan government halted the negotiations then underway with the foreign oil companies for new concessions, blaming the companies for these protests. Shortly thereafter, however, the government agreed to resume negotiations. Royal Dutch-Shell's enlargement of its facilities on Curaçao was a potential source of friction with Venezuela, but the additional equipment was ancillary to existing equipment, and even dependent upon it. The Armed Forces of the United States, moreover, had authorized the amplification, as winning the war took precedence over offending Venezuelan sensibilities.

## Reform under Betancourt, Reaction under Pérez Jiménez

In late 1945, after a group of young Army officers had overthrown the Medina Angarita government, a decided shift took place in Venezuelan oil policy. The old government was replaced with a revolutionary junta dominated by the left-wing Acción Democrática. The junta installed as provisional president Rómulo Betancourt, a thirty-seven-year-old liberal intellectual. Charging that the 1943 petroleum law had proven inadequate, one of Betancourt's first actions as chief executive was to decree a multimillion dollar extraordinary tax on the oil companies. Apparently the revolutionary junta periodically felt impelled to issue such decrees in an attempt to satisfy the strong nationalistic sentiment then emerging in the country.

The U.S. government was highly concerned about this turn of events. On January 7, 1946, Secretary of State James Byrnes wrote to the Chargé in Venezuela, Dawson, about ". . . the shocked surprise with which the extraordinary excess profits tax was received in both [government] and financial circles here and what a blow it might entail for the Junta's standing abroad."[9] It was the suddenness of the action that was most disturbing to American officials, who admitted its technical legality. In searching for a justification, the Venezuelan government took the position that it needed the extra money to finance new programs, including an Instituto de Fomento de la Producción, a merchant marine, and a housing program. Betancourt added that the tax revenues might be used to buy equipment in the United States. It was also during 1946 that Venezuela suspended the granting of new petroleum concessions to foreign interests.

While the American oil companies had taken huge earnings from Venezuela, they had made at least a modest contribution to the social and educational development of the nation. As of 1945 Creole and Mene Grande had spent a total of $36 million for this purpose. By 1947 illiteracy among the 15,000 workers employed by Creole had fallen to 12 percent, approximately one quarter of the national figure.

Despite this display of social consciousness on the part of the oil companies, with the left-wing government in power, labor agitation began to mount. Betancourt now informed the petroleum firms that improving the lot of the workers was a political rather than industrial issue. Then during June of 1946 Jersey Standard and Royal Dutch-Shell began to sign contracts with the unions. Remembering the disastrous series of events that had taken place in Mexico prior to World War II, the oil companies operating in Venezuela were most generous with labor, offering substantial raises, paid vacations, and assorted fringe benefits. The overall settlement increased the wages of 25,000 petroleum workers from 35 to 50 percent. Significantly, the U.S. government does not appear to have been upset at these terms. On June 17 Acting Secretary of State Dean Acheson wrote to Secretary of War Robert Patterson: "The terms of the agreement are not excessively out of line as compared with agreements reached in the domestic petroleum industry in recent years."[10]

The following year, in December 1947, the novelist Rómulo Gallegos became president in what has been characterized as Venezuela's first truly free election. Having won by a four to one majority, Gallegos took office promising a democratic future for his nation. But his ideas and ideals soon began to make enemies; among other things, he suggested an increase in government oil royalties, which alarmed business leaders. At the end of November an Army *coup d'état* drove Gallegos from power and into exile, along with Betancourt. The State Department felt it necessary to issue a press release on December 8, denying that U.S. petroleum companies had supported the overthrow of Gallegos or had participated in the revolution in any way whatsoever.

On November 12, 1948, twelve days before the Gallegos regime fell, a new income tax law had gone into effect, mandating a 50-50 profit split between Venezuela and foreign oil firms. Creole, moreover, agreed that it would pay back taxes to the Venezuelans that would "equalize" that government's income from Creole with the company's profits since 1946. This 50-50 profit sharing formula was to remain in effect during the four years that an Army junta was shaping Venezuela's destiny, when total oil industry taxes increased from $374.8 million in 1948 to $514.2 million in 1953.

Once the new regime was firmly in power, the American oil firms began to build additional refineries in accordance with the agreements signed while Medina Angarita had been president. Since Venezuela purchased more American exports than any other non-Marshall Plan country with the exception of Canada, it is little wonder that a group of leading Venezuelan businessmen visited America at this time. In 1950 the total Latin American imports of crude oil and residual fuel from Venezuela were not equal to the U.S. imports from that nation.

But despite the large scale commerce that was going on between America and Venezuela at this time, the State Department was hesitant to invite their military leaders to the United States for a number of reasons. Deputy Assistant Secretary of State Willard F. Barber claimed in a memorandum, dated May 15, 1950, that "the recent strikes in the oil fields and among other unions in Venezuela have revealed the deep and widespread dislike of the military government in Venezuela on the part of large sectors of the people."[11] Then on June 30 the State Department finalized a top secret policy statement on Venezuela that rather pessimistically evaluated the current relationship between the foreign oil firms and the country: "Both the companies and their foreign employees have been the objects of envy, distrust and dislike by the Venezuelans."[12] Among the obstacles to harmonious relations the report singled out the high standard of living enjoyed by foreign managers, their isolation in Americanized settlements, the conflict between the "hurry-up" Yankees and the "mañana" Venezuelans, and the foreign control of that nation's greatest natural reserve.

In 1950 through a secret poll which it conducted in Venezuela, Jersey Standard discovered that 70 percent of those interviewed favored the nationalization of foreign oil firms. During the years that followed Creole, which accounted for more than one-third of Jersey Standard's net income at that time, continued to attempt to integrate itself into the life of Venezuela in a number of ways. These included the employment of qualified Venezuelans where feasible; compulsory Spanish lessons for all non-Spanish speaking employees; and various cultural contributions, including relief maps for the schools and a large number of scholarships. In October 1951 Creole elevated to its board of directors Guillermo Zuloaga, an outstanding administrator and geologist who had obtained a Ph. D. at the Massachusetts Institute of Technology before joining Creole's staff in 1939. Zuloaga became perhaps the most important South American in the petroleum industry.

Despite the existing support for nationalization, however, it would appear that around 1952 most of the citizens of Venezuela were relatively satisfied with the behavior of the Jersey Standard subsidiary. A public opinion poll taken at this time revealed that the American firm received high marks in a number of areas, including governmental relations, labor attitudes, and financial payments. But given the repressive atmosphere in Venezuela, one must question the accuracy of any poll taken there.

On August 28, 1952 the American and Venezuelan governments signed a new trade agreement, amending that of 1939. Under the new understanding the duty on petroleum imports under 25° API gravity was reduced to 5.25 cents per barrel, while the 10.5 cent per barrel duty on imports above 25° remained in effect. At the same time that the United States granted new concessions on iron, Venezuelan concessions were doubled overall on a wide variety of American agricultural and industrial items. In a message dated August 29, President Harry Truman advised Congress: "The imported oils which would pay the lower tax [5¼¢ per barrel rather than 10½¢] are among those which are in relatively short supply in the United States and generally throughout the world."[13] This included only the asphalt crudes, the residual fuel oil, and some of the topped crude.

On October 11 the United States agreed to jettison the quota system for Venezuelan oil imports into America. Among those most disturbed by the recent developments was Democratic Representative Omar Burleson of Texas, who wrote to Truman that U.S. favoritism of Venezuela had affected seriously the domestic oil industry of America, especially the independent operator. In his reply, dated November 2, Truman pointed out that it was unlikely that Venezuela would try to glut the American market with petroleum; furthermore, the United States needed to have access to the oil resources of the Western Hemisphere, so as to be prepared in the event of a major struggle against Communist aggression. HST added that he was interested in developing the large, high grade iron ore deposits of Venezuela for the benefit of the United States.

It was during this month that Venezuela held its long delayed presidential election. When early returns indicated that the opposition was defeating the government, junta leader Marcos Pérez Jiménez imposed censorship on future vote tallies and proclaimed himself provisional president. In January of 1953 the Venezuelan Congress adopted a new constitution that awarded dictatorial powers to the president; it then granted Pérez Jiménez a five-year term as chief executive in April. Announcing a "New National Ideal" program, Pérez Jiménez used the money derived from petroleum to build a wide variety of public works. Although he was looked upon by the U.S. government with favor, Pérez Jiménez inaugurated half a decade of flagrant corruption and widespread repression that were without precedent in Venezuelan history.

During these years perhaps the leading foe of oil imports in Congress was Republican Representative Richard Simpson of Pennsylvania, who, following the termination of the quota system, introduced a bill sharply reducing American imports of foreign petroleum. Although the bill was sent back to committee, it aroused the ire of the Venezuelans who felt that its passage might cost them several hundred million dollars annually. In defending his measure, Simpson observed, "For every Venezuelan who buys a refrigerator, I can show you five or six Americans who are on unemployment relief and can't afford to buy one."[14]

Seizing the counteroffensive in April of 1954 the Venezuelan Chamber of Commerce of the United States published a 115 page study attacking the Simpson bill. According to the report no less than 450 American cities stretching from Gloucester, Massachusetts (glue), to San Gabriel, California (adding machines), benefitted from American trade with Venezuela. Simpson himself admitted that such Pennsylvania towns lying within his eighteenth congressional district as Mt. Union (firebrick) and Waynesboro (machinery) profited from their dealings with that Latin American nation. In criticizing the Simpson bill, the Venezuelans noted that there had been a transition from coal to oil in the American economy and that in its attempt to remedy this situation, the United States was ready to sacrifice the working class of a friendly and loyal country.

Throughout the Pérez Jiménez regime the official Venezuelan attitude towards

American petroleum operations in Venezuela was supportive. When Creole President Arthur T. Proudfit retired in 1954 after a quarter of a century in Venezuela, the nation awarded him its highest nonpolitical honor to be given to a foreign citizen, the rank of Knight Commander of the Order of the Liberator. Proudfit also was praised by Foreign Minister Aureliano Otañez as the best ambassador that the United States had ever sent there. Under Proudfit, Creole had become the largest overseas investment by a single U.S. company.

The National Planning Association offered a favorable assessment of Proudfit's firm in a 1955 study entitled *U.S. Business Performance Abroad.* Aside from pioneering the 50-50 split, Creole paid top wages, "lived in a fishbowl," employed numerous Venezuelans, and required its non-Venezuelan employees to speak Spanish. But Creole had brought an additional innovation to Venezuela: community integration. Under this concept the American oil firm turned over the operation of towns, schools, hospitals, and commissaries to other organizations, such as religious orders; the net result was a break with paternalism, the Latin American tradition of centralized government, and the company store concept. Unfortunately, many Creole workers did not like this new departure.

In the early part of 1956 President Marcos Pérez Jiménez decided "to open the door to offers" for new petroleum concessions for the first time since the end of World War II. But in declaring that Venezuela would grant concessions to those firms that offered the most in royalties and "advantages," Pérez Jiménez indicated that his nation was moving away from the existing 50-50 formula. As a result of taxes on the petroleum companies, the split in fact already had widened to 56-44. In addition, Pérez Jiménez imposed three major conditions on future concessionaires: refining of newly discovered crude oil was to take place in Venezuela; the petroleum firms would build "open cities" rather than company towns; and there would be provision for the conservation of natural gas. Disturbed by the high tariff bloc in the U.S. Congress that was constantly threatening Venezuela with import restrictions, Pérez Jiménez planned to give potential European concessionaires preferential treatment. Creole, however, announced that it would bid for new concessions and would undertake a $200 million expansion program there.

Creole's positive attitude apparently paid off, for during that year it purchased 24,700 acres of underwater land in Lake Maracaibo from the Venezuelan government. This concession was to last for twenty years and was to be renewable for another forty years. Creole also obtained 98,800 acres of exploration territory for three years in areas where oil potential was uncertain. The total cost of these concessions was $25 million. Other U.S. oil firms in Venezuela were also active. Mene Grande (the Gulf subsidiary) spent $121 million on 126,000 acres of Lake Maracaibo concessions and 98,800 acres of exploration areas, while Signal Oil, Superior Oil, and Sun Oil obtained concessions with a total value of $79 million. Non-American petroleum firms, too, joined in the effort for concessions; Royal Dutch-Shell, for example, expended $65 million on lake and exploration concessions.

By the end of 1957, despite the mass granting of concessions, the Venezuelan petroleum bonanza had begun to go sour. *Barron's* subheaded its December 30 article, "Gusher in Trouble: U.S. Import Curbs, Dwindling Demand Hit Venezuelan Oil Producers." During this month independent U.S. petroleum producers and others finally forced the Eisenhower administration into requesting eighteen West Coast importers, who had been specifically exempted from the restrictions that had been imposed in July, to voluntarily reduce their imports for the first six months of 1958 by approximately 37 percent or face mandatory controls. The action was a serious blow to Venezuela, whose daily petroleum output had fallen from 2.9 to 2.6 million barrels in a depressed world market, even though it had been shipping 30,000 barrels of crude oil daily to California via the Panama Canal. Under these circumstances, it was only to be expected that the Venezuelan government would make some countermove.

## The Movement Toward Nationalization

The countermove, though, was not to come from Pérez Jiménez, who held a plebiscite in lieu of an election in December 1957. Although the plebiscite gave him another term as chief executive, he already had alienated most, if not all, of the major interest groups in the country. After the military turned against him and a general strike ensued, he was forced to flee in January 1958. A five-man junta then took over, but rather than holding power indefinitely, it agreed to a new presidential election, which was to take place that December. The election saw the return of Rómulo Betancourt, the Acción Democrática candidate, as chief executive.

Before Betancourt even assumed office, however, the provisional government had decreed a change in the petroleum taxes without consulting the oil firms. On January 1, 1959, a new tax structure went into effect under which Venezuela would obtain approximately 60 percent of the profits. According to journalist Tad Szulc, the ruling junta planned to use the tax increase to pay the debts Pérez Jiménez had left behind. It was estimated that this would cost the U.S. petroleum companies $170 million initially, $90 million of which would come from Creole alone.

Among the most vocal critics of the higher tax rate was H. W. Haight, the new president of Jersey Standard's subsidiary in Venezuela. According to Haight, "Venezuela becomes the first country in the world to break the 50-50 principle [for existing concessions], completely disregarding acquired rights and ignoring moral if not legal obligations to negotiate with interested parties. . . ."[15] The oil companies stood to lose no matter what position they took. If they claimed that Venezuela had a contractual obligation not to alter the 50-50 formula, they might become liable for several billion dollars in U.S. taxes on past profits; if they gave their consent to the Venezuelan tax increase, they might encourage various oil-rich countries in the Middle East to increase their taxes. After thinking it over, the petroleum firms apparently concluded that consenting to the tax increase was marginally safer.

When Jersey Standard later sent Arthur Proudfit back to Venezuela to assume the presidency of Creole, replacing Haight, Creole proceeded to cooperate with the Betancourt administration, an unquestionably wise decision on its part. In assessing Haight's remarks the Venezuelan ambassador to the United States, Marco Falcón Briceño, had observed that his tirade was "offensive to the national dignity, prejudicial to [Venezuela's] international relations and. . . deliberately misleading."[16] Falcón Briceño pointed out that, as iron ore also was affected, the petroleum firms were not the specific target of the new income tax rates. Venezuela, he went on to state, no more needed the permission of Standard Oil to raise its taxes than the American government did. His hard-line stance found favor in Caracas, where the idea that a private business would dare to challenge a sovereign government was widely resented.

Most interesting of all was the reaction of the incoming president, Rómulo Betancourt. In 1956 Betancourt had published a book in Mexico entitled *Venezuela: Política y Petróleo*. Although he congratulated the provisional government publicly for its "vigilant patriotism," in private he expressed shock at its action; that many members of that group were politically hostile toward him may have been a factor in his reaction. While Betancourt had imposed a multimillion dollar extraordinary tax as provisional president in 1945, he later acknowledged that the tax was a mistake. Yet the mutual distrust between Betancourt and the U.S. petroleum firms continued throughout his second term as president (1959-64). In 1960 several American oil companies repatriated large amounts of capital from Venezuela, while *Time* magazine had reported as early as January 5, 1959, that foreign investors were withdrawing their capital from the country at the rate of $30 million per month.

In the spring of 1959 Venezuela asked the United States to alter its import quota program, leaving to Venezuela the responsibility for limiting its oil exports to America. The Venezuelan government was particularly upset at the company-by-company quotas, since it preferred an overall country quota. According to Juan Pablo Pérez Alfonzo, the Minister of Mines and Hydrocarbons, the present American import quota arrangement was "complicated and cumbersome, and is upsetting the industry's whole operations, including those in Venezuela."[17] Pérez Alfonzo conceded the right of the United States to limit oil imports, but he objected to Canada's exemption from American import curbs, unless Venezuela received the same treatment. In December the American government agreed to allow 1,530,000 barrels of petroleum products to enter the United States daily for the first six months of 1960, an 80,000 barrel increase over the last six months of 1959. As most of the extra 80,000 barrels daily was residual fuel oil, Venezuela was the major beneficiary of the agreement.

Such concessions as these by the U.S. government, though, did not cause the Venezuelan government to relinquish its growing stranglehold over the nation's petroleum industry. During 1960 Venezuela established the Corporación Venezolana del Petróleo (CVP), which was to operate on a limited basis only for the next decade; it nevertheless served as an irritant and threat to the foreign oil firms. The CVP was created to meet "a legitimate aspiration of the Venezuelan people," but as *The Magazine of Wall Street* pointed out, it did not diversify the country's economy, nor did it confront Venezuela's main problem at that time, which was marketing rather than production.

The Betancourt regime dealt another severe blow to the petroleum companies during that same year, when it announced it would not grant future concessions to them. Creole became so concerned about its position in Venezuela that it successfully opposed an attempt by Exxon (Jersey Standard) directors in New York City to cut the posted price there. In the most spectacular Venezuelan oil maneuver of all, Juan Pablo Pérez Alfonzo joined forced with Abdullah Tariki of Saudi Arabia in September of 1960 to form the Organization of Petroleum Exporting Countries in Baghdad. Other charter members were Iran, Kuwait, and Iraq.

Having participated in the birth of OPEC, Venezuela now turned to organizing the state-owned oil firms of Latin America. In this connection the Venezuelan Minister of Mines and Hydrocarbons convened a gathering in Maracay, Venezuela in June of 1961. Upon assembling, the delegates established the Conferencia de Empresas Estatales Petroleras de America (CEEPA), using the term "America" in the organization's title to "provide for subsequent entry of any future state-owned petroleum company of the United States or Canada."[18] In the opinion of CEEPA, state-owned oil firms should be financially self-supporting, autonomous, and self-sufficient.

It was perhaps not coincidental that, shortly before this meeting, the Kennedy administration formally renounced the policy of withholding foreign aid from state-owned petroleum companies. In the opinion of the New York *Times*, this action was "a result of recognition that the Latin American oil monopolies are a fact of life, and that Soviet bloc suppliers are eager to conquer the Latin American market through barter deals and financing plans."[19]

Early in 1961 Harry A. Jarvis assumed the presidency of the Creole Petroleum Corporation. Fortunately, his relations with the Betancourt government proved to be generally harmonious. Jarvis took two steps to improve relations between Creole and Venezuela: the Creole Investment Corporation invested $10 million in new Venezuelan business ventures, and Creole itself purchased $16 million in government bonds. Creole, whose policy was to acquire up to 49 percent of the stock in new and expanding businesses, even invested in a firm that grew mushrooms which was headed by a psychiatrist; neither this investment nor any of its other investments—including a 50,000 acre cattle ranch—were giving signs of failure by the end of 1962. At this time Creole enjoyed net assets of $1.2 billion and gross revenues for 1961 of over $1 billion. It was the world's only billion dollar petroleum firm with a pretax profit of over 50 percent of sales.

By 1964 *Time* was reporting that Creole, faced with Communist sabotage of its facilities, had become so skilled at repairing dynamited pipeline that the Marxists were taking a defeatist attitude toward their prospects for success. Under its new "Dividend for the Community" program Creole and other large foreign companies in Venezuela were devoting several percent of their profits to new hospitals and social projects. The number of foreigners working for Creole, moreover, declined under its "Venezuelanizing" policy. Thus it is not surprising that the Asociación Venezolana de Ejecutivos (The Venezuelan Executives' Association) was of the opinion that the Americans working for Creole did not fall into the "ugly American" category.

Despite these efforts on the part of Creole, U.S. oil companies continued to repatriate large amounts of capital from Venezuela during the presidency of Rául Leoni (1964-69), who was the first freely elected president ever to succeed another freely elected president in Venezuelan history. In 1966 Venezuela enacted a Hydrocarbons Law, stating that more than 70 percent of the concessions held by foreign oil companies would be subject to relinquishment no later than 1983. This measure established the basic terms for service contracts, and forced the foreign petroleum firms to coordinate their efforts with those of the Venezuelan state oil company, the CVP. A year later, in 1967, Venezuela unveiled a "tax reference" price scheme in lieu of the previous "posted price" concept.

In 1969 the administrations in both Caracas and Washington changed. Rafael Caldera became President of Venezuela, and Richard Nixon assumed the presidency of the United States. Caldera had been the candidate of the principal opposition party, the Comité de Organización Política Electoral Independiente (COPEI) and had taken office in still another constitutional transfer of power. Unfortunately, relations between Venezuela and the United States worsened as soon as Nixon selected a protectionist Texas oil man as the American ambassador. He, in fact, was forced to withdraw the nomination after opposition began to mount both in the U.S. Senate and in Venezuela. Caldera himself later visited Washington in the summer of 1970, but his discussions over oil with American government leaders did not prove fruitful.

Then in the summer of 1971 the Venezuelan Congress enacted a Hydrocarbons Revision Act. Under the act, as of 1983, all improvements previously made by a concessionaire would revert to the state when the concession lapsed. Foreign oil firms protested the act in the Venezuelan courts, as they would be forced to surrender not only the concession land and all the equipment on it but also such items as bowling alleys, art on the walls, and the company president's car. Still another provision of the act stipulated that the petroleum companies had to contribute hundreds of millions of dollars into a special government fund; their contributions would be returned to them upon the lapsing of their concessions if their equipment was not worn out or run down at the time of the reversion. Shortly thereafter the Venezuelan Congress passed another law nationalizing natural gas.

That Christmas, moreover, Venezuela imposed a tax increase of 32 cents a barrel on the oil companies. Then a week later it announced that it would fine the petroleum firms, if their quarterly production varied more than two percent from the 1970 output levels. Because of its heavy sulphur content, there had been a decline in the demand for Venezuelan crude oil, and as a result, the oil companies faced fines in the hundreds of million dollars range, if this trend continued. Leo E. Lowry, then the chief executive of Creole, observed about this series of developments: "The most important longer-range challenge confronting us is to provide for the profitable continuing of our company beyond 1984, when most of Creole's concessions expire."[20]

By this time, though, the American government had come to realize that it could no longer rely on an endless flow of petroleum from the Middle East. The Nixon administration therefore sent its "energy diplomat," James Akins, to visit Venezuela. Talks between the American and Venezuelan governments reached a climax in the winter of 1972-73, following the lapsing of a twenty-year-old commercial treaty between the two countries. Rather than sign a new treaty, Venezuela decided to continue its "most favored nation" treatment of U.S. imports, and at the same time the American government reaffirmed its preferential tariff stance toward Venezuelan oil.

The trend toward the nationalization of petroleum in Venezuela reached its inexorable climax during the presidency of Carlos Andrés Pérez, who took office early in 1974. Calling for a reversion of the foreign oil concessions to the state, Pérez set up a commission that reported to him in December of that year. On August 29, 1975, the Venezuelan Congress passed a measure providing for the state operation of the $5 billion petroleum industry, effective as of January 1, 1976. Summing up this trend of events eight years earlier, Rómulo Betancourt had observed: "Venezuela's oil policy after 1959 was neither capricious nor inconsistent, but was realistically planned and applied in a flexible but irreversible manner, and followed patriotic and nationalistic principles."[21]

## Colombia: From de Mares and Barco to the Anglo-Persian Contract

Like Venezuela, during the twentieth century Colombia has struggled to establish a democracy. It has elected to public office such enlightened statesmen as Alberto Lleras Camargo (1958-64), who held office at approximately the same time as Betancourt. But here, too, dictators have seized power, including the contemporary of Pérez Jiménez: Gustavo Rojas Pinilla (1953-57). Since 1958 Liberals and Conservatives have alternated the presidency between them in a National Front government.

Although oil long has been a major Colombian export, it has not played as dominant role in the country as it has in Venezuela. The modern Colombian petroleum industry began in 1905, when General Virgilio Barco obtained a concession in the Catatumbo region of Norte de Santander. In that same year Don Roberto de Mares signed a contract with the Colombian government to exploit oil and asphalt near Barrancabermeja. As World War I approached, foreign investors became more active. In 1913 the British Cowdray, or Pearson, interests obtained a concession to exploit 10,000 square kilometers of petroleum land anywhere in Colombia, including that part of the nation bordering on the Panama Canal. The contract in fact gave the British interests the right to build another canal, as well as communications systems. In the United States, President Woodrow Wilson worried about a possible challenge to the Monroe Doctrine, and William Jennings Bryan again unleashed a verbal onslaught against foreign concessionaires.

As a result of these actions, in January 1914 Ambassador Walter Hines Page wrote from London that the British government had taken up "the dangers that lurked in the Government's contract with Cowdray for oil; and they pulled Cowdray out of Colombia and Costa Rica—granting the application of the Monroe Doctrine to concessions that might imperil a country's autonomy."[22] Lord Cowdray, according to the *Independent*, believed that one reason for the withdrawal of the contract was its use elsewhere to stir up opposition to the Pearson interests. The British, however, did not emerge from this episode as total losers. Although there was no direct trade-off, the American government satisfied a British request to repeal the law that had granted U.S. ships exemption from Panama Canal tolls. With the start of World War I only six months away and with the worsening conditions in Europe, officials in London obviously were interested in maintaining friendly relations with Washington. The last, but not least reason for the withdrawal of the contract was the hostility of the Colombian Congress to the Cowdray concession.

By 1916 Standard Oil had spent $1.75 million in Colombia, only to discontinue its investments out of frustration over the country's petroleum regulations. Articles 939 and 1126 of the Fiscal Code of 1873 had declared that mines on public lands were the property of Colombia. Independent U.S. oil men, though, were not deterred by Standard's pullout. In that year an American group that included wildcatters M. L. Benedum and J. C. Trees set up the Tropical Oil Company, which obtained an option on the French-owned de Mares concession. Whereas early in the Wilson administration the State Department had opposed "monopolistic concessions" by American firms in Colombia, with the United States on the verge of entering World War I, it was now viewing "with approbation the participation of capitalists of the United States in the development of natural resources of South America."[23]

In June 1919, though, the President of Colombia issued a decree declaring that the subsoil hydrocarbons of the country belonged to the government. The decree went on to require a governmental permit for oil exploration and prevented the surface owners from executing petroleum leases. This action predictably offended the American petroleum companies. Back in the United States, Republican Senator Henry Cabot Lodge of Massachusetts used it to force the recommittal to committee of a pending treaty with Colombia that had been designed to pacify Colombia for the U.S. role in the Panamanian Revolt of 1903. But the Colombian Supreme Court then invalidated the oil decree, and the Colombian government informed the United States that it was prepared to give a "full guarantee" in protecting American interests there.

The U.S. Senate was not satisfied with this pledge. Lodge's Foreign Relations Committee insisted on including an article in the Colombian treaty that pledged each party never to nullify the rights of the citizens of the other country. This action, however, was not acceptable to Colombia. Fortunately for the treaty negotiators, the Colombian Congress enacted a new oil law in December that eliminated government ownership of subsoil hydrocarbons. Six months later, in June of 1920, the U.S. Senate Foreign Relations Committee, having been reassured by that nation's latest petroleum measure, favorably reported the Colombian treaty.

Less pacified was the American Consul at Barranquilla, who felt that the law still contained many of the worst features of the earlier decree. Pure Oil and Sinclair, moreover, were so discouraged that they withdrew from Colombia entirely. On the other hand, in 1920 Jersey Standard bought into the Tropical Oil Company. Since it now controlled the de Mares concession, which had been granted fifteen years earlier, it could legally bypass the recent Colombian legislation, even though a 1921 ruling by the Colombian Supreme Court had held that no new private mineral or petroleum rights had been acquired since the Fiscal Code of 1873. By 1927 Tropical Oil was outproducing any other foreign holding of Jersey Standard, despite the problems caused by labor friction and government inspections.

In 1921, with Theodore Roosevelt now dead for two years, the United States Senate—apparently lured by the petroleum deposits of that nation—finally approved the treaty with Colombia. This granted to the latter country $25 million in "conscience money" for damages suffered by Colombia during the Panamanian revolt of 1903, when Roosevelt encouraged the creation of a separate nation. Especially interesting in this connection was the reversal in position of TR's friend and one-time treaty foe Henry Cabot Lodge, who had released to the press in 1917 this ringing indictment: "Even if I favored the treaty I would not support it now because I am not willing to have my country blackmailed."[24] Four years later Lodge gave his endorsement, pointing out that petroleum was a necessity for a great maritime nation, and dwelling on the dangers inherent in a British oil monopoly.

Not every member of the Senate, though, was as enthusiastic as Senator Lodge. Democratic Senator James Reed of Missouri attacked the idea that the treaty would prevent British oil penetration in Colombia as "unsound," affirming that "it never should have been uttered upon the floor of the Senate."[25] Then there was Republican Senator George Norris of Nebraska, who wondered what the cost would be of ousting British petroleum interests from Mesopotamia, South America, and Mexico. Still another unsuccessful protestor was Republican Senator William E. Borah of Idaho, who complained: "It would not be so bad. . . if this treaty actually settled the oil question."[26]

Among the supporters of the Colombian treaty was Jersey Standard, which needed permission from the Colombian government before it could construct a pipeline from the de Mares petroleum fields to the coast. During July and August of 1920, the operating manager of Standard's pipeline affiliate in Colombia had engineered a meeting between the Colombian Minister to the United States, Carlos Urueta, and Albert Fall, then a member of the Senate Foreign Relations Committee. James Flanagan, the operating manager of the company, also talked with such key Senators as Lodge, Gilbert Hitchcock, and Oscar W. Underwood and even had obtained support for the treaty from President-elect Warren Harding.

By September 1922 the Andian National Corporation (a Jersey Standard subsidiary) had obtained a fifty-year concession to construct and operate the needed pipeline. While its production for the years 1922-25 was only 500,000 barrels annually, the construction of the 360 mile pipeline to the coast enabled Jersey Standard's output in Colombia to reach the 20 million barrel mark by 1928. Aside from the pipeline, the company built factories, refineries, boats, harbors, railways, roads, and cities; it also undertook mosquito and sanitary campaigns and furnished free medical and hospital service to all of its employees.

Because of accomplishments such as these, the Colombian government began to adopt a more sympathetic attitude toward foreign oil firms. In 1923 and again in 1925 the Colombian Congress modified its petroleum legislation. By 1925 approximately three dozen American companies and a dozen British ones were exploring for oil in Colombia.

Assessing the reaction of the leading U.S. petroleum subsidiary in Colombia to the 1925 oil measure, Assistant Trade Commissioner C. Reed Hill wrote to the Director of the Bureau of Foreign and Domestic Commerce on November 16, 1925:

> The Tropical Oil Co., the only producer on a considerable commercial scale, does not consider this law as particularly dangerous to them as their concessions held at present time are under an arrangement previous to the extant national laws on hydrocarbons. But they judge it as even more severe than previous rigorous Colombian laws on this subject.[27]

Let us now shift our attention from the de Mares to the Barco concession, which Henry L. Doherty and Company had obtained in 1919 and then explored. In January 1926 Gulf Oil purchased the Barco concession from the latter firm. This proved to be a most unwise move, since in February the Colombian government declared that the Barco concession had been cancelled, even before Gulf Oil had made its final payment to Doherty. The result was a bitter dispute. As H. Freeman Matthews, the Assistant Chief of the Latin American Division of the State Department wrote at the end of the Hoover administration, "from 1926 on the Barco concession played a role in Colombian politics which it would be impossible to exaggerate; it became a 'Muscle Shoals' of Colombia."[28]

In 1927 the Colombian Congress approved a new petroleum law that was less favorable to foreign oil interests than earlier measures had been. Under the new law proof of ownership and lease contracts were to be presented within six months, consideration of concession applications were postponed indefinitely, and the tax due the Colombian government on petroleum production from private lands was doubled. Even more restrictive was Executive Regulation Number 150, which was issued two months later in January 1928. This not only reduced the six-month period to thirty days but also demanded permits for drilling and forbade any appeal to the courts over the rulings of the Minister of Industries.

Persistent American pressure led the Colombian government to suspend the executive regulation temporarily on June 1, pending a Supreme Court and Council of State ruling on its constitutionality, but the November 1927 petroleum legislation remained in effect.

During this period U.S. oil firms also confronted the presence in Colombia of two agents of the Anglo-Persian Oil Company, Colonel Henry Yates and Lieutenant Colonel Sir Arnold T. Wilson. Yates and Wilson were interested in acquiring a concession in the northwestern part of Colombia next to Panama, and apparently were gaining the favor of the Colombian government, which at this time wished to achieve greater diversification of direct foreign investment in the petroleum industry. As George Eder, who was Chief of the Latin American Section in the Division of Regional Information, pointed out to the Bureau of Foreign and Domestic Commerce, "the presence of English capital in Colombia would offset alleged Yankee imperialism."[29] The American government was disturbed by the British presence not only because of the threat Anglo-Persian posed to U.S. petroleum firms in Colombia but also because of the nearness of the proposed concession to the Canal Zone. Secretary of State Frank Kellogg in particular feared American exclusion. Ludwell Denny reflected the alarmist point of view in the American press, when he wrote: "Colombia probably will be the scene of the next international oil explosion. Grave consequences are threatened by the efforts of the Anglo-Persian Oil Company. . . ."[30]

Since the U.S. government was applying diplomatic pressure on both Mexico and Nicaragua at this time, it was hardly the most appropriate moment to wave the "Big Stick" at Colombia. Fortunately, there were certain Colombians who had mutual interests with the United States. These individuals protested to their government about the Anglo-Persian menace, complaining that the Constitution and laws of Colombia forbade the acquisition of petroleum rights by a foreign government. The secrecy surrounding the contract further militated against its final approval. Yates, thrown on the defensive, countered with the claim that he was acting as a private citizen, not as an agent of the British government, but when the Colombian Congress rejected the Anglo-Persian application he left Colombia without acquiring the concession. Earlier his partner, Wilson, had confidently proclaimed:

> The august shadow of the illustrious Mr. Monroe has not yet fallen over the province of Uraba and there is no reason to suppose that the North American companies or the State Department are opposed to the legitimate development of the natural resources of the Colombian Nation by any company whatever, national or foreign.[31]

## From the Voided Barco Title to the Age of Ecopetrol

The bitter dispute over the voided Barco title continued for many years. When the American Minister to Colombia protested the 1926 cancellation early in 1928, the Colombian Foreign Minister observed, "The Secretary of State of the

United States has committed an error in initiating this intervention in respect to an affair which, since it deals with the judicial relations between the Government and a national entity, pertains exclusively to the tribunals of the country."[32] In August the Colombian government reaffirmed its cancellation, charging that the Americans had failed to exploit its resources between 1923 and 1926, since they had marketed no coal, asphalt, or petroleum from there.

These protests on the part of the U.S. government and the American oil firms, as well as the rather belligerent attitude of Secretary of State Kellogg, aroused resentment not only in Colombia but also in Argentina and other parts of Latin America. While at the same time criticizing Colombia's attitude, Kellogg eventually found it necessary to state publicly that he had not asked the Colombian government to reconsider the cancellation of the Barco concession. The Secretary of State also announced that he would continue to support the activities of American citizens abroad.

Although the Barco concession continued to be a bone of contention between Colombia and the United States, other developments in the petroleum area tended in the direction of harmony. On June 3, 1928, the Colombian government voided the Executive Regulation Number 150, a move that immensely pleased foreign petroleum firms. Then in December of that year the government, after two years of negotiations, finally reached an agreement with the Tropical Oil Company over some disputed royalty payments.

Two years later the people of Colombia elected a new President, Enrique Olaya, a Liberal whose victory ended twenty-six years of Conservative rule. Because of a collapse in the price of coffee, the Colombians were suffering the ill effects of the worldwide depression at the time that he took office. Olaya reversed the position of the Colombian government on the Barco concession, reconceding a part of it to Gulf Oil for fifty years as an exclusive grant. Although the Colombian Congress and public opinion were rather hostile toward Gulf, Olaya maneuvered the new contract through Congress in 1931.

At this time he received a $4 million loan from the National City Bank to counteract the financial crisis the country was experiencing. The fact that Andrew W. Mellon was still Secretary of the Treasury did not escape the attention of oil company critics. Democratic Representative Wright Patman of Texas in particular attempted to link the concession with the loan; Patman pointed out that just prior to the settlement, Mellon and the Colombian President had met at a dinner in Washington. Republican Senator Hiram Johnson of California was also critical of this loan.

Jefferson Caffery, the U.S. Minister to Colombia, believed that the Colombian government would not have showed much interest in revalidating the Barco concession, if it had not thought that the State Department favored these negotiations. President Olaya's primary concern appears to have been the suspicions of his Congress that Gulf Oil would not actually exploit its concession, but instead create additional reserves for itself. Gulf, moreover, was at odds with the government over the royalty rate and remained strongly opposed to accepting a time limit on exploitation.

While Olaya was coming to terms with Gulf, the Colombian Congress was in the process of passing a new petroleum measure. The new law was framed with the assistance of George Rublee, the former oil advisor to the U.S. Ambassador to Mexico, Dwight Morrow. The March 4, 1931, measure declared the petroleum industry of Colombia a public utility, gave preference to qualified Colombian nationals, and referred all disputes over oil to the Colombian courts. Under the law, the subsoil was the property of the state; there were to be no petroleum export duties for thirty years after exploitation began, but no individual was to acquire a concession larger than 50,000 hectares. Undoubtedly influenced by the Anglo-Persian episode of a few years earlier, the new law denied grants of petroleum lands to foreign governments. Royalty payments were to vary on private and public lands, and concessionaires were required to operate all wells at a minimum 25 percent capacity, despite worldwide overproduction. The oil companies found this last provision, and the requirement that they had to prove the title to their concessions, quite distasteful. But while the Texas Company announced that it would stop developing its 800,000 acre holding and Sinclair proclaimed that it would not continue operating under the new law, Jersey Standard stated that the petroleum measure was satisfactory to it, thus mirroring the opinion of Rublee who found the statute "workable."

Unfortunately, by the mid-1930s labor problems began to poison Colombian-American petroleum relations, but unlike Mexico these did not lead to expropriation. In 1934 Alfonso López assumed the presidency for a four-year term. Unrest and conflict on Tropical Oil property during 1935 led to widespread demands in the Colombian Congress for action; on November 16 the Chamber of Representatives set up a committee to investigate labor radicalism in the petroleum industry. After touring the facilities of both Tropical Oil and the Andian National Corporation, the congressional committee issued a report early in 1936, criticizing the U.S. petroleum firms and pointing especially to social and hygienic inadequacies at Barrancabermeja.

When the Colombian Congress passed still another piece of petroleum legislation in 1936—Law 160 of November 14—it did not consult the American oil companies or the U.S. legation, as it had in 1931. The new measure, which was technically more sophisticated than its predecessor, was nevertheless more favorable to foreign investors. There was a substantial reduction of taxes on those properties that were privately owned and more liberal royalty payments. Furthermore, there was no provision calling for a Colombian-owned refinery.

During this period, President Alfonso López also engineered drastic changes in the Constitution of 1886. One amendment, disturbing to both Colombian and foreign business interests, stated that the nation could expropriate property "for motives of public utility and social interest." A more active reformer than Olaya,

López attempted to implement what Hubert Herring has described as a Colombian version of the New Deal.

The complex Barco story, which seemed to have ended in 1931, began again in 1936, when the Texas Company and Socony-Vacuum acquired control of the concession from Gulf, after agreeing to split expenses and profits on a 50-50 basis. To acquire control of the concession, the Texas Company and Socony-Vacuum purchased all of the stock in a Gulf subsidiary that held 79 percent of the stock in Colombian Petroleum. The other 21 percent was held by the Carib Syndicate, which sold out at the end of the year, after Texaco and Socony-Vacuum had announced plans to construct an expensive new pipeline from the concession over the mountains to the sea. At that moment not a single barrel of Barco oil had yet gone to market. The construction of the 263 mile long pipeline, which began in February 1938, was one of the most difficult engineering feats in the annals of the oil industry, but by the fall of 1939, 25,000 barrels of Barco petroleum flowed to the coast each day.

Summarizing the attitude of the various Colombian presidents toward the U.S. oil companies between the Great Depression and World War II, Stephen J. Randall has concluded that:

> Official pro-Americanism was most in evidence during the Liberal presidencies of Enrique Olaya (1930-1934) and Eduardo Santos (1938-1942). Yet the desire to attract foreign investment was apparent throughout the Conservative-dominated nineteen-twenties as well as the Leftist-Liberal administration of Alfonso López (1934-1938).[33]

By the end of World War II the Tropical Oil Company and the Andian National Corporation had spent over $18 million for social and educational purposes in Colombia. Yet as the decade neared its end, U.S. petroleum operations were on the decline. Within a six month period seven U.S. oil companies, as well as Royal Dutch-Shell, suspended drilling operations, leaving only three big American firms, Jersey Standard, the Texas Company, and New York Standard. This trio had invested too much money in profitable wells to withdraw precipitously. A number of factors were responsible for the flight of the foreign petroleum firms, of which the most important was the negative impact of Colombia's frequently revised oil legislation. Not only did it take ten years for a petroleum company to obtain an exploration concession, but the country even taxed a new well when no oil was found. Since it also imposed royalties, the Colombian government might take over half the value of the net revenue of a foreign petroleum firm. These conditions led one entrepreneur in Bogotá to observe, "Colombia is the graveyard of oil profits from other countries."[34] Equally important, and discouraging to foreign firms, was the establishment of Empresa Colombiana de Petróleos (Ecopetrol) in 1948 under Law 165. Although at first limited in its functions, this state enterprise eventually so widened its activities that it was involved in every phase of the petroleum industry, thus challenging the private sector.

A new wave of labor discontent occurred during the 1940s. On August 31, 1948, the Ambassador to Colombia, Willard L. Beulac, complained to Secretary of State George Marshall that members, or former members, of the Communist party had infiltrated the unions, leading to frequent costly and illegal strikes. President-elect Laureano Gómez admitted to Ambassador Beulac in January of 1950 that the labor situation confronting Tropical Oil was a national disgrace, since that firm had to maintain hundreds of idle workers on its payroll and even increase their wages in compliance with a presidential decree.

Like Mariano Ospina, the chief executive of Colombia from 1946 to 1950, Gómez was a Conservative. In fact, he was more so than his predecessor; not only had Gómez once admired Adolf Hitler, he also was a friend of Francisco Franco. Given these circumstances, it is most thought provoking that the creation of Ecopetrol and the reversion of the de Mares concession to the state three years later took place during Conservative presidencies rather than Liberal ones. Economic considerations rather than political ones obviously were the motivating factor; in the words of one Colombian petroleum ministry bureaucrat, "What will Venezuela have to show for lying supine before the drillers? Holes, that's all."[35] Although the neighboring state was oil rich, its petroleum had not greatly helped it in developing its agriculture and its industry.

When Colombian Petroleum returned its de Mares concession to Colombia in 1951, a band played the Colombian national anthem, "The Horrible Night Has Ended." The Colombian Supreme Court had ruled in 1944 that the concession would expire in seven years. Rather than fight its decision, the oil company cooperated with the government and worked out an agreement that entailed surrendering the concession without compensation, but retaining a slice of the profits for at least a decade. As a part of the agreement, Colombian Petroleum was to supply refining technicians for five years and handle distribution and sales for ten.

With the expiration of the agreement Ecopetrol took over the operation of the de Mares concession and the Barrancabermeja refinery. In this year (1951) a new petroleum law went into effect that did less to stimulate exploration and development than earlier legislation. As a result, Mobil and Royal Dutch-Shell began to restrict themselves to depleting their existing fields. As the decade of the 1950s progressed, moreover, the government of Colombia slowed down its processing of new applications for concessions. A dozen years later, in 1963, labor unrest again erupted in the form of a crippling forty-day strike, during which saboteurs rendered as many as 500 oil wells nonoperative. Backed by the Communists, this attempt at disruption eventually failed, and Colombian petroleum exports continued.

By the mid-1960s the government of Colombia, in another of its seemingly endless shifts in oil policy, began to cultivate the foreign petroleum firms. While the production of Ecopetrol declined between 1962 and 1969, that of the private companies increased; Texaco and Gulf opened up the Orito field in southern Colombia. On the eve of the Arab oil embargo the two largest refineries in

Colombia were operated by Intercol (a Jersey Standard affiliate) and Ecopetrol, with most of the petroleum marketing still remaining in private hands. Despite the creation of Ecopetrol and the reversion of the de Mares concession, as of 1973 private oil companies continued to play an important role in the Colombian economy.

## Bolivia: The Pre-World War II Years and the Standard Expropriation

"Politically," Hubert Herring wrote, "Bolivia is an abyss. Parties are mythical."[36] With the lowest per capita income in Latin America and a longstanding and widespread hostility toward large landowners, tin barons, and foreign capitalists, Bolivia has evolved in a more totalitarian direction than Venezuela and Colombia. In 1952 Victor Paz Estenssoro and the Movimiento Nacional Revolucionario (MNR) overthrew a military junta and instituted various reforms. Since then Bolivian relations with the United States have been far more harmonious, even though friction between the Bolivian government and Jersey Standard during and after the Chaco war had culminated in 1937 with the expropriation of Standard Oil. In fact, more than $400 million in loans and grants were given by the United States to Bolivia up to 1967. No other Latin American nation has obtained as much. Even so, the nationalization of key industries has continued: the tin mines in 1952, Gulf Oil in 1969.

Although approximately 15 to 20 percent of the world's tin comes from Bolivia and has provided that nation's government with the major share of its revenue, since World War II petroleum has been exported on a greater scale. The state-owned Yacimientos Petrolíferos Fiscales Bolivianos (YPFB) has played a key role in this development. Unfortunately, the unstable political climate, as well as a shortage of capital and technological knowhow, has prevented Bolivia from upgrading its economic status.

American interest in Bolivian oil dates back to 1920, when the Bolivian government negotiated a rather stringent understanding with the Richmond Levering Company of New York. Under this agreement, the U.S. firm obtained the right to explore 4 million hectares of oil lands in the departments of Chuquisaca, Tarija, and Santa Cruz. Although Richmond Levering obtained the privilege of exploiting 1 million acres without taxes for fifty years, it had to pay Bolivia a fifteen percent royalty on all crude and finished oil products and begin drilling within a one-to-three-year period. Bolivia had the option of participating in the company up to fifteen percent of the total capital investment.

The contract also contained a "Calvo clause": foreign companies should not have the right to appeal decisions by the courts of the contracting government. Article 20 of the contract stated that the Bolivian Supreme Court should handle all petroleum disputes, but the most controversial section of the contract was Article 18 with its "failure of performance" provision:

> If during the execution of the contract, the Government should claim as to anything which in its opinion involves failure of performance, it shall give notice thereof to the

capitalists (i.e., Company) who from this moment shall have a maximum term of six months in order to correct the failure which forms the motive of the claim. In case the capitalists should not do so, upon the expiration of said term, the Government may declare the forfeiture, rescission or modification of the contract. . . .[37]

Then during the following year, the Bolivian government considered a proposal that would have raised the tax on oil lands. Protecting the interests of the U.S. petroleum firms, S. Abbot Maginnis, the American Minister to Bolivia, complained that such a measure would be prejudicial to the development of the Bolivian oil fields. While he received assurances that the legislation would be "softened," the new law limited concessions to 100,000 hectares and continued to assert state ownership of subsoil oil holdings. Although there was a reduction in the state petroleum royalty from 15 to 11 percent, there was an increasingly steep tax on all concession lands.

In March of 1922 Jersey Standard acquired the Richmond Levering Company concession and later obtained the Spruille Braden holdings. Employing the strategy it had used in Colombia, it was able to bypass the 1921 Bolivian petroleum law by obtaining concessions that had been granted prior to its passage. These maneuverings incurred the wrath of Bolivian Senator Abel Iturralde, who compared Standard Oil to an octopus that not only dominated Tammany Hall in New York City but also had arranged the assassination of President Carranza of Mexico in 1920. Such attacks hardly restrained the American petroleum giant. In November 1921, it formed the Standard Oil Company of Bolivia in the United States with a $5 million capitalization and then proceeded to incorporate it in La Paz.

Seeking to expand its operations even further, on July 27, 1922, Jersey Standard obtained a new concession in southeastern Bolivia, which was a continuation of the Richmond Levering concession with certain modifications. While Article 28 of the new contract omitted the arbitrary right of the government to declare forfeiture in the event of fraud, under Article 15 Jersey Standard was held liable for any internal intervention:

The capitalists pledge and obligate themselves not to intervene in the internal affairs of the country, in such manner as not to put their elements, capital, rents, employees, dependents and workers at the electoral or political service of a party or individual. The breach of this clause will give place to the retiring or suspension of those who violate it.[38]

Between 1922 and 1937 Jersey Standard invested approximately $17 million in Bolivia. Other firms also obtained Bolivian concessions, but only Jersey Standard actually developed its concession. Although it had stopped drilling new wells by the time of expropriation, it had already built refineries, roads, camps, and sanitation facilities. But even with the right to exploit 2.5 million acres of petroleum lands, the fact that Bolivia was landlocked made it difficult for Jersey Standard to compete in the global crude oil market. Transportation costs inevitably raised crude oil prices.

A new diplomatic controversy involving petroleum erupted in 1925, when Bolivia introduced a new clause into its oil contracts with private firms. According to the clause, contracts could be transferred only to European entrepreneurs, ostensibly to prevent an American monopoly over the Bolivian oil fields. Although Jersey Standard was not overly concerned by this move, since it felt the clause would never actually go into effect, the U.S. Embassy at La Paz took a hard line against the clause as a matter of principle: "The fact that the Standard Oil Company is not greatly interested in the discriminatory features of these concessions. . . does not alter the position of the Department in this matter."[39] The American Minister to Bolivia found the clause to be not only discriminatory but also in violation of the second article of the 1862 treaty of friendship, navigation, and commerce between the United States and Bolivia. As a result of these protests, the Bolivian government decided in the beginning of 1926 to remove the controversial clause from its concession contracts.

If Jersey Standard had not been overly concerned about the new clause, it did disagree with the Bolivian government over the date on which the American firm's concessions had reached the production stage. After promulgating a resolution in 1928 that officially set the date as occurring in 1930, in 1931 the Bolivian government moved the date back to 1924. From 1931 on this case was before the courts of Bolivia, but the Supreme Court never handed down a final ruling.

In 1932 with the Great Depression still gripping the country, Bolivia, in a search for access to the Atlantic Ocean, entered into a three years' war with Paraguay over the Gran Chaco. On paper Bolivia possessed a decided advantage, but it did not win the war; the 1938 peace treaty left most of the disputed area in the possession of Paraguay. A complicating factor was the undetermined extent of petroleum reserves in the Gran Chaco. In 1975 E. J. Hobshawn rather extravagantly characterized the war as one in which "Standard Oil and Royal Dutch-Shell fought each other for oil deposits. . . ."[40] His interpretation holds that Standard Oil supported Bolivia and Royal-Dutch Shell, Paraguay in a war over petroleum, but as a matter of fact Paraguay never claimed the oil lands of the Chaco. Petroleum only became an issue after the Bolivian armies had gone on the defensive, and the Paraguayans neared their oil fields.

On the other hand, various other critics of Jersey Standard charged that the firm was actually undercutting Bolivia. According to one report, Standard Oil bought sixty bombers in Italy for the use of the Paraguayan army. When the firm refused to lend Bolivia several million dollars to help in the conflict, it merely added fuel to the fire. Standard Oil's enemies charged that it had been uncooperative and had not provided the La Paz regime with a full measure of aid and assistance. This charge even found supporters in the United States. Senator Huey P. Long of Louisiana observed on the floor of the Senate: "As is usually the case, it is the forces of imperialistic finance which are today responsible for war between Bolivia and Paraguay. . . ." Long also charged that it was "the Standard

Oil Company of the United States and other affiliated interests which have been guilty of promoting this war. . . ."[41]

Jersey Standard's practice of transferring some of its equipment to Argentina, both before the war and during that conflict, further irritated the Bolivian government. It added insult to injury by its refusal to refine aviation gasoline on the grounds that it lacked the technical ability to do so. This apparently was the final straw; Bolivia seized the Standard refineries, and proceeded to produce aviation gasoline for its armies.

But it was not just equipment that Jersey Standard had been shipping to Argentina. In 1925 the U.S. company had transferred a limited quantity of crude oil to Argentina to be used in drilling operations there; not only had the Argentine government given its permission for this operation, but Bolivian customs officials had issued certificates of origin. A decade later, in 1935, a special commission in Argentina ruled that Jersey Standard had not imported Bolivian petroleum illegally, but still, Bolivia used this operation as an excuse to seize the American oil properties in 1937. Although the U.S. Embassy was aware of the petroleum shipments to Argentina, it did not advise Jersey Standard to report these oil exports to officials in La Paz.

In 1936, with the Gran Chaco war now ended, Bolivia created a government bureau to handle the future exploitation of its petroleum reserves. The bureau was modelled on the Argentine Yacimientos Petrolíferos Fiscales (YPF) and in fact was called the Yacimientos Petrolíferos Fiscales Bolivianos (YPFB). Placed directly under the president, the corporation was to enjoy freedom from taxation and import duties, and had the right to enter into partnership with private capital in mixed companies. A five-man directorate, which was to draw up an organic law and a table of organization during the next three months, was to supervise the YPFB.

Throughout 1937 the YPFB expanded its operations in Bolivia. In January the American Chargé Muccio informed the Secretary of State that the corporation was contemplating entry into the production field: it apparently was going to explore for oil on concessions Jersey Standard had found unproductive. On January 16 the YPFB received various concessions throughout Bolivia, covering almost all of the known oil lands other than those held by Jersey Standard. In March, with a few minor exceptions, it obtained full control over the importation of petroleum and petroleum products.

The guiding force behind the YPFB was Colonel David Toro, who had begun the drive toward "socialization" as soon as he had seized power in a bloodless coup during May 1936. Toro promised to develop the resources of Bolivia for the benefit of its citizens. His campaign reached its logical conclusion on March 13, 1937, when the Bolivian government annulled the oil concession of the Jersey Standard subsidiary there, and seized its properties. By taking this step Bolivia became the first Latin American nation to take over a foreign oil company.

To justify its seizure, the Toro regime charged that Jersey Standard had not only avoided the payment of taxes and thus defrauded the Bolivian government, but that it also had exported petroleum illegally to Argentina. Yet it might be technically inaccurate to describe this seizure as an "expropriation." Bryce Wood has observed in this connection:

> The action of the Bolivian government was not an expropriation, for which compensation might have been offered. Instead, relying on its claim of fraud, the government took the position that it could cancel the concession and take title to the properties without making any compensation whatever.[42]

There were other factors involved in the seizure. For one thing, the physical and cultural isolation of the oil camps from the population centers discouraged harmonious relations between the foreign companies and the Bolivian populace. In addition, there was the matter of saving face. In the aftermath of the seizure Foreign Minister Enrique Finot complained to the U.S. Minister to Bolivia, Raymond Henry Norweb, that Jersey Standard had been uncooperative at the time of the Chaco War. Finot pointed out that it was necessary for his nation "to dispel the impression current throughout the world that the weak and impoverished Bolivia had been merely an instrument of the all-powerful, imperialistic world monster, the Standard Oil properties."[43]

The Toro regime hoped to continue in power by appeasing nationalists on both the Right and Left. The greatest pressure came from the war hero Germán Busch, whose discontent with the rate of social change led him to hand in his resignation as Chief of the General Staff—which Toro rejected—in early March, just prior to the takeover. Under these circumstances, it is not surprising that the Bolivian Foreign Minister informed his American counterpart in May that Jersey Standard had been expelled for political reasons, and that it would never be allowed to return. Unfortunately, at this point Toro decided to raise the taxes on tin, causing an even greater furor. On July 14 the powerful Busch ousted him and set up a new government with Enrique Baldivieso as Foreign Minister.

Baldivieso and another Bolivian official quickly arranged for talks with the U.S. Ambassador to Argentina, Alexander W. Weddell. During this conversation Weddell suggested referring the expropriation dispute to a commission of reputable citizens. He went on to suggest that Bolivia make a deal with Jersey Standard in order to create the proper climate for private investments. By September the new U.S. Minister to Bolivia, Robert G. Caldwell, had concluded that the failure of Jersey Standard to develop its oil properties had been the main reason for their seizure:

> . . . the real grievance of the Bolivian Government lay in the fact that the Standard Oil Company of Bolivia was supposed to be treating its oil concessions as a potential reserve and that, however sound the reasons might be from an economic point of view, it had proved unwilling to proceed to the development of these oil fields even to the extent of supplying the ordinary necessities of the country.[44]

On October 22 the Bolivian government demanded that Jersey Standard file suit against that nation in the Bolivian courts within ninety days, thus revoking earlier legislation which had allowed a period of no less than thirty years. The State Department complained about this drastic change in procedure, but was only able to obtain a two months' extension for Jersey Standard. By November 17 Minister Caldwell had concluded that ready adjustment or compromise of the oil controversy was impossible. Ten days later the Bolivian government gave credence to his assessment by deporting to Argentina Jersey Standard's Bolivian attorney, Dr. Carlos Calvo. The State Department did not protest the deportation, as it was obviously an internal matter over which it had no control.

Back in Washington, Under Secretary of State Sumner Welles mirrored Caldwell's pessimism, opining that further talks between Jersey Standard and Bolivia would be a waste of time. Welles felt that the best hope for a solution to the oil controversy would be international arbitration. Speaking as Under Secretary of State, he informed Bolivian Minister Luis Fernando Guachalla early in 1938 that "the only way in which public opinion in this country was going to support the 'Good Neighbor' policy as a permanent part of our foreign policy would be for the policy to be recognized throughout the continent as a completely reciprocal policy and not one of a purely unilateral character."[45] The Bolivian government, however, rejected arbitration even after Guachalla had endorsed it. Moreover, Secretary of State Cordell Hull at first apparently preferred to wait for the case to pass through the Bolivian courts. In the interim he proposed exerting pressure on Bolivia to compensate Jersey Standard for its loss.

On March 21, 1938, twenty-four hours before the Bolivian deadline, Jersey Standard filed its legal briefs under protest. During the next year the Bolivian Supreme Court deliberated the case under the most difficult conditions imaginable. In mid-February of 1939, for example, the Director General of Police verbally threatened the judges in a radio address:

> It is just at this moment that we should make the members of the Supreme Court understand our decision to tear out their entrails and burn their blood, in case they should decide against the sacred interests of the country, and in favor of Standard Oil. How long are they going to delay their decision? Are they afraid, perhaps that we, the Bolivian people, veterans of the war, will destroy them with our own hands?[46]

After Justice Placido Molina had resigned on March 5, citing ill health as an excuse, the remaining members of the Court ruled unanimously three days later that Jersey Standard did not have the right to enter a suit against the state as it lacked legal status in Bolivia. Thirteen years later, Samuel Flagg Bemis, a noted U.S. diplomatic historian, criticized this Bolivian judicial ruling: "In 1937 a *de facto* government in Bolivia outrageously confiscated, on the flimsiest pretexts, property belonging to a Bolivian subsidiary owned by the Standard Oil Company of New Jersey and then packed the courts, and stirred up public opinion so as to deny justice to the claimants in the tribunals of the country."[47] Despite widespread popular support for the court's action at the time, the Bolivian Minister for Mines and Petroleum, Dionisio Foianini, told Caldwell on March 16 that his government was disappointed in the Supreme Court decision, and that he personally had favored a friendly settlement of the dispute.

A month later, on April 24, Germán Busch with the support of labor assumed complete control over Bolivia. Bolivia then adopted a new constitution that paralleled in many ways the Mexican Constitution of 1917; emphasizing human rights and the nationalization of subsoil rights, this document constituted an attack on foreign capital, tin barons, and great landowners. Those political leaders who were critical of Busch were imprisoned on an island located in the middle of Lake Titicaca.

That May Guachalla, still the Bolivian Minister to the United States, noted that his government had nothing further to say to the American oil company and suggested that the two nations attempt to work out an agreement. By June the State Department had formulated the draft of an understanding between Bolivia and Jersey Standard, which Jersey Standard found acceptable. The Bolivian Minister, though, informed Lawrence Duggan, the Chief of the Division of the American Republics in the State Department, that his country could only accept a settlement that fixed the amount due Bolivia as well as that due Jersey Standard. In addition, he stated his government could only admit that Jersey Standard should receive a payment in equity; it would never admit that it owed Jersey Standard anything as compensation. As for Jersey Standard, it was only with reluctance that it was willing to accept an arbitral tribunal ruling that fixed how much it owed Bolivia in addition to what it had already paid.

The tumultuous course of events in Bolivia took another dramatic turn late in 1939, when Germán Busch was found shot in the head. Even today it is not known whether he committed suicide or was murdered, perhaps at the instigation of the tin magnates. A year later, in 1940, General Enrique Peñaranda won the presidential election. In sharp contrast to Busch, who had imported Nazi technicians to work the petroleum fields of Bolivia, Peñaranda expelled the German minister in 1941 and replaced the German airline operators with Pan American Airways.

By the summer of 1940, Bolivia had come to the conclusion that it was necessary to reach an agreement on the oil issue with Jersey Standard. Accordingly, during the autumn the executive branch approached the Bolivian Congress, requesting authority to negotiate with Standard Oil. Although the executive branch asserted that it still rejected both arbitration and the return of company properties, the Chamber of Deputies failed to take action before it adjourned on April 30, 1941. The Senate had granted this authority at an earlier date.

While Bolivia had come to realize that a settlement with the United States was necessary, the outbreak of World War II had impressed upon the American government the need to establish harmonious relations with its Latin American neighbors. Having backed down on arbitration in the Mexican oil controversy, the United States could hardly insist on arbitration in Bolivia's case. After the

Bolivian government discovered and thwarted a planned coup in July 1941, the United States sent a military mission there. It also proposed greater economic collaboration between itself and Bolivia. At the end of the year a group headed by Merwin L. Bohan arrived in the Bolivian capital; among the four major goals recommended by Bohan were increased petroleum output and the growth of pipeline and refinery capacity. But it was Bolivian tin that the United States especially wanted, now that the Japanese had cut off its Malayan supply.

The Japanese attack on Pearl Harbor provided an extra incentive for the settlement of the Bolivian oil controversy. At the Rio de Janeiro Foreign Ministers Conference in January of 1942 the Foreign Minister of Bolivia, Eduardo Anze Matienzo, proposed that Jersey Standard accept an indemnity of $1 million. The American petroleum firm, however, rejected the proposal, requesting an outright sale of the Standard Oil property, which the company felt was worth as much as $3 million. (In 1937 Jersey Standard had valued its installations and equipment in Bolivia at $17 million.) A compromise agreement, signed on January 27, 1942, provided for the payment by Bolivia of $1.5 million plus interest to Jersey Standard within ninety days "for the sale of all of its rights, interests, and properties in Bolivia" as of March 13, 1937. Included in the settlement were Jersey Standard's highly valuable geological studies and exploration maps.

Assessing the Bolivian reaction, U.S. Chargé Allan Dawson confided to Secretary of State Cordell Hull that

> the press had accepted the idea of a Standard Oil settlement without too much opposition and public opinion also seems to be inclined to the same sense, although the picture will undoubtedly change somewhat when it is found that the terms are not as favorable as the Bolivian government has irresponsibly announced.[48]

Significantly, the Bolivian Congress was not in session at the time the government of Bolivia announced its acceptance of the Standard Oil settlement. In fact, it was not scheduled to meet again until August, when the company already would have received full payment for its holdings. There was considerable opposition to the agreement in the cabinet, which only approved the contract by a narrow majority. Although a motion of censure later in the year failed to pass the lower house of the Bolivian Congress by a single vote, the petroleum controversy was finally at an end.

With the dispute ended, the U.S. government promptly agreed to finance a $25 million economic development program for Bolivia. There were to be credits for mining, agriculture, and highway construction, as well as a $5.5 million loan from the Export-Import Bank, which would enable Bolivia to develop its petroleum resources by drilling new wells and building refineries. Having broken off relations with the Axis powers early in 1942, Peñaranda persuaded the Bolivian Congress in 1943 to declare war on Germany and Japan.

Like a number of his predecessors—as well as successors—Peñaranda fell victim to a coup at the end of 1943. The coup, led by the National Revolutionary Movement, installed as president a hero of the Chaco War, Major Gualberto Villarroel. The American government, suspecting that Villarroel might be another Juan Perón, withheld recognition for six months. Once in power, Villarroel unveiled a series of economic and social reforms that alienated both the business interests and the mining magnates. As a result, the final years of World War II in Bolivia were chaotic for the people of the country as well as for foreign investors.

## The Post-World War II Era and the Nationalization of Gulf

Villarroel's presidency ended violently in July 1946, when he was lynched by a mob. The right wing then assumed power for six years. Earlier, in the spring of 1946, the Banco Central de Bolivia had negotiated a $5 million loan from the Bankers Trust Company of New York. Secretary of the Treasury Fred M. Vinson stated that he had no objection to the loan, as it would be used to construct a pipeline that would be highly valuable to Bolivia. In June Jorge Lavadenz, President of the YPFB, arrived in New York to hold a series of interviews with North American oil men; Bolivia wished to increase its exports and to promote industrial growth. The Bolivian government, which had approved a new pipeline, announced a preference for small U. S. petroleum operators "who will actually get the oil out of the ground for us."[49]

That fall the Export-Import Bank extended a $5.5 million loan to the Bolivian Development Corporation and through it to the YPFB, to develop the oil resources of Bolivia. The money was to be used for drilling for petroleum, building a pipeline, and constructing a refinery. Assistant Secretary of State for American Republic Affairs Spruille Braden severely criticized the loan, feeling that it was unsound and ran counter to the policy of the American government in Chile. (On March 26 a group of officials at a State Department meeting had come out against a Chilean loan.)

During the next several years, though, the intellectual climate of Bolivia became more and more friendly to U.S. capital. On December 28, 1950, the U.S. Ambassador to Bolivia, Irving Florman, reported: "Since my arrival here, I have worked diligently on the project of throwing Bolivia's petroleum industry wide open to American private enterprise. . . the whole land is now wide open for free American enterprise. Bolivia is, therefore, the world's first country to denationalize. . . ."[50] Although Bolivia terminated its monopoly over oil exploration in 1951, major U.S. firms remained reluctant to reenter the country; Jersey Standard in particular was unenthusiastic.

An election held during this year saw the MNR candidate, Víctor Paz Estenssoro, poll 45 percent of the vote. The Bolivian army promptly installed a military junta, only to see a revolution overthrow it in April 1952, and place in the presidency the one-time economics professor Paz. As chief executive, Paz Estenssoro reduced the power of the army and nationalized the tin mines, inaugurating a new era of reform.

Between 1952 and 1955 the production of the YPFB increased fivefold. Nevertheless, the Paz Estenssoro administration felt that it was in the national interest to encourage foreign oil companies, as it lacked the financial resources to develop its petroleum resources fully. For this reason the Bolivian Ambassador to the United States, Victor Andrade, asked his friend Nelson Rockefeller to intervene with Jersey Standard Rockefeller persuaded Jersey Standard not to discourage the entry of other petroleum firms into Bolivia. In October 1955 a new oil code, drawn up with the help of U. S. experts and more favorable to foreign investors, went into effect. Although a small Texas concern had taken a concession in 1954, it was not until March 1956 that Bolivia entered into an agreement with a large firm: the Gulf Oil Company.

Bolivia's hostility toward foreign oil concerns had begun to lessen by the mid-1950s because the YPFB failed to make any spectacular petroleum finds, even though the government had invested $140 million into its operations. Although there had been rumors that prior to its expropriation Jersey Standard had discovered oil in large quantities but had postponed its exploitation, government technicians soon discovered that such rumors were not true. A dozen other companies that had begun exploring for petroleum after 1957 experienced a similar lack of success; having lost over $50 million, they had abandoned their drillings by 1962. The only exception was Gulf Oil, which found petroleum in the Santa Cruz area. Discouraged by these results, in 1964 Paz Estenssoro drastically limited those areas available for foreign oil concessions.

Paz Estenssoro was succeeded as president by Hernán Siles in 1956, who operated in a less repressive manner. But four years later, in 1960, Paz Estenssoro again won the presidency, and obtained still another term in 1964, after an election that was rigged when his support had begun to crumble. He remained in office from that June until November 4, when the Commander in Chief of the Armed Forces, General Alfredo Ovando Candia, told him that he was through as president. After Ovando Candia and General René Barrientos had shared presidential powers for slightly less than a year, Barrientos won the presidency on his own in an election held during July of 1966. Barrientos remained in office until his death in a helicopter crash in April 1969. Luis Adolfo Siles then became president, but a junta headed by General Ovando Candia seized power in September.

One month later, Ovando Candia expropriated the holdings of the Gulf Oil Company, or Bogoc, after cancelling the Bolivian petroleum code. At that time Gulf had an investment worth approximately $160 million, although it had only been exporting oil since 1967. Gulf reportedly offered to split its profits with the new regime, but the offer was turned down. In the preceding months Gulf had agreed to increase its local tax payments voluntarily, had undertaken a public relations campaign, and had offered to supply the department of Santa Cruz with free natural gas for a decade. After government troops had moved to occupy Gulf's La Paz headquarters and its Santa Cruz facilities, the Bolivian Minister of Mines and Petroleum declared that his government would reimburse Gulf only for its machinery and real property, which amounted to about one-third of the firm's total investment.

Summarizing the reasons for the Bolivian seizure, George M. Ingram has concluded: "The expropriation of the Bolivian subsidiary of the Gulf Oil Corporation arose from discontent with the terms of its operating contract, from concern over the size and area of its operations—and more important, from Bolivian political events during 1969."[51] One prominent critic of Bogoc was the leftist intellectual Sergio Almaráz, who held that the possession of its oil reserves was the key to success for Bolivia. Ironically, by offering to supply large amounts of natural gas free to Santa Cruz, Gulf became caught in the cross fire between its inhabitants and Bolivians living elsewhere, causing the Bolivian government to accuse it of interfering in the nation's domestic affairs.

Three years later, in the summer of 1972 an American journalist, Selden Rodman, visited Bolivia to investigate conditions there and to determine why Gulf had been disliked, despite the economic benefits it had brought to the nation. In the opinion of Rodman:

> Their public relations were terrible. They staffed the headquarters with people from the mountains, ignoring the intense hostility. They never joined the Santa Cruz Chamber of Commerce, though invited to do so repeatedly. The executives were not friendly types, like your AID officials, for example; at parties they didn't mix with the Bolivians. Finally, Gulf was too efficient, too automated; perhaps it should have padded its payroll a little to include more Bolivians.[52]

By September of 1970 the Bolivian government had reached a settlement with Gulf under which it agreed to pay the company a net compensation of $78.6 million, after having deducted $22.5 million as a special "single consolidated tax" from a gross compensation of $101.1 million. Bolivia also pledged itself to repay those Gulf loans that Gulf had made to the YPFB, as well as the advance tax payments made by the company totalling approximately $16 million.

A year later Ovando Candia resigned, and the leftist Juan José Torres took his place after a short civil war. In 1971, however, Colonel Hugo Banzer Suárez overthrew Torres in still another coup. Banzer had a new hydrocarbon law adopted the following year which outlawed foreign concessions in the future and instead provided for service contracts with foreign companies. The contracts themselves mandated a 50-50 split of profits following the recovery of the risk capital. If a petroleum company from abroad were to operate in Bolivia, it was not to be as an integrated operation but rather to provide a single service: exploration, transport, or extraction. More recently, the 1975 exposé of cash payments by Gulf Oil to Bolivian politicians prior to 1969 has not led the Bolivians to look favorably upon foreign petroleum firms.

## Peruvian Oil: 1913-1963

During the twentieth century the political development of Peru has resembled that of Bolivia. The leading figure in Peruvian politics in the first three decades of the twentieth century was Augosto B. Leguía, who was president from 1908 to

1912 and from 1919 to 1930. Given the feudal nature of Peruvian society and its domination by the so-called forty families, it is not surprising to find a totalitarian tendency there. In 1924 Víctor Raúl de la Torre set up the Alianza Popular Revolucionaria Americana, or APRA; the Apristas favored the nationalization of land and industry and opposed alleged U.S. imperialism. But because of intermittent government oppression, APRA was never able to gain the presidency for its own candidate. Instead, there were non-Aprista reformers such as Fernando Belaúnde Terry, who as president from 1963 to 1968 instituted land redistribution.

Economically Peru is not tied to a single product or export. The nation remains agricultural, with the mining industry providing employment for only 2 or 3 percent of the people, despite heavy exports of copper. Oil also has been a major industry for many years. As early as 1870 the Peruvian Refining Company, whose president was from New York City, had begun to export petroleum, but it was not until the second Leguía presidency that official American interest in Peruvian oil manifested itself.

Following the close of World War I the rivalry between American petroleum men and Royal Dutch-Shell intensified in Peru. Because the U.S. government was concerned that Peru might award an exclusive concession to Royal Dutch-Shell, the Chargé d'Affairs ad interim reported to Secretary of State Bainbridge Colby on March 25, 1920, that he had met with President Leguía and had informed him of "the attitude of the Department of State that petroleum lands not be granted to nationals of other countries so as to exclude future entrance by American nationals."[53] Leguía had told him at this time that no concession would be granted that would be detrimental to U.S. interests. It would appear, however, that Leguía had considered the possibility, for when Royal Dutch-Shell did not obtain an exclusive concession from Peru, one of its representatives complained that the president had gone back on his word: "the privilege had been solemnly promised to his company by Mr. Leguía while he was yet in London last year [1919], and while he was still a filibusterer rather than President, that promise being given in return for the help being given him by the company in his revolution."[54]

Once in office, Leguía apparently had concluded that it would be advantageous for him to stimulate rivalry between the Jersey Standard subsidiary in Peru, the International Petroleum Company (IPC), and Royal Dutch-Shell. In this connection he suggested that the International Petroleum Company spend no less than $25 million on civic and sanitary improvements. Since Jersey Standard was experiencing tax problems with the Peruvian government, Leguía recommended that the export tax International Petroleum owed Peru should be capitalized, and pledged in connection with other taxes to the servicing of a loan to Peru. If this were done, Leguía claimed, the Peruvian government would be ready to settle the tax dispute between itself and the IPC.

Jersey Standard became increasingly sympathetic to a tax deal with Leguía in 1921. It agreed to pay its taxes for several months in advance to help the

Peruvian government in meeting its current expenses. The actual amount that it lent—$1 million—was smaller than Peru had hoped, but Royal Dutch-Shell still did not benefit, since in the years that followed Leguía grew more and more favorably disposed to the American oil interests.

It was also during 1921 that Great Britain signed a treaty with Peru, arbitrating the status of the La Brea y Pariñas concession. As early as the 1870s Henry Meiggs had drilled for oil in the coastal desert of the far northwest; a British company, London and Pacific Petroleum, had obtained Meiggs' property in 1889. But in 1913 Jersey Standard had leased the La Brea y Pariñas holdings from London and Pacific, using the International Petroleum Company as its agent. Two years later, however, a Peruvian decree subjected the U.S. firm's holdings to a mining tax. When Peru was unable to obtain an international loan in 1916, it blamed Jersey Standard, which was accused of not meeting its financial obligations. The Peruvian Congress now debated the status of the leased La Brea concession before Peru signed the 1921 arbitration treaty.

Early in 1922 Great Britain and Peru exercised the option of determining the status of La Brea y Pariñas on their own, and in April the arbitral tribunal reaffirmed their understanding. The award stipulated that the International Petroleum Company would pay $1 million to Peru in settlement of outstanding claims, while Peru pledged itself to revoke the 1915 tax decree that had precipitated the original controversy. In 1924, once the dispute had been resolved, the International Petroleum Company purchased the La Brea y Pariñas oil fields outright. Although the Peruvians attempted in 1932 to have the World Court nullify the award, the court refused to do so. Writing fifty years later, in 1974, Jessica P. Einhorn claimed that the IPC had placed great pressure on the Peruvian government to force an award favorable to the company:

> IPC conspired with the Canadian government to requisition the steamer *Azof*, one of the two steamers carrying oil from Talara to the metropolitan area of Lima-Callao. The warning could not have been more explicit—either the Peruvian government reached a compromise acceptable to the Company, or else it would face first a shortage and possibly a complete lack of petroleum with resulting hardship to the economy of the country. The Peruvian government was brought to its knees.[55]

During the years that followed, the International Petroleum Company made a major effort to improve the sanitary conditions in those areas of Peru under its jurisdiction. Existing native camps were torn down and new quarters were built; a modern water plant was constructed; and doctors vaccinated the entire population of its settlements against smallpox and typhoid fever. As a result, there was a sharp decrease there in both infant sickness and mortality rates, which at one time had been extremely high. Moreover, IPC constructed new school buildings and playgrounds for children, as well as recreational facilities and clubhouses for adults.

In January of 1927 the American oil firms in Peru persuaded President Leguía to issue an executive decree softening the current petroleum legislation. Among other things, Leguía agreed to extend the length of oil concessions to forty years. By the time of the Great Depression Jersey Standard was producing three-fourths of Peru's petroleum and faced two only competitors, Zorritos (Peruvian and Italian) and Lobitos (British). Peru had become one of the three most important oil-producing countries in Latin America, Venezuela and Colombia being its leading rivals.

As the depression worsened, however, hostility toward the International Petroleum Company began to grow. The criticisms sometimes took the form of wild accusations. Thus Acting Commercial Attaché Julian D. Smith reported from Lima to the Bureau of Foreign and Domestic Commerce on July 9, 1931, that it was rumored that "the International Petroleum [administrators] have a hidden underground pipe by means of which they surreptitiously load their tankers at Talara and avoid the payment of export duties."[56] In an attempt to quiet rumors, the IPC measured the storage tanks before and after loading, but this action in itself was insufficient to disprove the charge. Moreover, according to contemporary Peruvian newspaper accounts, the company had been avoiding payment of the production tax ever since it was first enacted.

The petroleum industry of Peru, though, was to recover rather quickly from the depression. The growth of production in the 1930s was due in part to political pressures and in part to an increase in exports. American oil companies preferred the administration of Oscar Benavides (1933-39) to a regime under the direction of the Apristas, who remained highly critical of the large IPC profits in Peru. By the eve of World War II the La Brea y Pariñas oil field had reached its peak, and the need to explore for oil had become apparent.

For this reason, Benavides set up a new state enterprise, a petroleum department, to drill for petroleum in 1934. In 1939, before the presidential election, the Benavides regime proposed that IPC enter into a joint enterprise with it to explore and develop the Sechura desert. But nothing came of this proposal at the time, and the IPC began a futile search for oil in Ecuador.

Manuel Prado won the presidential election of 1939, and he held this office for the duration of World War II. Prado was pro-Allies, breaking relations with the Axis in 1943 and declaring war on them in 1945. The United States made funds available, not only for the modernization of the Peruvian military, but also to stimulate the production of minerals. Yet during the war, the state petroleum venture began to lose its momentum. The initial version of Prado's new oil tax law was unpalatable to the oil companies, but even the final version more than doubled the export tax receipt on petroleum. Fortunately, it also allowed those companies that paid royalties to offset these against their export tax liabilities. The target of this provision was the International Petroleum Company, which had enjoyed an exemption from the payment of royalties on production since 1922. In addition, because of pressure from APRA, Prado was hesitant to grant exploratory concessions to foreign oil firms.

José Luis Bustamente, who held the presidency of Peru from 1945 to 1948, began by cooperating with the Apristas, but then tried to abandon them before General Manuel Odría deposed him in a coup. Anxious to revive the Sechura Desert project, the Bustamente government drafted a contract with the IPC, only to have the Peruvian Senate block passage of it in 1947, because of both left and right wing opposition. A year later with the support of the army General Odría took over as President, holding dictatorial power from 1948 to 1955.

The Odría regime promulgated an important new petroleum law in 1952. Abolishing the state oil reserves, the law allowed the government to grant new exploration and production concessions on liberal terms: five years for exploration and forty years for exploitation. In addition, the export taxes and production royalties were replaced with an income tax; the Peruvian government was to receive 50 percent of the oil companies' net profits. Unfortunately, the three years that followed the passage of this law witnessed much searching for petroleum in the Sechura but few oil discoveries. While petroleum companies operating offshore enjoyed greater success, technical and marketing problems delayed full production.

In 1957 the International Petroleum Company consolidated its dominant position in Peru by obtaining control of the British Lobitos petroleum fields, the second largest in the nation. When a flurry of articles appeared in the Peruvian press discussing the consequences of this purchase, and the dangers inherent in oil monopolies, IPC felt compelled to print a full-page advertisement in the Lima *Extra* for February 4, 1958. By that time the IPC, according to its own figures, had invested more than $180 million in Peru. It claimed, however, to have made little profit during the 1950s, largely as a result of its reinvestment of its earnings.

Among the leading journalistic critics of the American oil firms between 1959 and 1961 was *El Comercio*, the best known daily in Peru. Faced with its editorial barrage, the IPC stated in 1960 that it had pensioned more laborers than any other firm in the country. It also claimed that it had paid a great deal of attention to housing its workers; for all practical purposes it had built an entire company town at Talara. Nevertheless, the attacks on the IPC were continued not only by the press but also by the labor unions and the Peruvian Congress.

By this time, having been friendly for a number of years, Peruvian-American relations had begun to deteriorate. In 1958 the United States had placed a quota on the purchase of foreign zinc and lead, thus undermining the economy of Peru. Furthermore, grants and loans from the American government were not now as generous as they had been. As a result, every large American business enterprise in Peru was under attack, including the Cerro de Pesco, a copper firm, and the shipping firm of W. R. Grace and Company. Given this increasingly hostile atmosphere, it is not surprising that when Vice President Richard Nixon visited Lima in 1958, the students there greeted him with an angry demonstration.

## The International Petroleum Company Expropriation

A Peruvian presidential election took place in June 1962. The leading candidates were former dictator Odría, Aprista leader Haya de la Torre, and Fernando

Belaúnde Terry, an architect. Haya led the balloting, but he had less than the 33.3 percent necessary to win election; when he tried to make a deal with Odría that would place Odría in the presidency and Apristas in his cabinet, an army coup occurred. Surprisingly, a year later the military junta allowed another election, which this time Belaúnde won, with Haya de la Torre second, and Odría third.

When Belaúnde had visited the IPC complex at Talara in 1956, he had commented, "If this is foreign imperialism, what we need is more, not less of it."[57] Yet the changes that were taking place in Peru (which became a net importer of petroleum in 1962) were mirrored by changes in Belaúnde's attitudes. During the 1963 presidential campaign Belaúnde promised to expropriate the holdings of the IPC.

After assuming the presidency, however, Belaúnde began to talk in terms of merely revising the terms of the IPC concession, rather than nationalizing it. His more restrained attitude incurred the wrath of the opposition parties, which then persuaded the Peruvian Congress to nullify the 1922 international arbitration award, the legal underpinning for the IPC concession at Talara. Belaúnde then hardened his line toward the Jersey Standard subsidiary, offering it either tax and royalty rates at a new, confiscatory level, or expropriation in return for compensation from which Peru would deduct $50 million in alleged back taxes. The well-paid IPC workers were against immediate expropriation, but the company itself expressed a willingness to pay the $50 million over a twenty-year period, rather than the five-year period the government had proposed.

Between October 1963 and April 1964 the IPC and Peru did not discuss the petroleum controversy. During 1964 and 1965, though, the State Department reduced its loan commitments to Peru. Despite the fact that grants and actual AID expenditures continued at their former levels, there was a disobligation of part of the 1964 loan, a sharply reduced loan for 1965, and an even smaller loan for 1966.

Then in 1967 the Belaúnde regime, rather than adopting a more conciliatory approach, in the face of this obvious State Department pressure, began to claim that Peru owed the IPC as much as $144 million in back taxes. This sum, which was far in excess of the $50 million that had been under discussion even in 1963, represented the total profits of the IPC in Peru during the last fifteen years. Although Adalberto J. Pinelo has suggested that the Jersey Standard subsidiary could have saved the Belaúnde government and itself by making a generous settlement with Peruvian officials even as late as 1967, there were limits beyond which the oil firm was unwilling to go.

Negotiations between the IPC and Peru dragged on for another year, when the two parties finally agreed on the Talara Act of August 13, 1968. Under this settlement Peru became the actual owner of the oil fields; in return, the government abandoned its $144 million back taxes claim against the IPC, which was allowed to keep its refinery and pipelines. The settlement also authorized the IPC to expand its operations elsewhere, should it decide to do so.

As an integral part of the settlement, the IPC and Peru signed a crude oil sales contract. But the president of the state enterprise then charged that page eleven was missing from the contract; both radical leftists and the anti-IPC publication, *El Comercio*, used this accusation as grounds for urging the replacement of Belaúnde, who tried to save himself by purging his cabinet. The eleventh page was said to set forth the proposed minimum price for crude oil, if in fact it actually did exist.

Many Peruvians, including some of the country's military leaders, believed the Act of Talara was overly generous. On October 3 the armed forces launched a successful coup; President Belaúnde was forced from office, and the Peruvian Congress was shut down. General Juan Velasco Alvarado now exercised power through a military junta.

The day it seized power, the junta issued a manifesto, labelling the La Brea y Pariñas settlement as a "surrender." On October 4 it proclaimed Decree Law Number 3, which voided the August contract between the IPC and the Belaúnde regime. It seemed most probable that the new government would expropriate the IPC holdings, and it did on October 9 under Decree Law Number 4. On this *dia de la dignidad nacional*—"the day of national dignity"—Peruvian infantrymen occupied the oil fields, seizing property worth somewhere between $90 and $120 million. In assessing the events of October 9, Velasco complained to the nation in a radio broadcast on February 6, 1969, about the activities of the IPC, "which in a fraudulent manner—not having any legal title and using the most varied methods of pressure—exploited our oil wealth in the northern part of the country."[58]

In all fairness to the expropriated firm, one should note that whatever its faults might have been, this was one of the most socially minded companies in Peru, offering high wages and extending fringe benefits. It had created a superior company town at Talara, and had promoted Peruvians whom it had employed; the latter received extensive technical and managerial training. The IPC even had attempted to stimulate nationalistic art and literature, establishing FANAL in this connection, and had sponsored open essay competitions on topics from Peruvian history.

Ironically, the IPC's accusers leveled a number of charges against it that were outright lies—such as the claim that the firm had used double-bottom ships and had arranged for the murders of two newspaper editors—but often overlooked its actual transgressions. For example, IPC had helped to finance the Sánchez Cerro administration from 1931 to 1933. Hostility toward the IPC was so great that a Peruvian general declared after the 1968 expropriation, "They bribed ministers, corrupted governments, and promoted revolutions."[59]

Jersey Standard naturally believed that the Peruvian seizure was a clear violation of international law, but it did not ask the State Department to intervene directly. It hesitated for a number of reasons. Peru was only responsible for less than 1 percent of the firm's crude oil production; in addition, the increasingly grandiose demands of the Peruvian government militated against any immediate compromise settlement. On February 6, 1969, the Velasco regime claimed that the IPC owed Peru in excess of $600 million: the total value of the petroleum the firm had "illegally" extracted over four decades. The accusation apparently ruled out any future compensation to the IPC, although International Telephone and Telegraph, which was also nationalized, was compensated for its losses.

On November 27, 1968, the Foreign Office of Peru sent a message to the American government in which it pointed out that the IPC had not exhausted its avenue of appeal through the Peruvian courts and added that the oil firm was registered technically as a Canadian rather than American firm. Nevertheless, on December 13 the State Department informed the Peruvian Foreign Minister in Lima that his country's treatment of the IPC was of concern to the United States. On March 11, 1969, the newly inaugurated President, Richard Nixon, sent John N. Irwin to Peru as his personal emissary; between March 13 and April 3 Irwin conferred with key officials of the Velasco regime in Lima.

It is unquestionably significant that the Peruvian government did not nationalize a number of other foreign oil companies then operating within the country. Among these were Gulf Oil, Mobil Oil, Texas Petroleum, Belco Petroleum, Occidental Petroleum del Peru, and Conchán Chevron. The Velasco regime felt that these firms had acted in accordance with Peruvian laws. According to Velasco, "The postulates of the Revolution not only respected but also encourage foreign investment, provided it agrees with the legislation and interests of Peru."[60]

Despite the talks in Lima, by the fall of 1969 no settlement of the IPC expropriation had yet been reached. Although the American government did not apply the Hickenlooper Amendment to the 1962 foreign aid act, which would have penalized Peru for its act of nationalization in the absence of compensation, the foreign aid program for Peru died a quiet death. The Hickenlooper Amendment was not invoked because there was widespread opposition to it among American businessmen in Peru. As Charles W. Robinson, President of the Marcona Mining Company, observed, the amendment was "our 20th century version of the British gunboat diplomacy of the last century."[61] (His firm had not been expropriated.) Furthermore, the U.S. Secretary of Agriculture decided to grant to Peru an increase of 5 percent in its sugar exports to America, thus allowing it to become the United States' fifth leading supplier.

Moreover, Jersey Standard was not unanimously supported by the other American oil firms in Peru. Armand Hammer of Occidental, the world's fastest growing petroleum company at that time, flew to Lima and offered to run the operations of the expropriated U.S. company. Under these circumstances, it is not surprising that during the summer of 1971 Petroperú, the new state enterprise that had been created upon the seizure of the IPC holdings, signed a thirty-five-year contract with Occidental. Although Petroperú was technically the actual holder of the concession, under the contract Occidental was given the right to explore a 3 million acre tract in the Upper Amazon Basin over a seven-year period. It was expected that Occidental would invest approximately $50 million in its operations. Petroperú was to pay all the taxes, Occidental and Petroperú were to share production on a 50-50 basis, and if justified Occidental was eventually to build a pipeline. In the opinion of the Peruvian *Times*, the "Oxy" agreement was "the model for future deals in what could turn out to be Peru's first major exploration boom."[62] Between 1972 and 1974, moreover, the U.S. firm did make a series of discoveries.

Later in 1971, the Tenneco Oil Company and the Union Oil Company of California obtained both offshore and onshore exploration and development contracts from the Peruvian government. But American-Peruvian petroleum relations continued their erratic course in the years immediately preceding the Arab oil embargo of 1973. On March 21, 1973, the Peruvian government sold at auction to Petroperú the Conchán Chevron refinery at Lima, which was then processing 8,000 barrels of petroleum daily, after the California Standard subsidiary allegedly had refused to pay back taxes to the government.

In the long run, except for Occidental, the oil exploration activities of every firm in Peru were not particularly successful. By the end of 1976 Peru not only lacked proven export reserves of petroleum, it was also spending $600,000 per day on imported oil. Petroperú was faced with a series of almost insurmountable problems: a lack of qualified foreign petroleum geologists and engineers, an inadequate educational system in oil technology, and insufficient capital for extensive petroleum exploration. Having refrained from issuing any new exploration contracts to foreign firms since August 1973, the Peruvian government in November of 1976 again opened its doors to foreign capital.

### Argentina: From Irigoyen to Perón

Along with Venezuela Argentina enjoys one of the highest per capita incomes on the continent. While there has yet to be an effective evaluation of its coal reserves, there are oil deposits in the far north and in Patagonia. Over the years, though, Argentine commerce in primarily agricultural products has been directed more toward Europe than the United States, as beef and wheat from abroad are not officially welcomed in this country. American tariff barriers alone have tended to make Argentine relations with the United States less friendly than those, for example, between America and Brazil.

The move toward democracy in Argentina during the early part of the twentieth century culminated in the election of Hipólito Irigoyen as President in 1916. Since 1930, however, the Argentine military has played a leading role in the politics of the nation, and coups have become a regular feature of the political process. Of all the Latin American countries, Argentina has been the most sensitive to the right-wing political trends found on the European continent; in part this is due to the many Italian, Spanish, and even German immigrants who have settled there over the years. Perhaps the leader who best mirrored the country's fascistic climate of opinion was Juan Domingo Perón, President of Argentina from 1946 to 1955 and from 1973 to 1974.

Strained as Argentine-American relations have been at times, U.S. investors have long displayed an interest in the country. A governmental expedition searching for water first discovered oil in 1907. By 1911 Standard Oil had invested in an Austro-Hungarian refinery; by 1914, it had gained control of the firm. Two years later, however, Hipólito Irigoyen won election to a six-year term as president and ushered in an era of reform.

Far from welcoming the foreign oil entrepreneurs with open arms after World War I, the Irigoyen administration began to adopt a more critical attitude. It placed limitations on the size of foreign petroleum holdings, and it began to talk more frequently about the establishment of a state monopoly over oil. Such a possibility bothered the U.S. State Department, which maintained that the enterprise would be in violation of the principle of reciprocity, might retard commerce, and would drive out U.S. capital.

The United States also was disturbed by the growing British economic role in Argentina, most evident in the railroads, meat packing industry, and mass transit systems of the country. As early as 1919, the State Department had instructed its diplomatic representatives in Buenos Aires to investigate possible British petroleum activities in Patagonia. At this time, French and German oil entrepreneurs were also active in Argentina.

In 1922 Marcelo T. de Alvear won election as president, succeeding Irigoyen, and the Dirección General de Yacimientos Petrolíferos Fiscales (YPF) came into existence. YPF was a state oil exploration and exploitation monopoly, which at first was under the jurisdiction of the Department of Agriculture, but after 1923 became an autonomous agency. A year earlier Standard Oil's refusal to deliver gasoline for military planes without prior payment had aroused the hostility of the Argentine government. The Director of the Army Air Force, Enrique Mosconi, who led the attack on the foreign monopolies, served as the first head of YPF from 1922 to 1930. Although originally he had favored private oil companies, by 1929 he had concluded that a mixed firm (both governmental and private) was the best solution for Argentina's petroleum needs. It is highly significant that despite its problems with the U.S. oil firm, the Argentine government had allowed the continuation of the extensive Jersey Standard and Royal Dutch-Shell retail sales operations there.

Like his predecessor Irigoyen, Marcelo de Alvear was a strong proponent of government action in the petroleum sector. In 1924 de Alvear issued the so-called reserve decrees, which set aside for governmental use those lands that appeared to be the most promising for petroleum exploitation. Such provinces as Salta and Jujuy then followed suit, and new national and provincial reserve decrees continued to appear over the years. In 1925 the Argentine government took the additional step of constructing an oil refinery near Buenos Aires; in 1930, after eleven different prices had been charged the previous year, the YPF imposed a uniform price for gasoline. The establishment of a uniform price appears to have been less an economic necessity than a political move designed to win the support of the interior. As petroleum refining took place in the coastal cities, its cost in a free market was the coastal price, plus the transportation costs to the inland markets.

During the 1920s the Director General of the YPF, Enrique Mosconi, continued his attacks on Standard Oil: "Whenever this company obtains a foothold it becomes not only a Government within a Government but a Government over a Government."[63] Mosconi also attempted to annul the 1920 Standard Oil concession in the Salta Province, but this attempt failed. But even though in a 1928 Mexico City speech Mosconi had gone so far as to characterize Jersey Standard as a hempen rope and Royal Dutch-Shell as a silken rope, both of which might be used to hang Argentina, foreign petroleum firms continued to operate in Argentina as they had before the government began its intervention in the oil industry. For one thing, they had been able to perfect title to many of their holdings; for another, the reserve decrees did not exclude them totally from exploration. Finally, much to the displeasure of the YPF, several provinces decided to cooperate with the foreign oil companies because they needed the funds they would receive from royalty payments.

Despite the various restrictions imposed upon it, Standard Oil did a great deal to fight disease in Argentina during the 1920s. It not only built a fifty-bed hospital at Tartagal, it also carried out an extensive health campaign in its employee camps. In addition, Standard Oil helped to contain, and then wipe out, a plague epidemic that had erupted in the northern part of the country, and at a later date it established the first Argentine training school for nurses. These operations built up a reservoir of goodwill for the U.S. firm that helped to offset the attacks of the petroleum nationalizers.

In 1927 the Argentine Chamber of Deputies passed a bill that would have established a state monopoly over oil transportation and provided for exclusive state exploration. Anglo-Persian, Royal Dutch-Shell, and Jersey Standard joined forces in a successful attempt to block the passage of this bill by the Argentine Senate during 1928 and 1929. Despite the fact that Standard Oil was only one of a group of foreign petroleum companies then operating in Argentina, the attacks in the Argentine Congress focused primarily on it.

Hipólito Irigoyen, now seventy-eight years old, took office for a second term in 1928. The "father of the poor" had been responsible for the passage of social legislation and assistance to the labor unions, but he also had made many enemies during his years in politics. With Argentinian products selling at low prices on the world market, a group of rightists led by General José F. Uriburu drove Irigoyen from power in the fall of 1930. Uriburu then served as president for two years (1930-32), after which Agustín P. Justo won election to a six-year term as chief executive (1932-38).

One characteristic of this historical era was a tendency on the part of the national government to intervene in the affairs of the provinces. Even before Uriburu's 1930 coup there had been talk in the Argentine Congress of depriving the provinces of their ownership of oil wells, as some of them had entered into favorable agreements with foreign petroleum firms, thus undercutting the "united front" the Congress wished to present. One of these was Salta Province, which came to an understanding with the YPF on October 19, 1932; this agreement, if enforced, would not only have restricted the operations of the foreign oil companies there, but might even have terminated them completely. Soon after Jersey Standard contested in the courts the heavy operating royalties imposed by this agreement, Salta Province officials decided that the courts might decide in favor of the oil company, and therefore entered into negotiations with the firm. Salta Province signed a contract with Jersey Standard in 1933 which was to last for thirty years. Under the agreement royalty payments were substantially reduced, and the firm was exempted from various taxes. The YPF was angered by this pact, and members of the Unión Cívica Radical party in the Argentine Congress bitterly criticized the province. This in turn caused much resentment in Salta Province; its legislature approved the understanding with Jersey Standard over the objections of the Argentine Minister of the Interior. Fortunately for the province, it found an influential supporter in the newspaper *La Prensa*.

At this time there was some confusion over the ownership of subsoil rights. According to the *Código de Minería*, mineral deposits belonged to both the national and provincial governments. The Argentine Congress attempted to clarify this situation through the passage of a new petroleum law, which apparently vested in the Argentine provinces or states full rights to the subsoil petroleum deposits within their jurisdiction. The law reserved to the YPF those areas in which deposits of oil were most likely to exist, and restricted private foreign companies to the concessions they already held.

A year later, in 1936, the Argentine government prohibited the exportation of petroleum, and also limited the importation of oil. On May 9 President Justo set up a committee to investigate the petroleum industry, and its report accelerated the trend toward governmental regulation. According to the report, the American firms' rebates and price reductions had adversely affected the market the YPF normally dominated.

Shortly after the creation of the three-man committee, Ambassador to Argentina Alexander Weddell reported to Secretary of State Cordell Hull that one U.S. petroleum company official had summarized the deteriorating Argentine situation as follows: "We are actually being squeezed at both ends."[64] Nevertheless, Raymond Cox, the Chargé in Argentina, reported to Secretary Hull on July 29 that he could not find any unanimity among the foreign oil firms concerning the course of action they should follow, since each firm was affected differently. The Texas Company, for example, was not engaging in domestic production at all.

Under the executive decree of July 20, 1936, each foreign oil firm was to receive a quota proportional to its production and distribution capacity. If a company exceeded its quota, it faced a fine. The measure obviously favored the YPF; the total quota for foreign petroleum companies was determined by calculating the difference between the annual Argentine consumption of petroleum products and the yearly production of the YPF. By 1941 it was estimated that the YPF would control 40 to 50 percent of the oil business in Argentina, a 10 to 20 percent increase from 1936, at the expense of Standard Oil and the other foreign petroleum companies.

In the fall of 1936 the Argentine government and Jersey Standard discussed seriously the state's purchase of Jersey Standard's holdings. But the Argentine Congress failed to act, and Standard Oil rejected the offer. There was a temporary flurry of excitement when Buenos Aires approved a municipal bylaw that established a gasoline monopoly for the YPF in the city. Jersey Standard brought suit, declaring that the bylaw was unconstitutional, but following a dispute between the mayor and the municipal council, the bylaw was abrogated.

Neither Jersey Standard nor Royal Dutch-Shell signed the agreement that was approved by the other foreign oil companies on September 29, which set forth marketing quotas in compliance with the July executive decree. These two firms charged that this understanding favored the YPF in the federal districts. As together they controlled over half of the Argentine gasoline market, their action severely restricted the impact of the September 29th agreement. On December 29, though, Jersey Standard and Royal Dutch-Shell did sign a temporary, or standstill, understanding with Argentina; then on June 28, 1937, they reached an agreement with the government that was to run through 1940. As a result, relations between the foreign oil firms and the government became more tranquil.

The tranquility, however, was only temporary. Following the outbreak of World War II on September 1, 1939, Argentine-American economic relations became more and more shaky. Irritated by the U.S. tariff on Argentine grain and its quarantine on meat, Argentina placed a temporary embargo on all American imports during the latter part of 1940. But the two nations signed an important trade agreement a year later, in October, demonstrating that Argentine-American economic relations were subject to abrupt shifts during this era.

Then in June of 1943 a major revolution broke out in Argentina, when the military seized power. As a result of the coup Juan Domingo Perón eventually came to lead the nation. By January 1944 Argentina had broken relations with the Axis, but the American government remained unconvinced about Argentine intentions. Not only did the United States freeze Argentine gold stocks, it also tightened shipping regulations. Finally, in 1945, after the Act of Chapultepac had been drawn up, Argentina declared war against Germany and Japan.

It was against this stormy background that American-Argentine petroleum relations evolved during World War II. In 1941 the U.S. government pressured the Phillips Petroleum Company into withholding information and technical assistance for the manufacture of gasoline from the YPF, and persuaded Intava (a Jersey Standard subsidiary) to restrict its aviation and fuel oil deliveries. At this time, too, the State Department strongly opposed Argentina's attempt to import petroleum from Standard Oil tanks in Brazil.

During the summer of 1943, the new Argentine government concluded a series of petroleum negotiations with the United States. Under the final terms of the agreement the United States would make 36,000 tons of oil industrial materials available to Argentina, in return for which Argentina would export 360,000 cubic meters of petroleum to Uruguay, Paraguay, and southern Brazil. Although the Argentine government expressed its willingness to begin exporting oil immediately, Secretary of State Cordell Hull advised the Ambassador to Argentina,

Norman Armour, on June 16, 1943, that the proposed agreement was unsatisfactory for a number of reasons. Among other things the United States was not prepared to provide the needed oil equipment and felt the saving in tankers would be inconsequential.

Ambassador Armour was especially disturbed by this rejection. On June 26 he wrote Hull:

> Until the installation of the new Government, the Embassy could have withdrawn from the discussions gracefully, and without damage to any existing situation. Conditions today are very different. The Embassy's telegram 1329 June 14, 6 p.m. informed the Department of [Minister of Foreign Affairs Segundo] Storni's premature announcement that an agreement was about to be signed. This was instantly given favorable publicity by the press in Argentina as well as in Brazil and Uruguay and the general tenor of the comment was approval of this positive step by Argentina's new Government toward closer Pan-American cooperation and in particular a closer approach to the United States which might lead to more important cooperative steps later.[65]

Two years later, on May 9, 1945, Argentina did sign a fuel and vegetable oil agreement with the United States. The understanding, however, was criticized in the American press, and Argentina proved to be lackadaisical in the implementation of it. According to this agreement, Argentina was to receive fuel oil from the Caribbean on a "heat equivalent basis" and in return was to supply vegetable oil cake and cattle feed to the dairy herds of Belgium, the Netherlands, and Denmark, as well as vegetable oils to the United Nations.

In 1946 Juan Perón was elected president of Argentina, thanks largely to labor support, having emerged as the strong man of the ruling military regime. He then began to implement an economic policy that stressed Argentine emancipation from control by American, British, and foreign capital in general. During 1946 Argentina acquired a national telephone system from International Telephone and Telegraph, and two years later it took over the British-owned railways. Although Perón retired the entire foreign debt, he also accepted a $125 million credit from the Export-Import Bank in 1950.

Perón preached the ambiguous doctrine of *justicialismo*: a middle way between capitalism and communism. It is significant that his strongest European supporter was Spain's General Francisco Franco. On this side of the Atlantic, Perón attempted to play the role of protector of the weaker Latin American republics, but his grandiose schemes fell far short of total realization. Because of his attempts to seize the initiative in Latin America and his anti-American propaganda campaign, relations between the United States and Argentina were shaky at times, with a representative example of friction being the hostility between Perón and American Ambassador to Argentina Spruille Braden at the end of World War II.

During the immediate postwar period, the United States experienced problems with Argentina involving petroleum. In March 1947 the Perón Administration incensed Jersey Standard by asking that it extend wage increases to petroleum workers before it approved compensatory price increases. In this connection Ambassador George Messersmith wrote Secretary of State George Marshall on May 9, 1947, "the situation of the American and foreign companies in the Argentine has not been on a very satisfactory basis for some years. . . ."[66] So serious was the conflict between them and Argentina that Standard Oil felt noncompliance with Perón's demands might lead to expropriation; this fear proved to be totally unjustified, but conditions did not improve. At one point the foreign petroleum firms felt impelled to counter a slowdown strike by closing their refineries, a decision that Ambassador Messersmith regarded as unfortunate. The shutdown led to a gasoline and fuel shortage that was hardly popular with the Argentine public. Contrary to their better judgment, the companies did agree to accept a single union for the oil industry, only to have the workers reject the wage settlement that had been negotiated between the heads of the syndicate and the companies.

The Perón administration proved more reasonable. Although he did ask that the wage dispute be settled first, Perón recognized that price increases were necessary. Thus the Jersey Standard office in Argentina regarded it as unnecessary and undesirable for the main office in New York City to request a formal protest by the U.S. government. Nevertheless, Standard Oil officials in New York did take this step.

In 1949 Argentina approved a new constitution that some commentators have felt was as important as the Mexican Constitution of 1917. Article 40 of the new Argentine constitution stated that all underground resources, including oil, were the property of the nation. The government also was granted the right to intervene in economic affairs, including foreign commerce.

It was during 1949, too, that the Perón administration signed an agreement with Great Britain, under which Great Britain would provide Argentina with its import requirements of oil in return for payments in sterling. Five months later the British announced that they would have an additional 75,000 barrels of petroleum available daily in 1950. The losers under this arrangement were the American oil companies (Jersey Standard, Socony-Vacuum, and Texaco) which had been supplying Argentina with approximately 40,000 barrels of petroleum each day. Although the effective date of the November announcement was postponed from January 1 to February 15, 1950, the end result of the understanding was to eliminate the fuel shipments the U.S. firms had been making, thus dealing the dollar a severe blow in international trade.

Although Argentina had discriminated against foreign companies wishing to invest in the country for many years, in 1952 the Perón administration adopted a new five-year plan that called for assistance from private capital to aid in the development of energy. A new foreign investment law enacted in August of 1953 opened the door to foreign capitalists. A year later, dissatisfied with the work of the YPF, which was producing only half the oil Argentina needed, Perón extended a hand of welcome to foreign oil companies.

The first American capitalist to benefit from Perón's new stand was Floyd Odlum, a New York financier who headed a Texas petroleum group. Under the terms of a twenty-five year contract, Odlum and his associates were authorized to engage in crude oil development on a limited scale and to build a pipeline 400 miles long at a cost of approximately $40 million.

Having set the precedent with this contract, Perón then signed an agreement early in 1955 with the Compañía California Argentina de Petróleo, a California Standard subsidiary incorporated in Delaware with both Argentine and American shareholders. The U.S. controlled firm was to spend $13.5 million on petroleum prospecting in the Santa Cruz territory. While the Argentine government was to refine and market the oil the California Standard subsidiary located, and was to receive half the profits, the company had the right to export any oil Argentina did not need. At the expiration of the forty-year contract, Argentina was to take over all the installations and equipment.

By July, though, political opposition to the Compañía California Argentina de Petróleo contract had begun to mount. It was not just the opposition that disliked the contract; some ultra-nationalistic Peronistas complained of their leader's policy of *entreguismo*: the giving away of precious natural resources. As a result, the Industry Committee of the Chamber of Deputies decreed that it would be necessary to modify the proposed contract in Argentina's favor.

Among the leading critics of the contract was Arturo Frondizi, the chairman of the national committee of the Radical Party, and future President of Argentina. Pointing out that the United States had sought military bases along the Strait of Magellan during World War II, Frondizi charged that American capitalists were planning to establish a sphere of interest in Argentina, the South Atlantic, and Antarctica. So intense was the feeling generated by the contract that a number of writers—including John H. Lind—have concluded it was the single most important reason for Perón's fall from power.

Perón's wife, Eva, had died of cancer in 1952. Widely popular, Eva Perón had transcended her humble beginnings to exercise political power to a degree that few women in Latin America have ever equalled, let alone surpassed. Following her death, Perón had to contend with worsening economic conditions and disintegrating national morale; there was growing opposition to him on the part of businessmen, landowners, the army, and the church. The Vatican even excommunicated him. After an abortive revolt in June 1955, his foes succeeded in driving him from power, and into exile three months later.

## The Recent Struggle over Foreign Concessions

Following three years of provisional government under Eduardo Lonardi (1955) and Pedro Eugenio Aramburu (1955-58), post-Perón Argentina alternated between civilian and military rule. The civilian government of Arturo Frondizi (1958-62) gave way to the military one of José María Guido (1962-63); then there was civilian rule again under Arturo Illia (1963-66), but the military again exercised power under Juan Carlos Onganía (1966-70), Roberto M. Levingston

(1970-71), and Alejandro Lanusse (1971-73). Torn between civilian and military candidates, the Argentine voters turned to the Perónist candidate, Hector Campóra, in a special election held during 1973. A mere stand-in, Campóra gave way to Perón himself by the end of the year. Upon Perón's death his wife, Isabel Perón, exercised presidential power until 1976, when a new military coup installed Jorge Videla as chief executive.

Although Arturo Frondizi, the candidate of the now splintered Unión Cívica Radical party, had promised during his campaign to implement a program of state control, nationalism, and social legislation, stating that he long had adhered to Perónist ideology, he swiftly began to move away from these principles once he assumed office. Frondizi himself noted, "When I came to power I encountered a reality which did not correspond to this theoretical posture. . . ."[67]

By 1958 Argentina's deficit in its balance of payments was threatening its foreign exchange reserves. On July 24 Frondizi, in a speech to the nation, pointed out that Argentina was importing approximately $300 million worth of oil annually. In an effort to obtain self-sufficiency, the YPF planned to drill almost 10,000 wells by 1964, twice the total number sunk in the preceding fifty years. As progress would be difficult, if not impossible, without foreign economic help, Frondizi informed the Argentine people that his government had concluded a purchase of petroleum for barter agreement with Colombia, and was negotiating deals for oil and petroleum machinery with Russia and other countries.

Frondizi, moreover, had signed various contracts, or letters of intent, with foreign oil firms, which hopefully would bring approximately $1 billion dollars into the country's petroleum industry, and lead to the drilling of a large number of wells around Mendoza, in Nequén, and in Patagonia. Frondizi also announced a search for petroleum in previously unexplored areas, mainly northern Argentina. Of the companies that were going to participate in this program, the most important were the Atlas Corporation, a consortium of U.S. and European firms; the Pan American Oil Company, an Indiana Standard subsidiary; and Carl M. Loeb, Rhoades, and Company, a New York investment concern. Each of these contracts, or letters of intent, differed somewhat from the others, but taken as a whole, they covered every aspect of the Argentine petroleum industry: equipment, machinery, exploration, drilling, pipelines, and plants.

The $800 million dollar Atlas agreement fell through by the end of 1958 because of doubts about financing and high costs. The Frondizi administration, however, had been able to persuade both Esso (a Jersey Standard affiliate) and Royal Dutch-Shell to sign long-term drilling contracts. These two agreements were closer to the traditional concessions than were the understandings Argentina had concluded with various oil firms earlier in the year. Esso and Shell were to invest varying amounts of money in prospecting, and were to market the petroleum they found with a minimum amount of interference from the YPF. After the initial prospecting period, both companies were to receive the equivalent of concessions for twenty years. While Frondizi's billion dollar oil program had now shrunk to several hundred million dollars, the participation of these two petroleum giants gave it greater respectability.

Despite the opposition of the Communists, Perónists, and nationalists to Frondizi's petroleum policies, Argentine oil production under Frondizi reached the point where there was an exportable surplus by 1961. His handling of his office won him the support of both the Eisenhower and Kennedy administrations. The Argentine military, though, was disturbed by Frondizi's failure to pursue a vigorous anti-Castro policy, and by Perónist victories in the 1962 elections for the Chamber of Deputies and provincial governors. As a result, the military ousted him on March 29, 1962.

During the summer of 1963, while Frondizi's successor, José María Guido, was still acting as president under the watchful eye of the military, the YPF signed a contract with Kerr McGee Oil of Oklahoma to drill an additional 350 wells. While Kerr McGee had admitted that its drilling fees were perhaps too high, and accordingly had reduced them from $26 to $16 per meter, other foreign oil companies, including Jersey Standard and Royal Dutch-Shell, complained that they already had spent $60 million in Argentina, which they had yet to recoup.

Shortly after the signing of the Kerr McGee oil contract, the Argentine electoral college chose Arturo Illia as president, since none of the three leading candidates had received a majority of the popular votes. During the campaign Illia had promised not only to terminate Argentina's connections with the World Bank and the International Monetary Fund but also to cancel the Frondizi contracts. By this time foreign petroleum firms had invested a total of $400 million in Argentina, and the nation owed them more than $100 million for oil purchases. So serious was the situation that President John F. Kennedy sent W. Averell Harriman on a special mission to Buenos Aires to warn President Illia that any hasty action not only might jeopardize Alliance for Progress funds, but also might scare private U.S. investment away from Argentina.

There was, though, extreme pressure on Illia to repudiate the contracts with the foreign petroleum companies, especially the controversial ones with Indiana Standard, the Tennessee Gas Transmission Company, Union Oil of California, and Cities Service Company. Thus in November Illia signed a series of decrees annulling as "unconstitutional" thc contracts with ninc U.S. pctrolcum companies operating in Argentina, as well as four other foreign firms. Technically Illia was not expropriating these companies but rather annulling the contracts in preparation for renegotiating them. The oil firms experienced a further blow at this time when Illia charged that they owed approximately $80 million in back taxes. A year later, in 1964, an Argentine congressional committee issued a report that claimed the foreign oil contracts were fraudulent.

Argentina's actions were not surprising, in view of a conversation between Argentine Foreign Minister Zavala Ortiz and the British Ambassador. Zavala Ortiz had observed that although Royal Dutch-Shell was in a different position, he could assure the British Ambassador that "most of American companies were "*tramposas* [swindlers],"[68] But back in the United States there were protests that the cancellations were illegal as well as threats in Congress to suspend American aid to Argentina. Less emotional observers pointed out that the Frondizi contracts had been negotiated without legislative sanction, and quite probably were illegal as well as nonbinding. The *Nation* went so far as to compare them to Secretary of the Interior Albert B. Fall's Teapot Dome transactions four decades earlier.

Yet another publication, the *New Republic*, hardly a tool of foreign petroleum firms, took note of various YPF deficiencies. While admitting that YPF had provided hospitals, schools, and houses for its workers, the *New Republic* feared that Illia's action would give the YPF "more opportunity to distort the Argentine economy. It already has debts of over $200 million; it is riddled with inefficiency and corruption. . . . It was almost impossible to break into the field of selling to Y. P. F. because there were so many palms to be greased on the way in."[69] As a result of the contract cancellations by the end of 1964 the YPF was importing wholesale quantities of petroleum for the first time in several years, and had even made a $9.3 million deal with the Soviet Union for both diesel fuel and crude oil.

Aside from the worsening petroleum situation, Illia's economic policy in general did not prove successful. Perónist candidates led the field in the congressional elections of March 1965; in November Illia appointed a Perón sympathizer as minister of war. When Illia refused to send Argentine troops to take part in the Inter-American Peace Force occupying the troubled Dominican Republic, military support for him began to disintegrate. On June 28, 1966, a three-man junta deposed him and installed General Juan Carlos Onganía as president. Onganía quickly instituted a program of repression, dismissed Congress and abolished all political parties.

Onganía's economic policy was diametrically opposite to that of Illia. Argentina obtained a $125 million credit from the International Monetary Fund and $400 million in loans from American, European, and Japanese banks. The new president even agreed to compensate U.S. petroleum firms. In a settlement announced at the end of the year, the Argentine government arranged to pay the Pan American Argentina Oil Company and the Argentina-Cities Service Development Company $56.3 million in full for past crude oil production. In return, Pan American was to drill 150 new wells and Cities Service 60 over the next several years. The Onganía regime also encouraged new oil contracts, but foreign companies were restricted to the less promising areas.

To aid in the creation of an economic climate more favorable to petroleum operations, a new oil law was passed in 1967. This measure in effect downgraded the role of the YPF, as it had become obvious that it could not fulfill Argentine petroleum needs by itself. Not only were more than 800,000 square kilometers made available for new concession agreements, but work in YPF production areas now could be contracted out to private companies. In the event that they should discover oil in the new concession areas, the foreign petroleum firms would have tax obligations to the Argentine government, not to the YPF.

Unfortunately, Argentina did not become self-sufficient in petroleum by the time of the second Perón presidency in 1973, and of the Arab oil embargo during the same year. As of 1972 Argentina still was importing $58 million worth of oil, and because of the sharp increase in the price of petroleum was to import $588 million in 1974. Given the rapid alternation in governmental oil policy between the encouragement of foreign oil interests and a reliance on the YPF, it is difficult to predict whether or not Argentina will be able to implement a long-term petroleum policy in the years ahead.

## Brazil: The National Petroleum Council and Petrobrás

Although Brazil, the fifth largest country in the world and the second largest in the Western Hemisphere, suffers from chronic inflation and widespread poverty, its industrial growth, which began in earnest after World War I, is perhaps the most impressive in Latin America. It possesses enormous deposits of iron ore and has its largest oil fields at Recôncavo, in Bahia, and along the Amazon River. Her leading trade partners are the United States, West Germany, and Argentina; unlike Argentine beef and wheat, Brazilian coffee is welcome in the United States.

Like Argentina, the democratic era came to an end in Brazil during 1930, when Getúlio Vargas seized power, following a coup. Vargas, whose ideal was the *Estado Novo*, served as president from 1930 to 1945 and from 1950 to 1954, when he committed suicide. In the early 1960's the brief presidencies of Jânio Quadros and João Goulart moved Brazil toward the left, until the military engineered a coup and installed a series of presidents who ruled by decree. Despite this politically repressive atmosphere, Brazil has progressed in a number of areas since the military took over in 1964, including the widely publicized development of the Amazon River Basin.

Official U.S. interest in Brazil and its oil dates as far back as the early 1890s. On January 10, 1894, Secretary of State Walter Q. Gresham wrote the American consulate in Rio de Janeiro that unless all foreign shipping suffered common restrictions, "no substantial interference with our vessels, however few, will be acquiesced in."[70] At this time, William Rockefeller, who was engaged in a worldwide competition with Russian petroleum, was seeking markets in Cuba and Brazil. Three years later Standard Oil built a crude oil refinery in Brazil, which it was forced to close after a year of operation when the Brazilian tariff was raised.

A quarter of a century later, on January 15, 1921, a new mining law went into effect, which the U.S. Chamber of Commerce praised for offering "ample protection to any one concerned in prospecting in Brazil."[71] Encouraged by the passage of this legislation, foreign oil companies entered the country but failed to discover much petroleum. While in 1928 the Brazilian Congress debated, but did not pass, a bill barring foreigners from oil exploration, the Constitution of 1937 did include such a provision; it also gave the state control over all natural wealth, and established new regulations for foreign enterprise. A decree law on April 29, 1938, then limited the exploitation of natural resources to Brazilian-born nationals.

Endeavoring to show why the Vargas regime had enacted the decree law, Commercial Attaché Walter J. Donnelly observed in a confidential memorandum on April 27 that he had learned from a Brazilian source that:

> the decree law was drafted and approved by the Federal Foreign Trade Council at the request of the Brazilian Army. . . . the Army pointed out to the Council that Brazil should nationalize the refining industry in anticipation of the development of crude petroleum resources in Brazil, and that the oil industry should be national in every respect. . . . the action was precipitated by the decision of the Standard Oil Company to build a refinery in São Paulo and by reports that the Texas Oil Company had completed plans for a plant in Pará. . . . the Standard Oil Company had proceeded with the construction of the refinery in São Paulo on the basis of a permit issued by the city of São Paulo, and . . . the company had not received or applied for a permit from the Federal Government.[72]

Brazil set up the *Conselho Nacional de Petróleo*, or National Petroleum Council, at the same time. As a result of the decree, one subsidiary of Standard Oil was forced to close its refinery, and another had to surrender various oil land leases. Even so, in the years that followed Americans continued to import and sell petroleum in Brazil, since neither transaction had been prohibited by law.

On May 12, 1940, after the start of World War II, the Brazilian government placed the entire oil industry under the National Petroleum Council, which was then two years old. Its objective was to stimulate exploration, and to impose import and price controls. Yet it was not so much the National Petroleum Council that caused friction between the United States and Brazil, as it was the contract between the Jersey Standard marketing subsidiary in Brazil and Condor, a German airline operating out of Brazil. By the terms of the contract, Jersey Standard was to supply oil to Condor, an unpopular move with the U.S. government during World War II.

Jersey Standard informed the State Department in October 1941 that there would be a law suit and fine if the contract were not honored. Although the State Department threatened to blacklist Standard's Brazilian subsidiary, the shipments continued until a substitute supplier could be found. The Truman Committee of the U.S. Senate, which investigated such matters, was displeased with the compromise, but halting the deliveries of oil entirely probably would have damaged the Brazilian economy.

By early 1942 the government of Brazil was able to inform the U.S. Ambassador, Jefferson Caffery, that Condor was now entirely Brazilian, and that the Brazilian army was requisitioning the remaining gasoline stocks of Condor, Lati, and Air France. Although President Vargas issued decrees on March 12 confiscating Axis properties in Brazil, he felt that American authorities were mistaken in their attitude toward Condor, which had discharged numerous German employees by this time. Vargas resented the dictatorial tone of the United States, since his nation, unlike Argentina and Chile, was supporting the Allied war effort.

It was during March of 1942 that the American and Brazilian governments signed the so-called Washington agreements, under which the United States would help to finance the industrial expansion of Brazil. Brazil wanted drilling rigs and a petroleum refinery, but the American government was not sympathetic to these requests. Standard Oil though, had offered on several occasions to construct a refinery at Niterói in return for concessions in the Amazon and Paraná River basins, and various modifications in Brazilian petroleum legislation. On two occasions the Brazilian government had given tentative approval to Standard's proposals, only to have the Director of the National Petroleum Council, General Júlio Caetano Horta Barbosa, block final acceptance, even after Brazil had declared war on the Axis powers.

Yet Horta Barbosa, having exercised this veto, was unable to obtain a clear government mandate for his state monopoly plan. Despite hostility toward foreign oil firms on the part of many army officers, professional men, civil servants, and journalists, there were influential people in the Brazilian government who questioned whether their country had either the capital or the technical and managerial skills to launch a state petroleum company. At this time President Vargas tended to prefer a mixed arrangement with both public and private participation.

At the beginning of the post-World War II era, there was a clash in American attitudes toward Brazil between Secretary of State James Byrnes and Ambassador to Brazil Adolf A. Berle. While Byrnes thought that Brazil should remove its restrictions on petroleum ownership and management and should halt the construction of two refineries, Berle complained that Standard Oil enjoyed a stranglehold over refining and marketing in Brazil, which Gulf then was attempting to break. In the opinion of Berle, Brazil should enter into refining as a partner of the Americans. Byrnes, however, challenged Berle's claim that Standard Oil was enjoying a monopoly.

Now that World War II was at an end and the Axis powers defeated, there was growing sentiment in Brazil for free elections. Fearful that Vargas would not allow them, the military deposed him in October; it chose as the new president one of its own: General Eurico Dutra. Under Dutra foreign capital was highly active in Brazil, yet there were also new government expenditures for railroads and highways, the expansion of Volta Redonda, the development of the Amazon, and the building of hydroelectric installations.

Under Article 153 of the new Brazilian constitution which was promulgated in September 1946, federal authorization was required before the exploration and exploitation of mineral resources could take place. Although the Chargé in Brazil, Paul C. Daniels, reported to Secretary of State Byrnes on October 8 that Brazil would have to pass a new oil law and cancel various presidential decrees before foreign oil companies could operate freely, he pointed out that all the representatives of the petroleum firms in Rio de Janeiro were pleased with the new constitution. They believed it had "opened the door" to foreign capital entering Brazil in the future.

But in February of 1947 President Dutra set up a commission under the auspices of the National Petroleum Council to investigate the oil situation. By the end of the year the commission had prepared a new petroleum bill that gave first priority to supplying the home market; the National Petroleum Council was to receive funds with which to stimulate exploration, production, and refining. Foreign oil firms were to be allowed to participate in the program on a concessionary basis.

The new oil bill did not please the foreign petroleum companies, which had hoped for legislation similar to that currently operative in Venezuela. Such prominent oil consultants as Herbert Hoover, Jr. and Arthur A. Curtice advised the Brazilian government that foreign petroleum firms would be unlikely to enter the country under these conditions. According to Hoover, the oil bill manifested an ignorance of the petroleum industry, a sentiment shared by the retiring president of Standard of Brazil, Wingate M. Anderson.

Just before Christmas 1947, under pressure from President Dutra, the president of the National Petroleum Council approached the American embassy with a request to purchase four fuel tankers. He also asked the United States to guarantee Brazil's minimum requirements for oil products during the first six months of 1948. The reaction of Secretary of State George Marshall was not encouraging; the supply of government-owned surplus war tankers for sale abroad had been exhausted, and there were petroleum shortages at that time in different parts of the United States. On January 6, 1948, Marshall wrote to the Brazilian embassy, "For your information it may be difficult to justify special measures to aid Brazil in present emergency unless it modifies present development criteria which will effectively prevent any real utilization [of] its own resources in [the] future."[73] It is perhaps not surprising that the controversial oil bill then before the Brazilian Congress was not adopted.

A year later, the Standard Oil Company of Brazil spent $140,000 on a series of advertisements in thirty Brazilian newspapers, pointing out that Brazil was importing 98 percent of the petroleum it consumed. It even flew a dozen Brazilian journalists to the United States to inspect its processing facilities. Yet their campaign did not dislodge the National Petroleum Council from its dominant position over Brazilian oil. In fact, Getúlio Vargas used the slogan, *O Petróleo e Nosso*, "The Petroleum is Ours," in his drive to return to power.

In 1950 Vargas won the presidency in a free national election, but he was no more capable of leading Brazil out of its economic quagmire than Dutra had been. Although he appointed Oswaldo Aranha as Minister of Finance and obtained a $300 million credit from the Export-Import Bank, Brazil's economy continued to suffer. Corruption multiplied at every level of government, which eventually contributed to Vargas' downfall. To make matters even worse, Brazilian relations with the United States began to disintegrate.

Vargas' second presidency, though, did witness a major achievement: the creation of the Brazilian Petroleum Corporation, or Petrobrás, in October 1953. Petrobrás was granted a monopoly over petroleum exploitation; foreigners—or Brazilians married to foreigners—could not obtain stock in the new company. But while national self-sufficiency in oil was a meritorious goal, if only because Brazil was spending $200 million each year in foreign exchange for petroleum products, the country lacked the hundreds of millions of dollars in venture capital needed to explore and drill.

A year later an unsuccessful assassination attempt on Carlos Lacerda, editor of the *Tribuna da Imprensa*, triggered the series of events leading to Vargas' suicide on August 24, 1954. In a signed note he proclaimed: "To the wrath of my enemies, I leave the legacy of my death"; an unsigned and perhaps spurious note attacked "international economic and financial groups."[74] Vice-president João Café Filho assumed the presidency until the presidential election in October 1955.

The victor in this contest was Juscelino Kubitschek of Minas Gerais, who employed the slogan "Power, Transportation, and Food." As chief executive, Kubitschek tried to reestablish friendly relations with the United States against the wishes of the nationalists and Communists. Unfortunately, inflation continued to undermine the Brazilian economy, and in 1959 the American government stopped extending credit to the country.

Let us now return to the early part of the decade of the 1950s, and trace petroleum developments in more detail. In a surprise move, Petrobrás hired as the chief of its exploration department Walter K. Link, who had been the head geologist for Jersey Standard between 1947 and 1953; Link enjoyed a lucrative contract with Petrobrás, which increased its daily production by almost 20 percent in three years. Even so, in 1955 Link was the subject of an investigation by the Brazilian Congress, in which it was suggested that he was attempting to undermine the petroleum development of the Amazon Basin.

In 1957 Petrobrás accepted a Texas Company offer of 15,000 barrels of crude oil daily, and a $6 million loan to help finance the construction of a petroleum refinery. It was hoped that these would help the country to balance its refining and consumption. Nevertheless, political agitation continued. In 1959 the new president of Petrobrás was interrogated by a Chamber of Deputies committee that was highly critical of a deal with the Esso Export Corporation, in which Brazilian high-paraffin crude oil (which the Brazilian refineries could not handle) was exchanged for Venezuelan crude oil.

When he resigned as the head of Petrobrás' exploration department in 1960, Walter Link drew up a report on the future of Brazilian petroleum. In his report Link portrayed the future of Brazilian oil in gloomy terms, even though Petrobrás had spent approximately $300 million over a six-year period on petroleum exploration alone. After his report had been leaked to the press, and his critics had charged that he was a saboteur in the pay of Standard Oil, the president of Petrobrás also attacked him, coining the term "Linkismo" to describe the "defeatist tendency, supported by the shirkers, the weaklings, and the cowards"[75] within Petrobrás. This was in reference to the fourteen Brazilian and foreign geologists whom Link had hired, and who concurred in his negative findings. Many oil firms in the United States, however, expressed relief that Brazilian nationalists had kept them out of what were apparently risky business ventures there.

Not everyone in America was a foe of Petrobrás. During 1960 Republican Senator George Aiken of Vermont had undertaken a fact-finding mission to Latin America. Aiken concluded that the U.S. government could stimulate Brazilian development by aiding Petrobrás, and stated, "The refusal to deal with Petrobrás simply because it is a government monopoly seems altogether too doctrinaire."[76] Democratic Senator Wayne Morse of Oregon expressed similar sentiments.

In October 1960 Jânio Quadros was elected president of Brazil with the largest plurality in Brazilian history, but he lasted only seven months as chief executive from January to August 1961. When Quadros attempted to institute economic reforms, the Brazilian Congress was not cooperative. While he irritated the United States by appearing too friendly toward Communist China and Castro's Cuba, the American government looked with greater approval on his domestic policies, and agreed to assume the lion's share of a $2 billion package of loans and grants. Unable to break the stalemate with Congress, on August 25 Quadros resigned.

His successor, Vice-President João "Jango" Goulart, had served as minister of labor under Vargas and also had flirted with the Communists. Unlike Quadros, his economic policies alienated the United States. In February of 1962 the governor of Rio Grande do Sul nationalized the International Telephone and Telegraph holdings there with inadequate compensation; in October the Brazilian Congress decreed that foreign companies could not remit more than 10 percent of their profits. But when Goulart announced a new program of economic reforms in January 1963, the United States, fearing the Communist threat, agreed conditionally to extend $398.5 million in credits to the country.

Under Quadros consumer subsidies on products derived from oil had ended in 1961. On Christmas Eve in 1963, while Goulart was chief executive, Petrobrás obtained a monopoly over the importing of crude oil; this was a severe blow to such petroleum importers as Gulf, Esso, and Shell. In March of 1964 the Brazilian government moved to nationalize the few private refineries that remained.

Assessing the history of Petrobrás between 1953 and 1964, Peter Seaborn Smith concluded that the politics of radical nationalism hampered the operations of the state enterprise. Nevertheless, Petrobrás had a major impact on Brazilian industrialization, as it stimulated a growing volume of manufacturing. The mere scope of Petrobrás' operations was impressive, moreover, even though it may have wasted a considerable amount of money.

By April 1964 João Goulart had so alienated the army that it staged a coup.

The Chamber of Deputies elected General Humberto Castello Branco as President; Castello Branco quickly established a military dictatorship that featured a witch-hunt for "subversives," "Communists," and "grafters," as well as the abolition of existing political parties. Although he nationalized the holdings of the foreign utility companies, he eased the regulations concerning the remittance of profits by foreign firms, and even allowed foreign mining companies to receive concessions. Castello Branco's anti-Castro stance and his dispatching of troops to the Dominican Republic so pleased the American government that Brazil received approximately $1 billion in foreign loans and credits, much of it from the United States.

Following the military takeover the profitability of Petrobrás increased significantly. The Castello Branco regime fired some of those Petrobrás employees who were either labor leaders or advocates of nationalization. In 1965 private refineries were reinstated under a July decree law; private investment was then encouraged in the petrochemical sector and in the development of oil shale reserves. By 1966 Brazil had achieved greater self-sufficiency in refined products than in production.

General Artur da Costa e Silva, who took office under a new Constitution only to suspend it on December 13, 1968, continued military rule in Brazil. When Costa e Silva died a year later, General Emilío Garrastazú Médici assumed the presidency, ruling by decree. The next in this line of military chief executives was General Ernesto Geisel, who once headed Petrobrás and who became president in 1974.

As of 1971 Petrobrás was the largest enterprise in Brazil. It entered into private and public oil exploration agreements with such countries as Iraq, Iran, Egypt, Libya, Algeria, the Malagasy Republic, and Colombia. Yet Petrobrás did not make its first major strike until 1974, when oil was discovered in the offshore Campos field; even then Brazilian domestic production was 175,000 barrels daily, while its consumption was 800,000. Confronted with numerous problems, Petrobrás nevertheless was proud of its many achievements. As John D. Wirth has noted, "In fact, Petrobrás took on traits associated with the majors: it advertised extensively in the press, it had a large publication and information department, and it was a strong political power."[77]

## Costa Rica and Cuba

The American government has taken an interest in petroleum relations with several other Latin American nations. In the case of these countries, however, there have been isolated episodes rather than a continuous concern on the part of the United States. At least two of these merit examination: the controversy over the Amory concession during the post-World War I era in Costa Rica, and the confiscation of foreign petroleum holdings in Cuba by the Castro regime in 1960.

At the time of World War I, foreign oil enterpreneurs were interested in the petroleum possibilities of Costa Rica. The Pearson interests had sought a concession there in 1913, only to have the Costa Rican Congress reject their application. They then tried to acquire various petroleum rights from a local group but lost them to a Dr. Gruelich of New York, who obtained the exclusive right to explore for oil in three provinces. Here, as in Colombia, the State Department took an intense interest in the activities of the British oil men; it told Minister E. J. Hale that the Costa Rican oil concession was of "unusual interest because of its relation to naval bases and the proximity of Costa Rica to the Panama Canal."[78] Even so, the State Department showed little confidence in Gruelich whose American citizenship was dubious. In addition, at this time President Wilson opposed monopolistic concessions. Even though he had signed the initial understanding, President Cletus González Viques of Costa Rica also turned against Gruelich, after listening to the British Minister to Costa Rica. In November 1916, while the State Department (which now suspected that Gruelich might have German ties) delayed taking action, the Congress of Costa Rica approved Gruelich's concession, only to have González veto it. A month later a subsidiary of Sinclair purchased the concession.

Costa Rica now experienced a revolution. The former Minister of War, Federico Tinoco, who had been educated in the United States and Europe, seized power. As chief executive, he ruled in an arbitrary fashion, placed restrictions on the press, and quickly incurred the ill-will of the people. But once established he recognized the Sinclair concession as valid, and in 1918 he granted a concession to Amory and Sons, a New York City based firm with British backing.

When it became known in the United States that Lord Cowdray and the Pearson interests were supporting the Amory concession, an uproar ensued. Among the leading congressional critics of the Amory interests was Republican Senator William E. Borah of Idaho, who proposed that Congress adopt a resolution demanding an inquiry into the circumstances surrounding the British attempt to obtain a share of Costa Rican petroleum. Borah's resolution carried, which indicates that he was not alone in his concern. In November 1918, during the off-year congressional elections, the Republicans made strong gains that resulted in a mounting chorus of GOP opposition to the Wilson administration's policy toward British investors, oil, and Costa Rica.

In the latter part of 1919 the Costa Rican public hostility toward the Tinoco regime approached revolutionary fervor. After mounted police had refused to disperse a mob of 3,000 to 4,000 rioting school teachers and children, Tinoco's brother was assassinated, and the besieged president fled into exile. The new government held elections and declared all acts of the Tinoco regime void. Approximately one year after the departure of Tinoco, in August 1920, the British government notified the United States that the Amory concession was in fact a British holding. According to the American Chargé in Costa Rica, Thurston, the president of the Costa Rican Congress told him that the concession "was obtained under American incorporation by direction of British backers and that tenor of British Minister's notes indicates this procedure to have been adopted as a wartime policy by Great Britain to avoid attracting American attention."[79]

By February of 1921 Great Britain and Costa Rica were on a collision course. It was reported that Great Britain, which was the principal market for Costa Rican coffee, had threatened Costa Rica. Refusing to be intimidated, the Costa Rican Congress cancelled the Amory concession on March 7 by a vote of twenty-four to ten. The American government's position remained unchanged; on February 26 Secretary of State Bainbridge Colby informed the British government that Tinoco had acted without proper authority when he made various concessions to foreign interests.

But within a week, there was to be a change of Presidential administration in the United States when Warren G. Harding replaced Woodrow Wilson. Arthur Millspaugh, the State Department's oil expert, then began to use his influence to persuade governmental officials in Washington to be more tolerant of the attempts by foreign nations to obtain petroleum concessions throughout Latin America, provided that they were not exclusive and did not violate the Monroe Doctrine. Millspaugh was opposed by Stewart Johnson and Sumner Welles of the State Department. In August Secretary of State Charles Evans Hughes informed the British Ambassador that although official support of the British position was not possible, he would not oppose arbitration of the Amory concession, thus rejecting the State Department's prior insistence on a settlement in the Costa Rican courts. Hughes then went on to advise American diplomatic officials in Latin America to respect the rights of all foreign nationals.

As the year progressed, relations between Great Britain and Costa Rica continued to deteriorate. In November the British government delivered a note that amounted to an ultimatum to the Costa Rican Minister for Foreign Affairs, and then sent a warship to Puerto Limón in a display of force. After two months of heightened tension, Great Britain and Costa Rica agreed in mid-January of 1922 to allow Chief Justice William Howard Taft of the U.S. Supreme Court to arbitrate the Amory concession.

During 1921 Costa Rica also had a boundary dispute with Panama, which it won with a display of military force. Chief Justice Edward White earlier had handed down a ruling that awarded the disputed territory to Costa Rica, which was authorized to occupy the area by 1921. After a border skirmish, the Costa Rican armed forces pushed into Panama without encountering opposition. The American government then persuaded Costa Rica to withdraw its troops, in return for which the United States would officially endorse the White ruling.

Despite the furor over the Amory concession, not every oil entrepreneur wanted to prospect for oil in Costa Rica. During 1922 the Costa Rican government asked George Moreno to approach Standard Oil with a liberal petroleum concession. Standard rejected the offer because a group of its experts had unfavorably reported on the oil prospects of Costa Rica six years earlier. Furthermore, the oil conflict in Costa Rica was not invariably between American and British interests. During the early 1920s, for example, there was friction between Sinclair, California Standard, and the United Fruit Company. In analyzing the squabble over the control of Costa Rican petroleum, it is interesting that Woodrow Wilson believed Minor Keith and the United Fruit Company were behind Tinoco.

On October 18, 1923, Chief Justice Taft handed down his long-awaited ruling on the Amory concession. Taft declared that the recognition or nonrecognition of the Tinoco regime by a foreign government was not a consequential factor in assessing the validity of the actions taken by Tinoco as President. Having made this point, he went on to say that the concession was invalid because the concessionaires had failed to fulfill their contractual obligations. Since the British government did not challenge the Taft ruling, Amory and Sons were removed from the Costa Rican petroleum scene, and Doheny and Sinclair moved in to fill the vacuum.

Let us now shift our attention from the years 1913-23 to the World War II era. Although the United States did not become an active participant in World War II immediately, under the Act of Havana (1940) the American government and the Latin American republics agreed to protect each other, should they be attacked from outside the Western Hemisphere. Latin America, though, proved to be more valuable to the United States as a furnisher of raw materials than as a supplier of troops. Only Brazil sent troops to Europe, but Latin America as a whole made available such items as tin, kapok, quinine, balsa wood, quartz, copper, sugar, vanadium, flax, mercury, tungsten, mica, and rubber. In return, the United States doubled its sale of civilian goods to Latin America between 1939 and 1944 and supplied a large number of loans to the republics there.

Of the West Indies islands, Cuba in particular sought additional petroleum from the United States during World War II. Although the Cuban government maintained that it would be used for essential agricultural purposes, the American government claimed that it was being diverted to other, less essential uses. In addition, Ambassador to Cuba Spruille Braden complained to Secretary of State Hull on May 27, 1944, that Cuba was obtaining gas oil from Mexican sources that were outside the regular inter-American pool. At this time, however, Cuba had one of the most pro-American heads of state in Latin America, Fulgencio Batista, and Batista had much-needed nickel shipped to the United States. In 1944 Batista retired to Florida after eleven years of dictatorial rule; once in Florida he established a multimillion dollar financial empire and bided his time until returning to power as a result of a coup in March 1952.

During Batista's second presidency (1952-59), the Cuban economy boomed, but at the expense of growing political repression. Although sugar was its dominant export, it derived some additional revenue from mineral products, but as of 1954 only thirty-three petroleum wells had been drilled. It was during 1954 that the Cuban government opened the country to foreign exploration capital by passing a law protecting the latter for the next two decades. The result was an influx of oil companies from abroad, including the newly chartered Cuban-Colombian Petroleum Company. Among the officers of this firm, which received a loan of $150,000 from the government of Cuba for drilling, were Board Chairman Joseph W. Frazer, formerly of Kaiser-Frazer, and Director John Roosevelt, the youngest of FDR's four sons. Cuban-Colombian agreed to drill six petroleum wells in the central Jatibonico Basin, where wildcatters earlier had located oil. In September 1955 Indiana Standard acquired exploration rights covering 13 million acres of Cuba. Unfortunately, there were no major strikes as a result of this new activity.

Opposition to Batista had erupted as early as 1953, when 165 youths attacked the Moncada army barracks. During 1957 and 1958 Fidel Castro successfully built up a revolutionary force in the Sierra Maestra jungle, and such powerful groups as the Roman Catholic bishops openly criticized the Batista government. While the United States embargoed shipments of arms to Batista, Castro was able to maintain an illegal flow of material to his revolutionary band. On January 1, 1959, Batista was forced to flee Cuba, thus ushering in a new era under Castro, which was to witness the eventual implementation of Marxist ideas and a growing deterioration in Cuban-American relations.

During his first year in office, Castro broke up the large estates, including the United Fruit Company's 270,000 acres, under the Agrarian Reform Law. He also nationalized the banks and seized the mining concessions belonging to American entrepreneurs. Other actions taken by the Castro regime affected oil producers, refiners, and marketers. In the words of *The Magazine of Wall Street*, the new petroleum law, which was approved late in 1959,

> a)...imposes a 60 per cent royalty on output; b)...established refinery production quotas; c)...limits exploration concessions to about 20,000 acres, far below the acreages held by many foreign firms; d)...cancels all requests for new exploration and exploitation rights; e)...provides for the loss of concessions if firms cannot prove that drilling is in progress; and f)...established the Cuban Petroleum Institute with authority to carry out all types of petroleum activities and—most significant of all—to which private concessionaires must turn over all operational and geological information, generally considered the best guarded secret of an oil company.[80]

Since foreign petroleum companies, which were primarily American, had been exploring for oil in Cuba for many years without much success, it was not difficult for Castro to use their failure as a propaganda weapon against the American firms. At the time of Batista's overthrow, the foreign petroleum companies had invested over $100 million in the island. Admittedly Jersey Standard, the Texas Company, and Royal Dutch-Shell, assisted by a much smaller Cuban-owned refinery, had refined oil in Cuba efficiently, but the revolutionaries still could challenge the prices they had charged and the impact they had had on the balance of payments.

It was only a matter of time until the Castro regime found some pretext to seize and nationalize foreign petroleum holdings, and by the following year (1960) it had an excuse: the reluctance of the three companies to refine crude oil shipped to Havana from the Soviet Union under a new trade agreement. Secretary of the Treasury Robert Anderson in particular encouraged the American petroleum companies to adopt a hard line against the Cuban request. There was, though, some other justification for the oil firms' stand, since the Cubans were $60 million in arrears in foreign exchange payments, and the refineries had been built specifically to process Venezuelan petroleum.

When the American Ambassador to Cuba, Philip W. Bonsal, delivered a note to the Castro regime on July 5 protesting the seizure, he noted that neither under the 1954 oil refining law nor the subsequent regulations were there any requirements for foreign refineries to process government crude oil. In fact, it was understood that the foreign companies would have the prerogative of supplying and refining their own crude oil. The U.S. Congress then authorized President Dwight Eisenhower to cancel the $150 million bonus America had been paying to Cuba for sugar imports, thus dealing a severe blow to the Cuban economy.

Although the U.S. government did protest the illegal seizure of the oil refineries by the Castro regime, Deputy Under Secretary of State Raymond A. Hare wrote a letter on August 16 to Chairman Clarence B. Randall of the Council on Foreign Economic Policy. In this he pointed out that the Eisenhower administration had not forced the U.S. oil companies to challenge the Cuban takeover, nor had it told them not to refine Soviet crude oil. In the case of Cuba, he wrote, the American companies "appear to have made their own decision on the basis of the information available to them and in the light of their own interests."[81] Yet Peter Odell, who was then a member of the Shell planning group, has written the author that the State Department called in the petroleum firms and pressured them into resisting the Cuban demands.

In the aftermath of the seizure Standard Oil of New Jersey had its Cuban affiliate, Esso Standard Oil S. A. (Essosa), release a document entitled "Esso in Cuba: 1882 to Expropriation" that set forth its side of the story. Among other things, the Essosa report pointed out that oil marketing in Cuba was highly competitive, that its average annual earnings in recent years had been only 4 percent of its total investment, and that the typical wage paid to its Cuban employees far exceeded the legal minimum monthly wage. (Essosa employed approximately 1,000 Cubans, 370 of whom had been with the firm for over twenty years.) In addition, according to the report, Jersey Standard's Cuban affiliate had been operating for seventy-eight years, so it was hardly a product of the Batista era. Castro nevertheless held firmly to his decision to continue confiscating "gringo" assets "until not even the nails of their shoes are left,"[82] even though expropriation made Cuba almost totally dependent on the Soviet Union for crude oil.

By the end of 1960 the Cuban government had nationalized approximately $1 billion in American-owned properties, leaving very little in U.S. hands. Just before he left the White House, President Eisenhower ordered the severing of diplomatic relations with Cuba. Conditions worsened even further during the Kennedy administration after the Bay of Pigs fiasco of April 1961 and the missile crisis of October 1962, but Cuban-American relations then entered a less volatile but still mutually hostile era. As for the present, due to massive Russian assistance, Castro remains in power, having held office now even longer than Batista did. Cuban-American diplomatic relations still continue broken, and private U.S. economic activity is a thing of the past.

# 5. The Middle East and North Africa

## The Middle East Since 1945: An Overview

Even as late as 1945, according to Thomas A. Bryson, the United States had not formulated a clear-cut, well-defined policy for the Middle East. As British influence declined there after World War II, the influence of the Soviet Union increased. As a result, the American government was forced to develop an overall policy toward the Middle East. In this connection Bryson also has observed that: "No longer would the missionary-philanthropic element dominate U.S. policy in the region, but oil interests and American Zionists would assume an increasingly larger role in the making of policy."[1]

While Israel does not challenge the right of the Arab countries to their oil, many Arabs believe that Israel should not even exist as a nation. Although the British through such pronouncements as the Balfour Declaration of 1917 had been the most enthusiastic supporters of a Jewish national state, their withdrawal from the Eastern Mediterranean following World War II necessitated the Truman Doctrine of 1947, which established a protective U.S. military umbrella over Greece and Turkey to counter a possible Communist takeover. During the last generation, moreover, the United States has so supported Israel that the Arab World has come to regard it as Israel's leading advocate; unfortunately this has had a highly adverse impact on Arab-American relations, even at times jeopardizing U.S. access to Arab petroleum.

During the immediate post-World War II years the proposed Trans-Arabian oil pipeline, or Tapline, was the focus of controversy. Built with American capital, the pipeline was scheduled to extend from the Persian Gulf area to the Eastern Mediterranean, a distance of 1,000 miles. Both Transjordan and Lebanon had approved its construction in 1946; during the following year Syria also gave its permission, but only after King Ibn Saud of Saudi Arabia had urged them to do so on behalf of the U.S. interests. In 1947, too, the U.S. Department of Commerce granted the American firms involved with the pipeline the right to ship 20,000 tons of pipe to Saudi Arabia. The favorable attitude of the Commerce Department contrasted sharply with the more critical stance of the Senate Small Business Committee, the Defense Department, and the Independent Petroleum Association.

At this point the General Assembly of the United Nations approved a proposal, supported by the United States, to partition Palestine between the Jews and the Arabs. As a result, Syria withdrew its support for the pipeline scheme in December; on February 22, 1948, the Arab League stated that no American pipeline would cross Arab lands, as long as the United States supported partition. Following a coup, however, Syria reversed its stand and signed the Tapline agreement in May 1949, after Israel had declared its independence and fought the first of a series of wars with Egypt and other Arab states. By September 25, 1950, the pipeline was complete, and on November 10 for the first time petroleum was moving through the pipeline.

The construction of Tapline occurred even as the controversy over Palestinian partition and Israeli independence continued to smoulder. At the beginning of 1948 Warren Austin, the American representative on the Security Council of the United Nations, had begun to shift his position on the partition of Palestine. At the end of March he proposed that the Security Council lay the partition plan aside, and ask the General Assembly to establish a temporary Palestinian trusteeship. Not unexpectedly, the American government was severely attacked for its change in position, both in the United Nations and in the United States itself; critics maintained that the oil companies had unduly influenced this policy switch. Secretary of State George Marshall claimed that an unexpected degree of opposition to partition had developed, but his explanation ignored the fact that the Arabs had never supported partition in the first place. Even more important, though, was the strength of pro-Jewish public opinion in the United States. On March 14, 1948, the Truman administration recognized Israel as a sovereign state, just minutes after it had declared itself an independent nation.

Throughout the debate over the partition of Palestine, U.S. oil firms had hoped the Arab states would distinguish between their position and that of the American government. On December 22, 1947, James Terry Duce, the Vice-president of Aramco, submitted a secret report to W. F. Moore, Aramco's President, recommending personal consultations with Arab rulers and political leaders to make known their opposition to the partition of Palestine. He also suggested that Aramco report the views of the Arab nations to the State Department.

So great was the hostility of the Arab states toward the new nation of Israel, that once it had become independent the armies of the Arab League (Egypt, Syria, Lebanon, Transjordan, Iraq, Saudi Arabia, and Yemen) went to war against it in 1948. During that summer, after the Israelis had repelled the Arab offensive, the United Nations sponsored a truce. Although Israel did enter into armistice settlements with the different Arab nations in 1949, there has been a cold war between them since then, except on those occasions in 1956, 1967, and 1973 when the cold war escalated into open conflict. In the aftermath of the first Arab-Israeli war, the Arab nations did agree that it would be in their own interest to accept Tapline but did so only on the condition that at no point it crossed Israeli territory.

Sensitive to the charge that the State Department had become a puppet of U.S. petroleum firms, Secretary of State James Byrnes wrote the American Chargé in the United Kingdom, Gallman, on March 16, 1946, that the "U.S. Govt. further feels that existing and future pipeline and refining concessions in Near and Middle East countries should reflect full recognition of principle that countries which contribute in any way to development and commercialization of petroleum resources should receive fair and reasonable compensation for such contribution."[2] As it expressed its growing concern for the host nation, the State Department began to offer encouragement to independent American petroleum firms. In August 1947 it supported the formation of the $100 million American Independent Oil Company, which was to engage in large-scale, long-term foreign petroleum operations, especially in the Middle East. The President of the new firm was Ralph Davies, once the senior Vice-President of California Standard and later Deputy Petroleum Administrator.

It was during this year that Jersey Standard and Socony-Vacuum reached an agreement with California Standard and the Texas Company. Under the terms of the agreement Jersey Standard and Socony-Vacuum would be allowed to participate in the Arabian-American Oil Company (Aramco), a step that would further strengthen the monopolistic position of the major U.S. oil firms in Saudi Arabia. The agreement was mutually satisfactory: Socal and Texaco needed more capital, and Jersey and Socony feared that their competitors might undercut their position in the world market with cheap Arabian oil.

Before Socony-Vacuum and Jersey Standard could buy into Aramco, however, the four firms had to maneuver around the Red Line Agreement of 1928. This restrictive covenant embraced all of the Middle Eastern oil-producing nations except Iran and Kuwait. The understanding already had been challenged under British law, since neither the Compagnie Française des Pétroles, nor C. S. Gulbenkian (who had been caught in France when it fell to Hitler), had been able to claim during World War II the petroleum to which they were entitled under the Red Line Agreement. After Jersey Standard and Socony-Vacuum made known their desire to buy into Aramco, the Compagnie Française des Pétroles with the support of the French government filed suit to enforce the understanding, and Gulbenkian too took an obstructionist position. It was not until November 1948 that the companies reached an out-of-court understanding, which terminated the Red Line Agreement. Jersey Standard was then able to complete its purchase of a 30 percent interest—and Socony, a 10 percent interest—in Aramco.

Summarizing developments with respect to Middle Eastern petroleum concessions during the first half of the Twentieth Century, Henry Cattan believes the most significant features of these to be:

(1) the large areas and long periods of oil concessions;
(2) the small number of operating companies;
(3) the relative uniformity and simplicity of concession terms;
(4) the royalty concept which constituted the principal financial basis of oil concessions;

(5) the comparative moderateness of the financial terms of concessions which was a consequence of the lower value of, and lesser demand for, crude oil during that early period;
(6) the slow evolution in the terms and conditions of concessionary agreements which was tantamount to stagnation and rigidity.[3]

On March 3, 1952, *Time* magazine noted that there were 900 oil wells operating in the Middle East, most of which were no more than ten to fifteen years old. Over the previous fourteen years the area's petroleum production had increased 650 times; whereas the average Venezuelan well produced only 200 barrels a day, the typical Middle Eastern well spouted forth 3,700 barrels. But, as *Time* pointed out, the social consequences of the petroleum boom were by no means uniformly desirable:

> Then came the oil companies, bringing industrialization and cities, and breaking down feudal tribal relationships. The outpouring of Western wealth (at the rate of $200 million a year) is destroying old values before new ones can be substituted. Oil money expanded the middle classes; it educated them, and created a class hostile to economic domination by foreigners. Oil money increased the gap between rich and poor, and gave the poor something to covet.[4]

Aside from the social ramifications of their presence, the U.S. petroleum firms had to face the fact that Middle Eastern governments were beginning to demand a larger share of the oil "pie." It will be recalled that the 50-50 formula had first gone into effect in Venezuela during the 1940's; during the 1950's Saudi Arabia, Kuwait, Iraq, Turkey, Bahrain and Qatar also began to implement it. Middle Eastern countries also set up a number of state petroleum corporations, beginning with the creation of the National Iranian Oil Company in 1951. Subsequent decades saw the establishment of other state petroleum corporations in Egypt, Kuwait, Saudi Arabia, Algeria, Iraq, and Libya.

Even more drastic than the imposition of the 50-50 formula and the creation of state oil companies was expropriation. In 1951 Anglo-Iranian was expropriated by the Iranian Mossadegh regime, triggering an international *cause célèbre*, which was only temporarily resolved by the consortium agreement of 1954. Then in the summer of 1956 after the U.S. government had withdrawn its offer of financial assistance for the construction of the Aswan Dam, Gamal Abdel Nasser seized the Suez Canal. While in neither case were the takeovers specifically directed against the United States, they obviously created a dangerous precedent for the future. Nor was the establishment of a leftist regime in Iraq following a 1958 coup a hopeful sign for Americans. Only the disunity of the pre-OPEC Arab world prevented these countries from using oil as an effective weapon against Western interests.

Although the United States was concerned about the fate of the Middle East, it was Great Britain that signed the Baghdad Pact of 1955 with Turkey, Iraq, Iran, and Pakistan. Secretary of State John Foster Dulles was opposed to his nation's ratifying it. The American government, moreover, did not support the British government when Great Britain and France joined the Israeli attack on Egypt shortly before the 1956 U.S. presidential election. Two years later, though, in January 1957, President Dwight Eisenhower asked authority from Congress to provide American economic and military assistance to any Middle Eastern nation that was threatened by the Communists, and which requested aid. In July 1958 the Eisenhower administration sent U.S. marines into Lebanon in an attempt to forestall a repetition of the Iraqi coup that had occurred earlier that month. In addition, since 1947 the American government had extended a considerable amount of foreign aid to such Middle Eastern nations as Turkey, Iran, and Saudi Arabia, thus intensifying U.S. involvement there.

Late in 1958 Secretary of State John Foster Dulles was interviewed for the Independent Television Network of the United Kingdom by William D. Clark. After Clark had asked Dulles about Eisenhower's plan for the economic development of the Middle East with U.S. assistance, Dulles raised the question of protecting the British and American petroleum interests there:

> . . . the important thing to recognize, I think, when we talk about this matter is that we are providing a market which provides the resources which can tremendously help the welfare of these countries. And there is a growing development there of plans to use oil royalties and so forth for economic welfare, and that it is not just a development of something that is of interest to the West. . . . A pool of oil is about the most worthless thing there is in the world unless you have the machinery for marketing it. And we provide that, and that is a joint enterprise between the West and the Arab countries.[5]

Despite Dulles' stated concern for the welfare of the Middle East, the region was becoming increasingly independent in its thinking. In September 1960, at a conference held in Baghdad under the leadership of Saudi oil minister Sheihk Abdullah Tariki, Saudi Arabia, Iran, Iraq, Kuwait, and Venezuela decided to establish the Organization of Petroleum Exporting Countries, or OPEC. Although OPEC usually is considered an Arab organization, Venezuela was a charter member, and in more recent years such non-Arab countries as Ecuador, Nigeria, and Indonesia have joined it.

At first, OPEC operated in a rather cautious manner, so that its actions did not generate as much concern as did those of such member states as Iraq, which seized 99.5 percent of the petroleum concession territory within its borders in 1961. As a result, it was not until the third Arab-Israeli war in 1967, and the Algerian expropriations of Jersey Standard, Mobil, Sinclair, Phillips, and Tidewater in that same year, that the basic fragility of the Middle Eastern petroleum supply became apparent to a growing number of Americans.

Throughout the period, however, American and other foreign petroleum activity in the Middle East continued to disturb the Soviet Union. Writing in the Moscow publication *International Affairs*, the Marxist author Victor Perlo complained that the "neo-colonialist" oil companies were mercilessly exploiting the region. "Thus," observed Perlo, "of every dollar's worth of petroleum products, only ten cents go to the Governments of the producing countries, and only one cent to the workers extracting the oil."[6] Perlo then went on to claim that it was the workers (and other "patriotic" elements) in such nations as Libya, Saudi Arabia, and Kuwait who had forced their governments temporarily to embargo shipments of petroleum to Great Britain and the United States.

Although the Arabs did try to stop petroleum exports to the West, the third Arab-Israeli war of 1967 failed to bring the oil-hungry Western world to its knees. After noting the diversion of American, Venezuelan, and Indonesian crude oil production to Europe, Naiem A. Sherbiny and Mark A. Tessler identified four conditions present in the Arab world itself that precluded a truly effective oil embargo at that time:

(a) lack of agreement among the Arab oil producers on the objectives of the embargo;
(b) the absence of preplanning the mechanics of the embargo;
(c) the loss of oil revenues to the Arabs;
(d) the desire of the boycotting countries to strengthen their economies and the economies of those Arab countries that had suffered from the war.[7]

A revolution in Libya two years later, though, had widespread consequences, not only for the Arab world but also for the oil-consuming nations of the West. During that year a group of radical young military officers led by Colonel Muammar el-Qaddafi overthrew King Idris and established a socialist republic. The movement toward the expropriation of foreign petroleum holdings then proceeded even more rapidly than it had in Iraq. In 1972 a major change occurred in the relations between the American petroleum firms and such Persian Gulf oil states as Saudi Arabia, Kuwait, Iraq, Abu Dhabi, and Qatar. After nine months of negotiations, the petroleum companies agreed to extend to these nations a 20 percent ownership in the oil operations within their borders; this percentage eventually was to rise to 51 percent. The price to be paid was to be a compromise between the book value of the petroleum installations—as the Arabs had suggested—and the "true market price" of them—as the oil firms had proposed.

Armed with additional funds from their petroleum bonanza, the Arab nations now faced the dilemma of what to do with their new income. Among the possibilities was to invest it abroad, even in the United States, in refining and distributing facilities. President Richard Nixon recognized this possibility early in 1973, when he sent his first international economic report to Congress in which he noted, "with huge and flexible capital markets open to foreign investors, [America] is likely to receive a large share of these funds."[8] Significantly, the first such investment was made by Iran rather than Saudi Arabia.

With the Arab oil embargo almost at hand, in September of 1973 Secretary of State Henry Kissinger—in one of his less perceptive judgments—stated that the United States had excellent relationships with such Middle Eastern countries as Saudi Arabia and Iran, and that there were no foreseeable circumstances in which they would cut off petroleum exports to America. A number of governmental officials at this time also misread the signs from the Middle East. In the crisis that followed the outbreak of the fourth Arab-Israeli war in October, the United States was deprived of petroleum by such tiny oil-producing nations as Kuwait as well as by its longtime friend Saudi Arabia, although Iran did continue to supply oil. Democratic Representative Lee H. Hamilton of Indiana lamented: "never before in the history of mankind have so many wealthy, industrialized, militarily powerful and large states been at the potential mercy of small, independent and potentially unstable states which will provide for the foreseeable future, the fuel of advanced societies."[9] The petroleum embargo came to a halt eventually, but the movement toward expropriation proceeded in Iran, Iraq, Saudi Arabia, Kuwait, Libya, and Algeria.

## The Ottoman Empire and the Chester Concession

As an ally of Germany and one of the Central Powers during World War I, Turkey suffered the loss of much of its Asiatic territory in the peace settlement. In the revolution that followed in 1923 General Kemal Ataturk seized power, inaugurated various reforms, and laid the foundations for present-day Turkey.

Unfortunately, Turkey bequeathed to the post-World War I Turkish republic a legacy of economic instability and insolvency.The absence of industry so militated against a balanced economy that only in recent years has there been a systematic attempt to exploit the mineral resources of the nation. Its petroleum potential,moreover, proved to be rather limited, once Iraq (formerly Mesopotamia) became independent of it, although there are large deposits of coal, iron, chrome, and copper in Turkey.

In the final decades before World War I Turkey underwent economic penetration by European capitalists. German involvement intensified after Wilhelm II came to power in 1888; German investors helped to finance railroad construction, which the Turks particularly favored, and commerce between the two countries boomed. The Deutsche Bank, for example, formed an international railroad syndicate based on German, Austrian, French, Swiss, and Italian capital.

Although today it is the United States that looks to the Middle East for oil, following the American Civil War large quantities of petroleum were exported from the United States to that area for use as an illuminant. By 1868 oil as an import was rivalled in Constantinople only by rum, and had become the leading American export to Syria and Egypt. In 1879 the American consul in Constantinople reported that "even the sacred lamps over the Prophet's tomb at Mecca are fed with oil from Pennsylvania."[10] In fact, during the late 1880s several Americans helped the Egyptian government in a futile search for petroleum.

American oil first aroused diplomatic friction in Turkey. In the summer of 1882 the Minister to Turkey, Lew Wallace, complained to the State Department that there was talk of imposing an additional 8 percent duty upon American petroleum; he coupled his protest to the Turkish government with a challenge to the extra charge that officials in Constantinople had placed on salted meat and on alcohol. To make matters worse, a Turkish subject obtained the right to store petroleum at such ports as Smyrna and to charge fees.

In January of the following year Democratic Senator J. N. Camden of West Virginia complained to Secretary of State Frederick T. Frelinghuysen: "These high expenses will work against American oil in favor of Russian oil, because the moment Russian oil arrives here it will be sold ex vessel and not be stored, *but this firman is against the international treaties*, and it is of great importance to American petroleum that it should be revoked."[11] While a commission was set up to resolve the controversy, it was not until April 21, 1884, that Wallace was able to report the Council of Ministers would undertake a compromise reduction in the extra duty.

Official U.S. concern for American oil interests in the Eastern Mediterranean again manifested itself several years later. John H. Flagg, an attorney for Standard Oil, reported to Secretary of State Thomas F. Bayard that the Russian government was attempting to obtain an exclusive concession in Cairo and at various other Egyptian ports for the construction of tanks to store Russian petroleum. Flagg suggested that the proposed concession would severely undermine the position of American oil in Egypt, but the Turkish government denied any knowledge of the Russian petroleum monopoly scheme, in response to official U.S. inquiries.

Five years afterwards, in July 1892, the American Chargé at Constantinople, H. R. Newberry, learned during an interview with the Grand Vizier that the people of Iraq were using petroleum as fuel. The Turkish government, which wanted to keep this oil out of Russian hands, asked for an American expert to survey these holdings and then report on them. "This Government," Newberry wrote, "is inclined to be liberal and there may be a chance to pipe oil to the sea and wipe the Russians out of all this Eastern trade."[12] Newberry then invited John D. Rockefeller to undertake the survey, using a possible future concession as a lure, but the head of Standard Oil declined the offer as insufficiently advantageous for his firm.

American and Russian oil interests were by no means the only foreign entrepreneurs active in Turkey. As early as 1903 Mr. Nichols, later the Managing Director of the Turkish Petroleum Company, obtained a concession for William D'Arcy, a Britisher also active in Australia, that covered petroleum rights in Mesopotamia. He also obtained a letter from the Grand Vizier that promised the Sultan would confirm this understanding. A year later German interests followed up the Kaiser's visit to Constantinople by obtaining a contract with the Civil List that encompassed oil rights in various parts of the Ottoman Empire. The contract required the Germans to execute certain geological and developmental work within a specific period of time, and agreed to reimburse expenses for the work should the concession become null and void.

By the end of the year a controversy had arisen when the Germans had failed to carry out the provisions of their contract, and Nichols pressured the Civil List into advising them that their rights under the concession had expired. As the

Civil List, however, refused to pay a £20,000 claim the Germans had filed with it to cover expenditures for geological and other work, the Germans took the position that the contract was still in force.

Between 1904 and 1912 D'Arcy and Nichols made intermittent attempts to obtain ratification of the rights that were originally granted to them in 1903. Finally in 1913 Sir Ernest Cassel, the Shell Company, and the Deutsche Bank set up the Turkish Petroleum Company, Ltd., which took over not only the German claims but also the Baghdad and Anatolia Railway grants. By this time, the Anglo-Persian Oil Company had obtained D'Arcy's interests. After the Turkish Petroleum Company and Anglo-Persian had succeeded only in blocking each other's efforts, talks between the British and German governments led the British to engineer Anglo-Persian's acquisition of a 50 percent interest in the Turkish Petroleum Company. Stockholders had wanted official British assistance in confirming their title at Constantinople.

These complications did not end here, however. Six years before this merger, in 1908, an American group led by Rear Admiral Colby M. Chester had also obtained a concession from the Sultan. In 1899 Chester had gone to Turkey aboard the U.S.S. *Kentucky* to give "moral support" to the American efforts to obtain an indemnity payment for the property damage suffered by U.S. missionaries during the Armenian massacre of 1896. The amount at stake was relatively small, only $95,000, but the case had become mixed up with international diplomatic intrigue.

Unlike many of his fellow admirals, Chester did not engage in gunboat diplomacy, thus making a favorable impression on Sultan Abdul Hamid II (1876-1909). As a result, the Turkish government approved a contract for the building of a cruiser by the Cramp Company of Philadelphia, which, according to the Sultan, was only the beginning of a fleet that was to be constructed in the United States. Equally important, Abdul Hamid agreed to extend the lease for Roberts College, which had been in operation since 1866.

The Turkish government, which regarded the imperialistic activities of various European countries with a rather jaundiced eye, was further impressed when the United States remitted its share ($25 million) of the indemnity which the Chinese were forced to pay in the aftermath of the anti-foreigner Boxer Rebellion. Officials in Constantinople began to make a positive effort to attract Americans to Turkey, while in the United States Rear Admiral Chester urged businessmen to extend their operations there. American businessmen responded favorably, and in May and June of 1908 Chester, having received a commission from the Chamber of Commerce and the Board of Trade and Transportation of New York, went to Turkey on behalf of American commerce. Lord Charles Beresford earlier had made a similar mission on behalf of the British Board of Trade to the Far East.

Before leaving for Turkey, Chester made application to Secretary of State Elihu Root in an attempt to obtain support for his activities from the American government. Root gave him his personal support, undoubtedly well aware that there was little U.S. trade with the Middle East at this time, while President Theodore Roosevelt also lent his encouragement. A year later, in 1909, the American Navy stopped off at Smyrna on its way home from a world cruise and took aboard twelve Turkish naval officers; it brought them to the United States, so as to generate some good will on behalf of Chester's activities. Then in 1910 President William Howard Taft publicly stated that the American government was supporting Chester in his plan to construct railroads in Turkey. During this year, too, the Assistant Secretary of State visited Constantinople with the covert objective of encouraging Chester's project. Eleven years later, in a letter to Secretary of the Navy Edwin Denby, Chester discussed his long-standing interest in obtaining an oil concession and stressed the governmental support his scheme had enjoyed:

> The basis of this concession was a report made by the U.S. Consul at Aleppo, Syria to the State Department in 1908. Secretary Root took up the matter with American capitalists and made arrangements to try and secure a concession from the Ottoman Government covering the situation. Having had some experience in diplomatic negotiations with the Sultan of Turkey, I, being then on duty connected with the Navy Department, was detailed by President Roosevelt to proceed to the Near East and work for the expansion of American trade in that section of the world.[13]

Chester's ability to obtain the concession in 1908 must be regarded as an even more impressive achievement, in view of the fact that there had been a renewal of Turkish-American diplomatic friction over petroleum. On May 27, 1905, the American Minister, John G. A. Leishman, complained to Washington about Turkish storage rates, which apparently were imposed to establish a new source of revenue. Shortly afterwards Turkish officials began to impose a monthly rather than an annual tax. Yet by this time American oil sales in Turkey had become somewhat spasmodic, since the Russians were dominating the market with their cheap oil; the only U.S. petroleum to enjoy regular sales in Turkey was that used by the lighthouse service.

Three years later Rear Admiral Chester returned to Turkey. In exchange for building most of the public works of Turkey, particularly the railroads, Chester obtained the right to exploit the Arghana copper mine and the Turkish oil fields which were in Mesopotamia and on the Persian border. The Sultan had thrown in other gold, silver, and copper mines for good measure. Chester was not expected to build the railroads until he had determined which petroleum fields and mines were profitable, so that he could then link them with his railroad system. The Ottoman government had also agreed to pay a large "kilometer guarantee" to Chester, which would cover half of the cost of building the railroads, as well as an annual subsidy. The concession became operative immediately, and was to extend for ninety-nine years, without any restriction or time limit on the actual inception of operations.

A year after the Sultan had signed the Chester concession, though, political reformers replaced Abdul Hamid with his brother, Mohammed V (1909-1918). Even with this change in government, the Young Turks and Mohammed V accepted and confirmed the Chester concession; in 1911 the Ottoman Parliament considered it for approval. German agents in Turkey had informed Berlin as early as May 30, 1910, that Chester was moving ahead with vigor, adding that it seemed certain the Ottoman Parliament would approve the concession. As a result, the German Ambassador received instructions to place obstructions in Chester's path. General European reaction to the concession was not as negative as it was in Germany, where the government regarded the concession "as a feeler on the part of the Standard Oil, Morgan and some others."[14]

The Chester concession was not ratified, however, not because of German intrigue but rather as a result of the deteriorating political situation in the eastern Mediterranean and the Balkans prior to the outbreak of World War I. After the favorably disposed Ottoman Parliament had voted to postpone formal approval until the end of 1911, Turkey and Italy went to war. The Parliament then again delayed action on the concession until a later session; with the outbreak of the two Balkan wars and World War I, the Ottoman Parliament continued to withhold its approval.

From the standpoint of international law, the Chester concession was not in jeopardy. Two centuries earlier the noted Swiss jurist, Emeric de Vattel, had stated that "the aliens' right of action is only suspended during the war";[15] the British government had upheld this position on a number of occasions prior to 1908. In addition, when war broke out between Italy and Turkey, article 6 of the Chester concession was invoked, and a Turkish messenger to the United States served formal notice of a stay of execution. According to article 6, "At all times, in case of interference by 'force majeure', duly proved, the delay fixed for the execution will be prolonged by one of equal duration to that of the interruption of the work."[16]

It should be noted, too, that every European government sent congratulatory messages to the Acting American Ambassador at Constantinople, John Ridgely Carter, after the Turkish Parliament had formally taken up the Chester concession. Sir Babington Smith, a British financial agent, also gave a banquet for Chester and his associates. But behind the scenes the intrigue continued. The Germans obtained an opinion from Jules Dietz, a French lawyer, that the Chester concession was in violation of an earlier Baghdad Railway concession. They even speculated that the American scheme was a Zionist plot, only to learn that neither Chester nor any of his associates were Jewish.

Writing in 1926, Carter observed: "The appearance of Admiral Chester at Constantinople in 1908 provoked one of the bitterest diplomatic struggles of pre-war days, and was the signal for a great reconciliation of German and British interests in the Near East."[17] By 1914 the deterioration in relations between Great Britain and Germany helped to inflate World War I into a major conflict, but until the very last moment there were occasions when the British and Germans worked hand in hand in the Middle East. As noted, there was co-operation between the British and German governments at the time that Anglo-Persian obtained a 50 percent interest in the Turkish Petroleum Company.

On June 28, 1914, the summer in which World War I erupted in Europe, the Turkish Grand Vizier, Said Halim Pasha, wrote a somewhat ambiguous letter to the British Ambassador at Constantinople. In this he stated "with respect to petroleum resources discovered, in the vilayets of Messoul [Mosul] and Bag[h]dad, [the Ministry of Finance] consents to lease these to the Turkish Petroleum Company, and reserves to itself the right to determine hereafter its participation, as well as the general conditions of the contract."[18] Said Pasha went on to observe—in apparent reference to the Chester concession—that if necessary the Turkish government would indemnify those third parties who were interested in the petroleum resources of the two vilayets. The German Ambassador at Constantinople received a similar letter. Seven years later, on November 17, 1921 the American Ambassador to Great Britain, George Harvey, wrote the British Secretary of State for Foreign Affairs, Lord Curzon, that Said Halim Pahsa's communication of June 28, 1914, "cannot well be considered a definite and binding agreement to lease," but rather that negotiations between the Turkish Petroleum Company and the Turkish government were "those of negotiators of an agreement in contemplation rather than those of parties to a contract."[19]

Once war had broken out, the British government seized the German share of the Turkish Petroleum Company, that 25 percent held by the Deutsche Bank group, and attempted to transfer it to Anglo-Persian. The action disturbed the Royal Dutch-Shell interests, who then held 25 percent of the Turkish Petroleum Company, and they warned the French government. The French then demanded the German share of the firm as a war indemnity, pointing out that under the Sykes-Picot Agreement of 1916 France had obtained title to Mosul. After extended negotiations with the British government, Lord Curzon did agree to the transfer of the German share to France, provided that the French renounced their claims to Mosul and agreed to the building of an oil pipeline through Syria.

By the fall of 1918 victorious British troops were in possession of both Mosul and Baghdad, and the American government became highly disturbed. It complained that the British were using military necessity as an excuse for building pipelines and refineries, operating oil wells, and prospecting for petroleum. Although Lord Curzon dismissed the American charges as false, as the war in Europe neared its end, such economic rivalry between the two leading Allied powers undercut their relations in the Middle East.

### The Uncertain Peace: Palestine and Mesopotamia

The years that followed the close of World War I were just as momentous for the Ottoman Empire as those before the conflict. Under the Treaty of Sèvres Turkey lost most of its enormous empire in Asia, but the treaty was never truly operative, as the progressive Turks united behind Kemal Ataturk, and drove out the foreign occupying forces, most notably those of Greece. In November 1922 the reform element in Turkey deposed Mohammed VI, who had ruled since

1918. Delegates from Turkey and various other European nations assembled at Lausanne, and by the following spring they had drawn up a new peace treaty.

Under the Treaty of Lausanne, Syria and Lebanon became French mandates; Palestine and Transjordan British ones. Great Britain also obtained the mandate over oil-rich Mesopotamia (later Iraq) and was to negotiate the final boundary with Turkey at a later date. The rather loose control the Turks exercised over the eastern and western coasts of the Arabian Peninsula also came to an end as British influence supplanted it. After a civil war in Arabia, Ibn Saud won recognition as a sovereign and independent ruler under the Treaty of Jiddah in 1927; Arabia then became Saudi Arabia in 1932.

Elsewhere in the former Ottoman Empire, Palestine emerged as a focal point of controversy. Restrictions were placed on the operations of American petroleum entrepreneurs under the British military administration. Great Britain not only refused to allow New York Standard geologists to investigate the claims which the firm had obtained before the war, it also barred U.S. geologists from looking for oil in Mesopotamia. When President Woodrow Wilson protested the Palestinian ban, the British government claimed that the restrictions were not directed specifically against American petroleum interests.

During the last two years of the Wilson administration, and the first year of the Harding administration, there was a continuous exchange of diplomatic correspondence between the United States and Great Britain over the British restrictions. In the summer of 1919 there ensued a minor furor, when British authorities examined certain maps and papers belonging to the Standard Oil Company's office in Jerusalem; the British stated that they had applied to the Spanish Consul, who was representing the U.S. interests there at the time, for permission to study these. The British wanted to determine what the Ottoman regime actually had granted to Standard Oil, which then held a total of sixty-four concessions in Palestine.

The State Department remained displeased with the British restrictions on American petroleum operations in the months ahead. On March 17, 1920, for example, the Acting Secretary of State, Frank L. Polk, wrote to the U.S. Chargé in Great Britain, Wright:

> [The State] Department has examined Standard Oil Company permits and concessions in Palestine and finds (1) that in respect to all its claims it has certain vested rights under the Turkish law; (2) that Great Britain in its temporary occupation of Palestine may legally enjoin the acquisition of further vested rights but must by virtue of its position as quasi trustee recognize and protect the vested rights of the Standard Oil Company already existing; and (3) Great Britain as trustee cannot legally allow discrimination in favor of her own or other nationals.[20]

Even with this firm stance, at the end of 1921 New York Standard was still complaining about the conditions which the British government had laid down relative to geological research in Palestine. Standard objected to the condition that delayed the development of its concessions until the peace treaty with Turkey, which among other things would settle the terms of the Palestinian mandate, went into force. In face, it was hesitant to supply a full and complete report of its investigations to the British government of Palestine, since it might lend support to any attempt by them to change the terms of the concessions that had been granted by Ottoman authorities. The State Department fully supported New York Standard on this point.

British Foreign Secretary Curzon replied that the Palestinian government did not wish to impose "unacceptable conditions" on New York Standard but it lacked the funds to establish a competent Geological Survey Department, and therefore was soliciting reports from the operators of concessions.He added that, given the state of political unrest in Palestine, it might be necessary in the future to restrict the movements of New York Standard representatives there, so as to safeguard their own personal safety.

It was not until April 1922 that Curzon notified the U.S. government that Great Britain had finally granted New York Standard the prospecting rights in Palestine that firm had been claiming. Since the British Foreign Secretary had admitted to Ambassador John W. Davis as early as 1919 that there were no counterclaims, he could not use this as an additional reason to keep Standard from proceeding with its operations. Nevertheless, Winston Churchill stated in the House of Commons on February 18, 1922: "the grant of this permission [to prospect] did not prejudice the question of the validity of the claims and that no actual exploitation would be permitted until the political status of Palestine had been 'regularized.'"[21] But lest one assume that the British indeed had treated New York Standard in a most unfair manner, it should be pointed out that Great Britain did not allow any concession holder in Palestine to conduct operations during the immediate postwar period.

Factors other than petroleum entered into British policy toward Palestine. According to a letter the Chief of the Fuels Division of the Department of Commerce, H. C. Morris, wrote to the Secretary of Commerce, Herbert Hoover, on April 24, 1922:

> The latest British note on the subject to the State Department says that oil concessions in Palestine may be handled quite differently from those of Mesopotamia because of the British desire to put no obstacles in the way of establishing the home land of the Jews in Palestine and their control over that country.[22]

There was, of course, far more petroleum in the Mosul area, hundreds of miles to the East, than in Palestine. American attempts to develop the oil resources of that area were not only complicated by the British mandate over Mesopotamia, but also by the bitter controversy that ensued over the boundary in the oil-rich Mosul area. It was not until 1926 that Turkey and Great Britain agreed on an official boundary that placed these petroleum fields in Mesopotamia, and gave a financial equivalent for its territorial sacrifice. Earlier the British themselves had admitted in the *Parliamentary Papers* for 1917 that the original British expedition to Mesopotamia during World War I had as its basic objective the safeguarding of the oil refineries, tanks, and pipelines of the Anglo-Persian Oil Company.

In 1920 Great Britain and France reached an understanding in San Remo, that gave the British and the French monopolistic control over Mesopotamian petroleum. Known as the San Remo Accord, it extended to France the German rights in the Turkish Petroleum Company and to Great Britain, the right to build a pipeline from Mosul across Syria. American oil entrepreneurs were excluded entirely, much to the displeasure of the United States.

By the end of that year, a number of anti-British editorials had appeared in the American press. In defending the U.S. petroleum companies and the concept of equal economic opportunity, the Boston *Globe* noted that "what the United States is asking on the subject of Mesopotamian oil is exactly what we have always been willing to grant to the whole world."[23] As one might expect, a focal point of attack was the San Remo agreement. The Indianapolis *News* complained that "there could hardly be a clearer violation of the spirit and indeed, as is believed, the letter of the League Covenant than this arrangement."[24] A further irritant to the U.S. press was Lord Curzon's claim that only the League of Nations could properly discuss the terms of the mandates. The Washington *Post* tartly observed that American oil was not acquired under pretense of a mandate and then transformed into a monopoly.

Although the Anglo-Persian, Royal Dutch-Shell, and French petroleum entrepreneurs operating in Mesopotamia bitterly opposed U.S. participation, American oil firms found a spokesman in the wily C. S. Gulbenkian. Gulbenkian advised Sir William Tyrrell, the British Under Secretary for Foreign Affairs, that it would be preferable for those firms to admit the U.S. petroleum entrepreneurs as partners in their operations than to compete with them for petroleum concessions, thus threatening their ascendant position in Mesopotamia. Available evidence indicates that Gulbenkian was at least partly responsible for bringing about U.S. participation in the development of Mesopotamian oil in the years ahead.

The U.S. entrepreneurs themselves debated their position on Mesopotamian petroleum, after the oil firms had joined forces in November 1921 under the leadership of Jersey Standard. At first the seven companies making up the American group tended to favor independent action, but later they came to feel that it would be preferable to cooperate with the Turkish Petroleum Company. In eventually opting for a share of the dividends in a multinational consortium, they abandoned their earlier desire for the actual division of the Mesopotamian oil territory. American petroleum firms also wanted the consortium only to produce petroleum; under this plan the oil dividends among the shareholders was to be directly proportionate to their stock participation.

As the Mesopotamian negotiations progressed, Secretary of State Charles Evans Hughes continued to insist on the "Open Door" principle. He felt that any interested U.S. firm should be allowed to participate in the American group, to prevent the monopolization of Mesopotamian oil by a few U.S. companies.

Hughes also was opposed to the self-denying ordinance, which stated that participants in any petroleum agreement could not act independently of their partners anywhere in the former Ottoman Empire. In view of the Red Line Agreement of 1928, it is apparent that Hughes' ideas did not prevail. Despite these strongly held beliefs, the Secretary of State did recognize the right of the American group to restrict its operations in Mesopotamia on business grounds.

Not unexpectedly, Great Britain's opinions were somewhat different. Among those who defended the British position was Sir John Cadman, a consulting petroleum advisor to the British government. Cadman pointed out that only 4 percent of world petroleum production was under the control of Great Britain in 1920; such an amount, he said, hardly constituted a monopoly over global oil reserves. Referring specifically to the Mesopotamian situation, he noted that certain British nationals had obtained concessions there from the Turkish government prior to World War I. Under these circumstances the British government could not reasonably deny its nationals these rights, and the exercise of these rights did not signal the inauguration of a "closed door" policy directed against American and other foreign oil entrepreneurs.

In contrast, Colonial Secretary Winston Churchill was far more favorable to the American position, if only because he doubted that the legality of the Turkish Petroleum Company concession would stand up under arbitration. In a letter to Foreign Secretary Curzon in February 1922, Churchill noted that as long as "Americans are excluded from participation in Iraq oil we shall never see the end of our difficulties in the Middle East."[25] Once U.S. petroleum interests were satisfied, he continued, it would be easier to obtain ratification of the mandates by the League of Nations, as well as to stabilize conditions in revolutionary Turkey.

While the Turkish Petroleum Company and the American group continued to maneuver for position, Rear Admiral Colby M. Chester was attempting to breathe life into the expiring concession he had obtained under rather questionable circumstances from the Turkish government before World War I. Hoping to strengthen his position, in 1921 Chester tried to obtain an appointment as a U.S. Commercial or Naval Attaché at Constantinople. Although in the past the State Department had shown great interest in Chester and his firm, the Ottoman-American Development Company of New York, its enthusiasm was now on the wane. Under Secretary of State Henry P. Fletcher noted:

> The situation has now completely changed and it appears to me that it would be almost impossible to determine to what government or governments the Ottoman Development Company should address itself for confirmation in its concession rights until at least the Treaty with Turkey and also the boundaries of Armenia, etc., have been definitely established.[26]

Yet Chester persisted in advancing his cause. In a letter to Secretary of the Navy Edwin Denby on March 30, suggesting his appointment as an Attaché,

Chester stressed the fact that he favored Anglo-American cooperation in the Middle East. "The incidental business likely to accrue to this country through this concession is enormous," observed Chester, "and aside from acquiring what is estimated to be over a billion dollars worth of oil it will be a strong factor in building up the American Merchant Marine."[27] He recommended that the Navy Department send a surveying expedition to explore in Turkey, Syria, and Mesopotamia.

Aside from his lessening State Department support, Chester faced other problems, not the least of which was a lack of capital. By the fall of 1922, moreover, a split had materialized in his ill-fated concern; at this time Major K. E. Clayton-Kennedy, a Canadian who had been in the British air service, had become the representative of the Ottoman-American Development Company in Turkey. After Chester cancelled Clayton-Kennedy's sole power-of-attorney without the approval of the firm's directors, Clayton-Kennedy informed Turkish officials that they should ignore Chester's opinions. Yet Clayton-Kennedy himself was widely suspected of being a British spy; U.S. Trade Representative in Constantinople, Julian Gillespie, noted, "I have heard on every hand unfavorable expressions from people who have come into contact with him."[28]

In June 1922, as the Chester comic opera proceeded, the Turkish Petroleum Company, acting in accordance with the recommendations of C. S. Gulbenkian, offered the American Group a 20 percent share in the company. To make this possible, it would be necessary to reduce the Anglo-Persian share to 40 percent, and the shares of Royal Dutch-Shell and the French group to 20 percent each. This proposal was objectionable to Royal Dutch-Shell, which felt that part of the U.S. share should be carved out of the Anglo-Persian share rather than its own, and the French, who wanted their share to be equal to that of the Americans.

W. C. Teagle, the president of Jersey Standard, wrote to Secretary of State Charles Evans Hughes on August 25 that the seven companies constituting the American group were the only ones that had expressed an interest in the petroleum possibilities of Mesopotamia. Although Teagle had indicated to Hughes in person that he expected the American interests to obtain a 20 percent participation in the Turkish Petroleum Company and an equal voice in its management, he failed to confront the monopoly issue. But Secretary of Commerce Herbert Hoover had written Hughes on August 19 that he felt that Teagle's group was fairly representative of the different oil interests, and that he believed the U.S. group should have at least a quarter share. "This," Hoover observed, "is perhaps more a matter of national pride than for intrinsic importance."[29]

That November an international conference met at Lausanne, where the major item on the agenda was the writing of a new peace treaty with Turkey. Lord Curzon announced that Great Britain was ready to jettison the San Remo Accord, and support the Open Door policy endorsed by the American government. By this time the United States had adopted a more favorable attitude toward British petroleum activities in Latin America. The Anglo-Persian Oil Company then offered to reduce its participation in the Turkish Petroleum Company, to allow both the American and French oil entrepreneurs a 24 percent share. Anglo-Persian, however, attached conditions to its offer that complicated matters; for one thing, it wanted a 10 percent royalty on the petroleum produced.

During the course of the Conference, the American delegation went to great lengths to advance the interests of the U.S. oil entrepreneurs in Mesopotamia. Ambassador to Italy Richard W. Child, the spokesman for the American delegation, adopted a hostile stance toward the Turkish Petroleum Company's concession in Mesopotamia, citing the time-honored Open Door principle. Even after these activities, four years later Secretary of State Frank Kellogg in a lengthy letter to Republican Senator William E. Borah of Idaho on December 29, 1926, flatly denied that the U.S. oil firms had manipulated the American delegation at Lausanne. According to Kellogg, "No member of the American Delegation at Lausanne was connected directly or indirectly as a representative or otherwise with any oil or tobacco interest before the Conference, during the Conference or subsequent to the Conference."[30]

After the delegates to the Lausanne Conference had adjourned the proceedings in February 1923 with plans to reassemble two months later, the Turkish government stunned the world by granting the Chester interests (the Ottoman-American Development Company) a lucrative new concession. Incorporated in Delaware with a maximum capital of $100 million, this firm had as its president General George W. Goethals of Panama Canal fame. The apparent reason for this stunning act of generosity was that the Turkish leader, Kemal Ataturk, hoped that Rear Admiral Chester could bring U.S. pressure to bear on behalf of the termination of capitulations in Turkey. Ataturk also appears to have believed that the Americans would not use economic concessions for political purposes to the same extent that the British and the French would.

By any standard the concession the Chester interests received from the Turkish government was a most generous one. The Ottoman-American Development Company obtained the right to build a railroad from Ankara to Mosul, which was included in the grant, along with a number of branch lines; it then was to develop the petroleum and mineral resources that lay along the railroad. But this was not all. The Ottoman-American Development Company was to operate public utilities, construct model villages and a new capital at Ankara, build harbors in the Mediterranean and the Black Sea, and even organize the working classes and "better the Turkish character"!

Although Great Britain declared that it would not recognize the Ottoman-American Development Company claim to Mosul, the British were less disturbed by the concession than the French were. Some of the Ottoman-American railroad rights under the 1923 concession conflicted with those found in a grant the Turkish government had made to French capitalists in 1914 in return for a loan of 1 billion gold francs, about half of which actually had been advanced. The French capitalists, too, had already begun to lay some of the tracks.

Back in the United States the Ottoman-American Development Company made a concerted attempt to win the support of the U.S. government for its abortive Turkish concession, but neither Wall Street nor Standard Oil were greatly interested in the Chester interests. When Premier Raoul Bey was approached by a petroleum company official who declared: "I wish to inform you that the Standard Oil Company is not behind Chester," the premier's retort was, "No, but the Turkish government is."[31] Bey supposedly issued a warning that American prestige in Turkey would rise or fall with the progress, or the lack thereof, of the Chester interests.

At the reconvened Lausanne Conference Great Britain attempted to persuade Turkey that it should confirm the validity of the Turkish Petroleum Company's concession in a protocol to the peace treaty. Not only did the American Minister to Turkey, Joseph C. Grew, inform the Turkish government that the United States was "unalterably opposed" to this scheme, but he also warned the Turkish delegate, Ismet Inönü, that its acceptance would lead to a lessening enthusiasm on the part of American investors for Turkish economic projects. By adopting this hard line position the State Department placed itself somewhat at odds with the American group, which was taking a more conciliatory stance toward its British and French rivals, even though on January 31 it had rejected the 24 percent participation offer.

On July 24 the Turkish government and representatives of the foreign powers involved signed a treaty that replaced the Treaty of Sèvres. Once the Treaty of Lausanne had been ratified, Turkish sovereignty over petroleum-rich Mesopotamia, or Iraq, and the Arabian peninsula ended. It should be noted, however, that in 1954 Turkey enacted a law that was supposed to encourage the participation of foreign capital in the petroleum industry there by terminating the government monopoly over the development of oil resources. In drawing up this measure the Democrat party government of Adnan Menderes relied on the advice of U.S. petroleum expert Max Ball. Unfortunately, despite the interest shown by various foreign companies, by the late 1960s only Mobil (New York Standard) and Royal Dutch-Shell had found petroleum.

### The Red Line Agreement and the Iraq Petroleum Company

Between 1923 and 1932 Iraq was a British mandate, but a 1925 constitution gave additional responsibilities to the Iraqis. In 1930 the British signed a treaty with Iraq that granted the country full independence in two years and allowed it then to join the League of Nations. Nevertheless, the British were to maintain a privileged position there from the standpoint of diplomacy and defense, as well as to enjoy power in such areas as finance, business, and education.

In September 1923, at the inception of the mandate, the Turkish Petroleum Company applied to the Iraqi government for a concession that would include Baghdad, Mosul, and Basra provinces. Out of these negotiations emerged a provision for an annual competition "without distinction of nationality" for twenty-four plots, which reassured the U.S. government. By this time Secretary of State Charles Evans Hughes realized that unless he softened his stand toward the Turkish Petroleum Company, the British and the French entrepreneurs might freeze out the American oil companies, if the Turkish Petroleum Company obtained a new concession from the Iraqi government. Accordingly, he informed President Calvin Coolidge that he was willing to give U.S. diplomatic support to the Turkish Petroleum Company application.

Other State Department officials had reached the same conclusion by 1923. Allen Dulles of the Near Eastern Division noted the difficulty in applying a theoretical principle (the Open Door) to a practical situation (Iraqi oil). Stanley Hornbeck of the Economic Advisor's Office pointed out that first comers should be given preference and that equal commercial opportunity could not continue indefinitely; Foreign Trade Advisor Wesley Frost had expressed similar opinions as early as 1920. Still another Economic Advisor, Arthur Young, citing the Chinese financial consortium, observed that the State Department could not oppose every monopoly.

It was now only a matter of time before the Turkish government withdrew its support for the Chester concession, and official notification came on December 18, 1923. One individual who was highly concerned about the activities of the Ottoman-American Development Company was Rear Admiral Mark L. Bristol, the High Commissioner at Constantinople. Bristol felt that its failure might handicap the future activities of U. S. entrepreneurs there:

> The present situation in Turkey is a standing invitation to the commercial adventurer and repels the conservative American business man. I foresee not only the failure of the Chester concession but also a number of "Chester fiascoes" of one kind or another for some years to come, unless the Department feels it can change its present policy regarding American business enterprises abroad, to the extent at least of exercising some effective direction and control over the activities of Americans who seek Turkish concessions. The Turkish Government will be seriously weakened by the failure of the Chester project, which will lead to a renewal of intrigues by rival concession seekers.[32]

With the Chester concession now apparently a dead issue, in March 1925 the government of Iraq approved a new Turkish Petroleum Company concession which was to extend until the year 2000. The shares in this firm were divided as follows: Anglo-Persian, 47.5 percent; Royal Dutch-Shell, 22.5 percent; the French group, 25 percent; and C. S. Gulbenkian, 5 percent. Significantly, no American individuals or companies participated in the new firm.

Finally, in April 1927 the Turkish Petroleum Company began drilling for oil in Mosul. But on October 15 Baba Gurgur Well Number One erupted, producing initially 50,000 to 90,000 barrels a day; it became more important than ever for the U.S. petroleum firms to come to an understanding with their European competitors. The result was the famous Red Line Agreement of July 31, 1928, which superseded the San Remo Agreement of eight years earlier.

The understanding covered a geographical area more or less coterminous with the old Ottoman Empire as it embraced everything to the east of Egypt and to the west of Persia (Iran) with the exception of Kuwait. An American group composed of the Atlantic Refining Company (16.67 percent), Gulf Oil (16.67), Pan American Petroleum and Transport (16.67), Jersey Standard (25), and New York Standard (25) were collectively to hold a 23.75 percent interest in the Turkish Petroleum Company. An equal share was to go to Royal Dutch-Shell, Anglo-Persian, and the French oil entrepreneurs, while C. S. Gulbenkian was to receive the remaining 5 percent. Anglo-Persian was to obtain an overriding 10 percent royalty on the total Turkish Petroleum Company output. Finally, under the agreement none of the partners could explore independently for oil outside the "Red Line," a stipulation that was to cause friction during the two decades the understanding remained in force.

While the American petroleum companies obviously were advancing their own interests in the Middle East, the U.S. government was supporting them through its insistence on the Open Door policy. The efforts of the government did not always win universal approval back home. Thus the Democratic Party included a provision in its 1924 national political platform, complaining that the Turkish-American treaty had traded "American rights, and betrays Armenia for the Chester Oil Concession."[33] Edward Mead Earle in an article in the *Political Science Quarterly* that same summer commented:

> Is it the business of the American Department of State so actively to concern itself with the ventures of American petroleum companies in the four corners of the earth? In particular, is it the business of the American Department of State so actively to paddle in Mesopotamian oil? Even more important, is there any advantage to be gained by the promotion of these so-called national interests which will not be more than offset by possible friction between the American and British peoples? How can the United States expect European Powers to recognize the economic implications of the Monroe Doctrine if the United States is unwilling to recognize that European Powers have their particular spheres of interest?[34]

Because of pressure from the U.S. government, the formulators of the Red Line Agreement had included a provision "that other oil companies would be given a chance to share in the development of the oil fields of Iraq."[35] Designed to further the Open Door principle, the provision became in effect a dead letter; in 1931 when the government of Iraq signed a new agreement with Turkish Petroleum (which was renamed the Iraq Petroleum Company in 1929), it did not include this important provision. The revised understanding also limited the company's concession to 32,000 square miles east of the Tigris River. Moreover, Atlantic Refining, Gulf, and Pan American withdrew from the firm during that year because of the Great Depression, leaving Jersey Standard and New York Standard as the only U.S. partners.

Between 1931 and 1934 the Iraq Petroleum Company began to build a $50 million pipeline from Kirkuk in the oil fields, which branched at Haditha on the way to two separate ports on the eastern coast of the Mediterranean Sea: Tripoli in Lebanon (the French terminus) and Haifa in Palestine (the British terminus). This was the first such pipeline built in the Middle East, but by no means the last;

since then these pipelines have played an increasingly significant role. In 1937 the Iraq Petroleum Company gained effective control over the British Oil Development Company with holdings to the north and east of the Tigris, which it assigned to the Mosul Petroleum Company. During 1938 the Iraq Petroleum Company also obtained a seventy-five year concession in southern Iraq and organized another affiliate, the Basra Petroleum Company, to operate it.

During the early part of World War II the British helped to defeat pro-Axis elements in Iraq. In return, Iraq then declared war on the Axis powers in January 1943, becoming the first Middle Eastern state to qualify for membership in the United Nations. Throughout the war American as well as British troops were stationed in the country, which had begun to obtain lend lease aid from the United States in 1942.

Early in 1943, however, the American government issued a protest to the Iraqi government, after it had accused the Basra Petroleum Company of having failed to execute certain requirements of its concession as of November 29, 1941, thus rendering the concession null and void. At the same time the Iraqi government asked that the Basra Petroleum Company pay ground rent for both 1942 and 1943. U.S. firms, it will be remembered, participated in the ownership of the parent Iraq Petroleum Company. It was the position of the U.S. government that the payment of the ground rent constituted recognition of the concession's validity. Although Prime Minister Nuri al-Said, who visited Washington in 1945, told American diplomats that he regretted the U.S. government's interest in a question that he felt was basically the concern of Iraq and the Basra Petroleum Company, by March of 1943 the Iraqi government had drawn up two alternate agreements which it hoped would settle the dispute. Basra Petroleum eventually accepted an agreement under which it would make a loan to the Iraqi government.

The post-World War II history of Iraq can be divided into two eras: the years before and the years after the revolution of 1958. During the first period there were frequent changes in the prime minister, although Nuri al-Said held this position on a half dozen occasions. In 1951 Iraq successfully approached the United States in a quest for Point IV aid and military assistance, but frustration and unrest within the country built up to such a point that in 1952 it became necessary for the government to declare martial law. In 1955 Iraq signed the Baghdad Pact with Turkey providing for military cooperation, with Iran and Pakistan ratifying this later; during the same year the British drew up a new treaty with Iraq, granting it greater freedom from British influence.

Important developments also took place in Iraq during these post-World War II years relative to oil. Under a 1950 law the government of Iraq was to devote 70 percent of its petroleum royalties to economic development, while under a new royalty schedule that the Iraq Petroleum Company agreed to in 1952 and made retroactive to 1951, the royalty rate was raised to fifty percent of its profits prior to taxes. By 1955 these royalties had reached the quarter billion dollar mark, in part because of world demand and in part because of a new pipeline from Kirkuk to Baniyas on the Mediterranean. The Iraqi government also opened a petroleum refinery in Basra in 1952, and during this same year, the Government Oil Refineries Administration took over both internal refining and distribution.

In 1958 Iraq decided to federate with King Hussein's Jordan, but on July 14 of that year General Abdel Karim Kassim engineered a coup, during which Faisal II and Nuri al-Said died. General Kassim then established a republic with himself as prime minister. Denouncing the recent Iraqi union with Jordan, he engineered the withdrawal of Iraq from the Baghdad Pact.

It was after Kassim had come to power that the New York *Times* headlined a story: "West to Keep Out of Iraq Unless Oil is Threatened." The new regime at first said it would honor the concession commitments, but in 1959 Kassim began taking a hard line against American and European concessionaires. "These concessions," he stated, "were given a time when Iraq was fettered, but we are now free and deal with the companies on a basis of mutual profit."[36] Although Kassim died four years later during a coup, the anticoncession forces continued to gain momentum.

By the end of 1961 the Iraq Petroleum Company was paying the Iraqi government $270 million in oil revenues. Even so, during that year Iraq seized 99 percent of the concession territory of the Iraq Petroleum Company and its affiliates under Public Law 80, while leaving the producing facilities of that firm undisturbed. The action ended investment and exploration there on the part of the Iraq Petroleum Company, which challenged the seizure; that firm earlier had offered voluntarily to surrender 75 percent of its holdings.

Three years later, in 1964, the State Department attempted to persuade large U.S. oil firms without Middle Eastern concessions to avoid Iraq. Although Indiana Standard, Union Oil, Continental, Pauley, Marathon, and Phillips adopted a cooperative stance, Sinclair pointed out that if it did not apply, the concessions might go to Japanese, German, Italian, French, and other foreign interests. It was also during 1964 that the Iraq National Oil Company was organized, although it did not actually begin operations until 1967.

The Iraqi seizure of the IPC concession territory remained an isolated act in that part of the world during the 1960s. Other Middle Eastern governments apparently concluded that it had been economically harmful to the Iraqi economy, and therefore did not follow suit. But Iraq still did not moderate its nationalizing stance. During 1967, the year of the third Arab-Israeli war, the Iraq National Oil Company obtained control of all territory within the country, except for that held by the Iraq Petroleum Company under Law 97.

Iraq also signed an agreement with ERAP of France during that year, giving it a concession that was located on territory taken over from the IPC. Concerning the ERAP agreement, Secretary of State Dean Rusk wrote the Japanese embassy in Iraq on September 2, 1967 that: "Department representations last year to France on behalf of Iraq Petroleum Company were notably unsuccessful."[37] Additional Iraqi oil understandings with Turkey were reached in 1968 and with Algeria in 1969, but a proposed deal between ENI (the Italian State petroleum firm) and Iraq fell through in 1967 because of differences over price, and because of the opposition of the American and other foreign governments.

On June 1, 1972, Iraq nationalized the Iraq Petroleum Company, after the company had cut its production. The government, through, showed its appreciation to the French for their anti-Israeli stance by offering them a separate agreement. A year later, in March 1973, the Iraq Petroleum Company reached a comprehensive settlement with the Iraqi government. Among the main provisions of the agreement was compensation in oil to the Iraq Petroleum Company for the seizure of its facilities; the Iraq Petroleum Company agreed in return to pay the taxes it had been withholding for approximately a decade to Iraq.

The Russians had advanced a $224 million loan to Iraq in 1971 for the construction of two petroleum pipelines and an oil refinery. The agreement, however, has failed to prove satisfactory to Iraq, with the result that the latter has again turned to the West for aid relating to petroleum. During recent years, too, the main customers for Iraq's experts have been Great Britain, France, and India, while her imports primarily come from Great Britain, the United States, and West Germany. Although since the 1958 revolution Iraq has leaned toward socialism and nationalism, these economic connections with the West have kept it from too close an identification with the Soviet bloc, especially in recent years.

## Saudi Arabia to World War II: The Rise of Aramco

In comparison with Iraq, Saudi Arabia is today, and always has been, a highly conservative country politically. During the mid-1920s, though, Saudi Arabia was the scene of a civil war that featured King Hussein of Mecca and Ibn Saud (Abd al-Aziz) of the Wahhabis. The British had supported both Arab rulers against Turkey during World War I, but after the war King Hussein became dissatisfied with the peace settlement. As a result, he was not able to take advantage of British assistance, which might have enabled him to repel the attack Ibn Saud launched against him in 1924, after he had taken the title of Caliph.

Hussein, who was already short of funds, was placed in an even more difficult position by the British decision in 1924 to halt subsidy payments to Arab rulers. At this time the pilgrimages to Mecca made by devout Moslems were an important source of income to King Hussein's government, and by occupying Mecca Ibn Saud seized these funds for himself. In 1925 Hussein withdrew to Cyprus, and in the following year Great Britain recognized Ibn Saud as the "King of the Hejaz and of Nejd and its dependencies." By 1932 the two territories had become a part of Saudi Arabia, along with other areas he had conquered.

Even before he had launched his revolt against Hussein, in 1923 Ibn Saud had granted a concession to Major Frank Holmes' Eastern and General Syndicate, a British group, to develop the suspected petroleum reserves of the Hasa province along the Persian Gulf. As early as 1915, the British government had signed a treaty of protection with Ibn Saud, by which, among other things, no economic concessions were to be granted to foreign entrepreneurs without the permission of Great Britain. In contrast, Hussein had looked with favor on U.S. petroleum entrepreneurs. On November 20, 1922, the Assistant Secretary of State wrote to Herbert Hoover, the Secretary of Commerce: "...it is reported that foreign interests are endeavoring to secure an oil concession from the King of the Hejaz. The Minister has indicated his view that the King of the Hejaz would be disposed rather to favor Americans as concessionaires than other foreigners."[38]

Three years later, in 1925, with his war against Ibn Saud largely lost, King Hussein had his financial agent in Alexandria approach Stewart Johnson, the American Chargé there. Hussein asked for U.S. intervention and an American loan of £1 million in return for which he would grant to U.S. interests a concession over petroleum and other minerals in the Hejaz. Fortunately for the Americans, Johnson showed no interest in this scheme; had the United States officially cooperated with Hussein, American oil firms might never have obtained their lucrative concessions from Ibn Saud at a later date.

Despite this rebuff to Hussein, the U.S. government did not embrace Ibn Saud immediately. A puritanical Wahhabi and an absolute monarch, Ibn Saud wished to keep Saudi Arabia protected from modernization, yet he also needed to obtain funds and to improve the lot of his people. He therefore had his financial minister, Sheikh Abdullah Sulaiman, ask President Herbert Hoover in January of 1930 to send an American expert in geology and mineralogy to Saudi Arabia to investigate the water resources and mining possibilities of the Hejaz. As a result, former Minister to China Charles R. Crane, who was interested in Yemen, paid a visit to Ibn Saud at Jiddah and recommended that he employ Karl Twitchell, a U.S. engineer who already had worked in Yemen.

Unfortunately, Twitchell was unable to stir up much enthusiasm among American businessmen over the development of the Hejaz even though he was optimistic about its gold mining possibilities. (One ancient mine was later reworked and yielded $20 million worth of gold and silver before being closed in 1954.) Twitchell also was unable to locate additional water resources in Saudi Arabia, much to the disappointment of Ibn Saud. Ibn Saud then asked Twitchell to investigate the suspected petroleum reserves of the Hasa province, where the British had failed to make any significant discoveries. Twitchell suggested that since Hasa was near the independent island state of Bahrain, it would be preferable to await the results of a Bahrain petroleum search before investing capital on exploration and drilling in the Hasa province.

In 1933 Standard Oil of California obtained a concession from Ibn Saud. The new Secretary of State, Cordell Hull, who did little to aid U.S. petroleum firms abroad during his early days in office, remarked, "The company obtained the concession on its own initiative....King Ibn Saud, suspicious of governmental diplomatic processes, had preferred to deal directly with company officials."[39] Under the terms of the concession Ibn Saud was to receive an advance payment of 30,000 gold sovereigns, as well as a royalty of four gold shillings per ton; the concession was to run for sixty years, until 1993. California Standard was to begin geological exploration on the 360,000 square mile concession by September 1933 and had to commence drilling operations within three years. It also was to employ Saudi Arabians "as far as practicable," and was to construct a refinery at the site after it had discovered oil to supply the needs of Saudi Arabia for refined products. According to Raymond F. Mikesell and Hollis B. Chenery, "this negotiation seems to have been practically the only case in middle eastern oil history in which the conditions approximated the standard of the 'open door.'"[40]

Having made its initial loan payment to Jiddah on August 25, the company executed a maneuver by which it technically assigned the concession to the California-Arabian Standard Oil Company (Casoc), a wholly owned subsidiary of the parent firm. This was incorporated in Delaware with a capitalization of $700,000. Then in an equally important development the Texas Company, or Texaco, became a partner of California Standard by purchasing a half-share in Casoc during 1936. While California Standard was in search of market outlets. Texaco wished to obtain additional sources of crude oil.

It was also during 1933 that the United States officially recognized the Ibn Saud regime, signing a treaty of amity and commerce with it. At this time Saudi Arabia was importing little (mostly automotive equipment) from the United States. The most curious American import was gasoline, which was provided primarily by New York Standard and Royal Dutch-Shell and was then sold in Jiddah for around 60 cents a gallon.

Five years later, in 1938, full-scale oil production began in Saudi Arabia, three years after drilling had begun. During the spring of the following year a pipeline of forty-three miles was opened, connecting the U.S. petroleum concession with the Ras Tanura terminal on the Persian Gulf; King Ibn Saud attended the celebration, performing the ceremonial function of turning the value of the pipeline, but despite the importance of the ceremony, there was no U.S. diplomatic representative present.

The flow of oil, however, would inevitably interest the American government in establishing more intimate contacts with Saudi Arabia. Some scholars have maintained that Ibn Saud granted the concession on political grounds, since he had opposed the French in Syria, the British in Palestine, and denied both Italy and Germany a political foothold in the country. Nevertheless, to quote New Left revisionist historian Lloyd C. Gardner, "Recent studies of the Middle East have pooh-poohcd that notion and instead insisted that he granted the concession to the Americans because they were the only ones who could pay what he demanded."[41]

## Lend Lease, Government Participation, and the Pipeline Scheme

The outbreak of World War II—and the eventual involvement of the United States in this conflict—drastically magnified the importance of Saudi Arabia in U.S. eyes. Unquestionably to officials of the American government the two most significant projects there were its possible participation in a Saudi Arabian oil company, and the planning of a pipeline from the Persian Gulf to the Mediterranean Sea. But the United States also extended Lend Lease to Ibn Saud during the conflict, and obtained permission from him to build an air base at Dhahran in

1944. Another key concern that emerged from these World War II negotiations with Ibn Saud was the American desire to prevent the escalation of British influence in Saudi Arabia.

During 1941 Ibn Saud demanded that the American oil firms pay an annual $6 million subsidy to the Saudi Arabian government, which at that time was in need of money. The Saudi budget at that time was approximately $10 million, and Great Britain had recently advanced it £400,000. Unfortunately the revenues from Meccan pilgrims had been declining and a recent drought had forced the king to feed some of his subjects. Although the petroleum companies probably could have raised the $6 million, they approached the U.S. government with a request that it grant Saudi Arabia a loan under the Lend Lease Act.

Although President Franklin Roosevelt was interested in the proposal, he doubted that he could supply Lend Lease to Saudi Arabia. An alternate approach was suggested by U.S. oil man James S. Moffett, who proposed that the U.S. Navy purchase $6 million worth of petroleum each year from Ibn Saud. If financial aid were not extended, Moffett warned there was a "grave danger that this independent Arab Kingdom cannot survive the present emergency."[42] FDR, though, doubted the legality of Moffett's scheme. Also unsatisfactory to the President was an alternate suggestion by his intimate advisor Harry Hopkins, who thought that instead of extending a loan the United States might send a shipment of food to Saudi Arabia under the Lend Lease program. Hopkins suggested using the royalties on the moneys Saudi Arabia expected to receive in the future from Meccan pilgrimages as collateral. But Hopkins was aware that even this approach had its problems: ". . . just how we could call that outfit a 'democracy' I don't know."[43]

Roosevelt did eventually evolve his own solution to this thorny dilemma, proposing that the British employ part of a $450 million Reconstruction Finance Corporation loan "to take care of the King's financial requirements," with the oil firms perhaps acting as middlemen between the British and Ibn Saud. The plan, however, did not appeal to the petroleum companies, which feared that it would enhance the prestige of Great Britain in Saudi Arabia at the expense of the United States. As the problems seemed insurmountable, the American government did not extend Lend Lease to Ibn Saud at that time.

Nevertheless, early in 1942 the Roosevelt administration offered to assist Saudi Arabia by sending a special agricultural mission which would be financed with the emergency funds available to FDR. It was also during 1942 that the American government set up a legation in Jidda. In 1943 the Chargé there, James S. Moose (who had served previously in Iran and spoke Arabic fluently), became a resident minister in accordance with the wishes of the oil companies.

One of the first problems Moose confronted as resident minister was what means the United States should employ to protect the petroleum fields of Saudi Arabia, thus denying the use of them to the Axis powers. The oil concessions were known to be vulnerable to air attacks; in 1940 the Italians had bombed both Dhahran and Bahrain but had only broken one pipeline. There was a real need for both antiaircraft guns and fighter planes. In Washington though, many government officials believed that the refinery on Bahrain Island was in greater danger than the oil fields of Saudi Arabia, and the War Department therefore had no immediate plans to send troops or equipment to the Arabian peninsula.

The planners also had to consider the possible demolition of the oil fields to prevent them from falling into Axis hands. The American oil firms could not undertake the demolition work themselves without threatening their concessions; furthermore, such a step would accomplish little unless it was accompanied by the destruction of the petroleum refinery on Bahrain Island. According to one California Standard official, the U.S. Army regarded Saudi Arabia as an area of British rather than American military responsibility. Fortunately, by the fall of 1942 Ibn Saud had agreed in principle to a defense program for his kingdom that involved the installation of antiaircraft facilities and the presence of U.S. troops.

But the oil companies apparently were as worried about the British threat to Saudi Arabia as they were about the German one. In February 1943 Texas Company President W. S. S. "Star" Rodgers and California Standard President Henry D. Collier drew up a memorandum after talking with certain government officials; among these were Interior Secretary and Petroleum Administrator Harold Ickes, Navy Secretary Frank Knox, War Secretary Henry L. Stimson, Under Secretary of State Sumner Welles, and Under Secretary of the Navy James Forrestal. In advocating direct U.S. government aid to Saudi Arabia, the two petroleum businessmen observed; "Concern is felt over the rapidly increasing British economic influence in [Saudi] Arabia because of the bearing it may have on the continuation of purely American enterprise there after the war."[44] Their memorandum favored extending Lend Lease and assuming Saudi Arabia's obligations to Great Britain. Apparently these talks and the memorandum had an effect on the Roosevelt administration, since on February 18, FDR issued an executive order that made Saudi Arabia eligible for Lend Lease.

By June 1943 Ibn Saud was telling Patrick J. Hurley, Roosevelt's personal representative, that Saudi Arabia looked primarily to the United States for support. In return Hurley remarked to the President that "Ibn Saud is the wisest and strongest of all the leaders I have met in the Arab States."[45] While admitting to Hurley that Saudi Arabia needed the friendship and assistance of a major power, Ibn Saud still feared foreign imperialism; he was prepared to deny entry to any business interest dominated by an imperialistic government. The petroleum resources of Saudi Arabia, therefore, would only be developed by American firms. Accusing the British of having violated the Atlantic Charter on at least two occasions and of scheming to impose imperialistic rule on the Arab states, Ibn Saud expressed confidence in the American commitment.

In his 1944 monograph *Petroleum and American Foreign Policy*, State Department Economic Advisor Herbert Feis minimized the threat that British financial assistance to Ibn Saud posed to American oil interests. The U.S. petroleum companies that had advanced $2,307,023 to Ibn Saud against future royalties and rentals in 1942 advanced him a mere $79,651 in 1943; conversely, the British government increased its advances from £403,000 in 1941 to £16,618,280 in 1943. By March and April of 1944 Cordell Hull had received reports of increasing British activity in Saudi Arabia that might jeopardize American interests.

As for developments in the United States, on June 30, 1943, President Franklin Roosevelt set up the Petroleum Reserves Corporation and had made $30 million available for its work by October. Interior Secretary Harold Ickes, who was the president of the Board of Directors, hoped the American government could acquire the stock of those U.S. petroleum companies active in Saudi Arabia. While the proposal was generally acceptable to the War and Navy departments, it met with the disapproval of both the American oil firms and the State Department, where Cordell Hull and Herbert Feis led the opposition.

Both Hull and Feis believed that Ickes' proposal would involve using strong-arm methods and power diplomacy, and neither California Standard nor the Texas Company would consent to the Petroleum Reserves Corporation even acquiring one-third of their stock, let alone a controlling share. After the State Department had adopted the position that the government should sign a contract with the American oil companies for the purchase of petroleum from Ibn Saud, Hull wrote to Ickes on November 13, suggesting that the Petroleum Reserves Corporation was hurting the United States both in the Near East and Saudi Arabia. Hull's eventual victory over Ickes on this government stock purchase question gains added luster retrospectively, when one realizes that FDR had originally tended to favor the American government's acquisition of all the U.S. petroleum companies' stock.

During August 1943 the California-Arabian Standard Oil Company (Casoc) placed an advertisement for men to work its properties in Saudi Arabia in the "help wanted" columns of a San Francisco newspaper. This single advertisement item brought no less than 3,000 responses. The petroleum firm also embarked on a long-range program for the economic and cultural development of Saudi Arabia. On July 27 Minister Alexander Kirk wrote to Secretary of State Hull, assessing the contributions of Casoc to date: ". . . building roads, planning irrigation projects, establishing schools, and even supplying and maintaining the amenities of civilized life for the King and other prominent Saudi Arabians."[46]

With the proposed governmental acquisition of stock now a dead issue, on February 6, 1944, Ickes reached an agreement with the American firms under which the U.S. government would construct a pipeline from the Persian Gulf to the Mediterranean Sea. The understanding was also to lead to the creation of a 1 billion barrel oil reserve in Saudi Arabia and Kuwait for the use of the American military. Gulf Oil, as well as California Standard and the Texas Company was a participant in this plan. In defending the undertaking, Ickes observed: "The primary purpose in that suggestion was to alert the British to the idea that we really meant business in the Middle East on oil."[47]

Although FDR approved the project, Secretary of State Cordell Hull was unhappy about its governmental ownership aspect, since he wanted the U.S. government neither to operate the pipeline nor to keep ownership of it after

World War II. The wrangling between Ickes and Hull apparently was so serious that FDR felt impelled to send the two secretaries a joint memorandum on January 10, in which he recommended that they come to some sort of an agreement. In his communiqué the President emphasized the importance of the time factor, noting that "after the war the American position will be greatly weaker than it is today."[48]

At first the British, who were busy planning the upcoming Normandy invasion and were miffed because they had not been given advance information about the plan, were reluctant to send high level negotiators to Washington to discuss the pipeline project and other matters concerning the Middle East. Although Foreign Secretary Anthony Eden favored the talks, there was much criticism of the plan in Parliament as well as in the British press. The *Economist* declared that the project heralded permanent American intervention in the affairs of the Middle East and noted that there had been no prior consultation with the British.

There were also varying reactions to the Saudi Arabian pipeline plan in the American press during the first half of 1944. On February 14 the *New Republic* complained that "we shall be lucky if we do not have later cause to regret this appeasement of the American oil companies" and that "we have a situation that looks dangerously like old-fashioned dollar diplomacy at its worst."[49] I. F. Stone compared the pipeline project in the February 26 issue of the *Nation* to the efforts of Standard Oil to encroach upon the British holdings in Iraq following World War I. Two days earlier the *Christian Century* had charged that "the principal consequences of this deal will be that the people of the United States will underwrite with the blood of their sons the investment of three oil companies."[50]

These journalistic critics found an ally on the floor of Congress in Democratic Representative John Coffee of Washington, who said that the American government should have followed the British example and insisted upon a partnership with the oil firms. Not every critic of the pipeline scheme was positioned to the left of center on the political spectrum. Oklahoma Republican Senator Edward Moore, who was an independent petroleum operator, denounced the project as a step in the direction of socialism. Both Moore and fellow Republican Senator Owen Brewster of Maine laid plans to introduce a bill abolishing the Petroleum Reserves Corporation, which was to be the financing agency for the pipeline.

Among the top oil men perhaps the leading critic of the pipeline scheme was James A. Moffett, who had only recently retired from California Standard and the Texas Company. The outspoken Moffett demanded that Ickes be fired—and possibly impeached—for his "Arabian adventure"; Moffett was highly displeased because "that obstructionist" Ickes wished to spend up to $165 million on a strictly commercial operation. The day after Moffett issued his press release, fifty-five members of the Petroleum Industry War Council, representing every major U.S. oil company except Gulf, the Texas Company, and California Standard, released a resolution stating that the pipeline scheme was not only unnecessary, but also a violation of the Atlantic Charter and a potentially dangerous commitment by the American government in an explosive area. Vice-president

James E. Pew of Sun Oil bluntly asserted that the pipeline was an invitation to World War III.

One of the most articulate supporters of the pipeline was *Newsweek* columnist Ernest K. Lindley. Pointing out that executive action had served the national interest well on several occasions throughout American history, Lindley noted that the prime mover in this scheme was not Ickes, but rather the U.S. Navy and the U.S. Army. Perhaps even more significant was the qualified endorsement of Everette L. De Golyer, one of the world's leading oil geologists. As De Golyer observed, "I hold no brief for the Government's Arabian pipeline [but] until some satisfactory substitute is found ... I am for the line."[51]

At the end of February 1944 the longstanding Anglo-American petroleum rivalry erupted again in an exchange of messages between President Franklin Roosevelt and British Prime Minister Winston Churchill. On February 20 Churchill cabled FDR that there were those in Great Britain who feared the United States wished to deprive the British of their Middle Eastern oil interests on which the British navy was dependent. Churchill added that the mere convening of an Anglo-American petroleum conference would result in a heated debate in Parliament. In answering the Prime Minister FDR made reference to possible British designs on the oil reserves of Saudi Arabia. Fortunately, however, Roosevelt and Churchill were able to reach an understanding on the Middle East rather quickly. On March 4 the Prime Minister summarized the meeting of minds in a message to the President:

> Thank you very much for your assurances about no sheeps eyes at our oilfields in Iran and Iraq. Let me reciprocate by giving you fullest assurance that we have no thought of trying to horn in upon your interests or property in Saudi Arabia. My position in this as in all matters is that Great Britain seeks no advantage, territorial or otherwise, as a result of the war. On the other hand she will not be deprived of anything which rightly belongs to her after having given her best services to the good cause—at least not so long as your humble servant is entrusted with the conduct of her affairs.[52]

During the subsequent months the prospects for the pipeline scheme grew progressively dimmer, although Ibn Saud had consented to a survey of his domain. At the beginning of the summer the U.S. Senate's special committee on national oil policy decided to postpone public hearings; the State Department feared they might interfere with the Anglo-American negotiations and there was strong sentiment in the Upper House against the U.S. government building the pipeline. It seemed unlikely moreover, that it could be completed by the end of the European war. The oil companies active in the Middle East continued to ship their petroleum by tanker until they could determine the cost of the pipeline. The petroleum firms were considering the possibility of approaching the American government for a loan to build the pipeline, if the transmission costs would not be lower than they were by tanker.

Yet pipeline or no pipeline, the Roosevelt administration continued to place great importance on harmonious relations with Saudi Arabia. Following the completion of the Yalta Conference in February of 1945, FDR and Ibn Saud met for talks on a warship. Roosevelt wished to obtain Ibn Saud's assent to the building of an air field in Saudi Arabia for the use of the U.S. military both during and after the war, but Ibn Saud had reservations over the nature of the arrangement and the length of the understanding. The Americans did begin to construct an air field at Dhahran before they had reached a final agreement with Ibn Saud, but Japan surrendered before the work on the air field was complete.

On February 22, 1945, the ad hoc committee of the State-War-Navy Coordinating Committee began to circulate a report on the "Recommended Procedure for the Extension of Financial Assistance to Saudi Arabia." The document concluded that "it is the wartime inability to develop the existing oil concessions in a normal commercial fashion which is the main source of the present budgetary deficits."[53] Up to this time the Saudi financial deficits had been met by grants from the British, Lend Lease, and royalty prepayments (which had been halted during the war) from the American oil firms.

Then in the late fall of 1945 the Export-Import Bank offered Ibn Saud a $25 million dollar advance, but the monarch refused it. Ibn Saud apparently objected to that provision he thought required him to pledge oil royalties to the repayment of the advance. The sponsors of the proposal, though, had merely wished to ensure that these royalties would not be extended to other lenders.

On August 8 Lord Beaverbrook, representing Great Britain, and Cordell Hull, representing the United States, had approved the Anglo-American petroleum agreement in its revised but still unacceptable form. Great Britain then gave its approval for U.S. oil interests to build a twenty-six inch pipeline from their holdings in Saudi Arabia to Haifa, Palestine. By this time the growing support in both Great Britain and the United States for the creation of a Jewish national state in Palestine had begun to incur the ill will of Ibn Saud and other Arab monarchs. Ibn Saud, however, expressed his satisfaction with the American petroleum firms, and went on to state that he differentiated between them and the American government. It seems unlikely that even under the circumstances, Ibn Saud would have substituted a British presence for that of the Americans.

While all of these international maneuverings were taking place during World War II, the Arabian-American Oil Company, as the California Standard-Texaco operation in Saudi Arabia became known in 1944, was significantly increasing its production. Prior to World War II Aramco had been able to market east of Suez only 12,000 to 15,000 barrels a day of its petroleum. During World War II, though, as a result of proliferating demand Aramco's production reached the 58,000 mark daily by 1945. Unfortunately, once the war was over Aramco's market began to shrink again, and California Standard and the Texas Company soon turned to Jersey Standard and New York Standard as new partners in their enterprise. At the same time the Middle East entered a tumultuous new era that saw the creation of a Jewish national state.

## Saudi Arabia Since World War II: The Nationalization of Aramco

In 1947 King Ibn Saud outlined an ambitious $270 million four-year plan for the development of Saudi Arabia. It involved the construction of railroads, highways, ports, airfields, schools, hospitals, electric power plants, and irrigation systems. By 1951 the railway between Dammam and Riyadh had begun operating. But the first Saudi budget had not been drawn up until 1948 and Ibn Saud had chosen to ignore it; then there was no published budget for another four years after that. In 1952, however, a U.S. financial mission reorganized the currency of Saudi Arabia.

Although a British army mission endeavored to modernize the Saudi military in 1947, it was again the United States that provided the bulk of this type of assistance, although Ibn Saud had rejected a $15 million Export-Import Bank loan in 1948 in protesting the favorable American policy toward Israel. The U.S. legation at Jiddah became an embassy in 1949; in 1951 the American government signed a mutual defense agreement with Saudi Arabia and in the same year it extended Point IV technical aid to the country. It was also in 1951 that the United States obtained a five-year renewed lease on the Dhahran air base.

Then in 1953 Idn Saud died, and Saud became king. Saudi relations with Great Britain worsened; there was a dispute over the Buraimi Oasis in 1953, and Saudi Arabia broke off diplomatic relations with both Great Britain and France at the time of the Suez Crisis three years later. The United States, however, in return for additional military assistance was able to obtain another five-year renewal of its Dhahran air base when King Saud visited Washington in 1957.

During the 1950s Saudi Arabia grew in a number of areas, but King Saud showed an inability to master the complicated world of finance. Deficits and irresponsible spending became the rule rather than the exception. As a result, Prince Faisal assumed in effect the leadership of Saudi Arabia as prime minister in 1958, although Saud technically remained king until he was deposed in 1964.

Examining petroleum developments in Saudi Arabia since World War II, between 1945 and 1947 the U.S. Navy bought $68 million worth of oil products from Aramco. But when rumors began to circulate in the United States that the Navy had paid too much for the oil, the Republicans, who had won control of Congress in the November 1946 elections, attempted to discredit the Truman administration by investigating these purchases. A special Senate committee headed by Owen Brewster of Maine began to make inquiries, but it discovered that the Navy had only paid $1.05 per barrel for the Saudi petroleum, which was well under the current world price. Aramco officials took the position that they in fact had saved the Navy as much as $26 million; according to Secretary of Defense James Forrestal, Aramco's product was "the cheapest oil, delivered, that the Navy ever bought."[54] Despite this claim, the British Admiralty was allegedly obtaining oil at $.40 cents a barrel from Anglo-Iranian, by virtue of the agreement under which the government of Great Britain obtained half of the Anglo-Persian stock in 1914.

Unfortunately, because of the continuing efforts of the American petroleum man James Moffett, this was not the end of the controversy. Early in 1947 Moffett, who had leveled the accusation of overcharging, filed a $6 million suit against Aramco, maintaining that the firm had not paid him for exerting his political influence with FDR in 1941, when Ibn Saud had asked the U.S. company for a similar amount. During the course of the trial the Aramco attorney observed, "I have yet to learn that friendship with the President is a saleable commodity,"[55] but in 1949 the jury awarded Moffett $1,150,000. Although he expressed his satisfaction with its decision, Moffett failed to get Ibn Saud to cancel the Aramco concession and then to award a new concession to him and his friends.

In August 1946 the Export-Import Bank had arranged a $10 million credit for Saudi Arabia to help that nation modernize its transportation and agriculture. Reassured of American goodwill by this gesture, Ibn Saud informed Aramco that he was not going to cancel its concession because of strong pro-Zionist sentiment in the United States, and that rumors to that effect were only "hot wind." Ibn Saud later made the same point to C. L. Sulzberger of the New York *Times*, stating that he had never even considered voiding the Aramco concession. With the U.S position in Saudi Arabia secure for at least the foreseeable future, Jersey Standard and New York Standard that December signed a contract with Aramco in what was then the biggest deal in the history of the American petroleum industry.

Under this agreement, which had the support of the State, War, and Navy departments, Aramco was to borrow $227.5 million from a group of banks. Jersey Standard and Socony were to guarantee $102 million of the loan. With this capital in hand, Aramco then was to construct a pipeline of 1,000 miles from eastern Saudi Arabia to the Mediterranean, build a deep water port at Dammam, lay the tracks of a short railroad, and install more refining equipment and connecting pipelines. The American oil firms hoped that these improvements would double, or even triple, Saudi petroleum output.

At the time Ibn Saud announced his $270 million four-year plan, Aramco already had been active on its own in helping Saudi Arabia to modernize. It had built houses at Dhahran for its Arabian laborers as well as for its American employees; it was supervising a 3,000 acre irrigation project. The U.S. firm labored under a handicap, since it was necessary to import almost everything that was to be used for development. Yet Aramco did more than just introduce American technology to Saudi Arabia. It demonstrated its respect for Moslem customs in making Friday, the Moslem Sabbath, the day of rest and in granting "prayer breaks" on work days, so that Arab workers could pray and bow toward Mecca. Aramco even encouraged its U.S. employees to study Arabic, although it is a difficult language to learn.

The growing petroleum receipts Saudi Arabia enjoyed, though, did not reduce friction between the Saudis and the Americans. In 1948 Ibn Saud began to demand his oil revenues in gold in accordance with the contractual agreement in the concession document. Aramco, which would be inconvenienced by the necessity of purchasing $20 million worth of gold annually from the Soviet Union or South Africa, offered to raise its petroleum royalty, but Ibn Saud (who distrusted paper money) remained adamant.

During this period the Department of Commerce suspended additional allotments of steel for the Trans Arabian pipeline, or Tapline. Concerned by the growing Middle Eastern tension, its Office of International Trade preferred to rely on tankers to transport Arabian oil from the Persian Gulf. In this decision, the Department of Commerce enjoyed the support of Secretary of State George Marshall, as well as Secretary of Defense James Forrestal, who had been an ardent Tapline supporter.

But the oil deposits of the Arabian Peninsula itself could also be found in the offshore areas. Therefore, in 1948 Aramco agreed to give up its rights in the Kuwait-Saudi Arabia Neutral Zone in return for the rights to offshore oil. The Saudi government, however, felt that the Aramco concession only covered the islands and territorial waters of the Persian Gulf. It did not believe that the concession extended to any rights Saudi Arabia might acquire in the event of a division of the continental shelf of the Persian Gulf among the riparian states.

So as to determine the extent of their rights, Aramco retained Judge Manley Hudson of the Harvard Law School, who sent his assistant Richard Young to Saudi Arabia to make an on-the-scene inspection; with the consent of Aramco the Saudis then consulted Young. These activities disturbed the British, who would have preferred to have an international convention divide the Persian Gulf. In the next year Saudi Arabia issued a proclamation defining the scope of its territorial waters and asserting its right of ownership to the minerals found therein. By December 29 Saudi markers had been placed on eighteen islands in the Persian Gulf. As a precedent for their actions, the Saudi government cited President Harry Truman's proclamation, dated September 28, 1945, which declared that United States enjoyed total jurisdiction over the continental shelf subsoil and seabed off Texas. The United States, though, considered the Saudi offshore claims too sweeping. But in the following year Saudi Arabia claimed most of the land area between Qatar and Buraimi, including part of the Sheikhdom of Abu Dhabi.

As Saudi Arabia was attemping to extend its boundaries, the Aramco territories within the country were shrinking. Beginning in 1947, when it surrendered its preferential rights to a large area in the western sector of the the country, Aramco relinquished its concession piece by piece. Two years later it made a similar withdrawal from the southwest. Even more relinquishments took place between 1955 and 1960, but Aramco retained 300,000 square miles of territory in the eastern region.

It was during 1950 that Aramco further altered its relationship with Saudi Arabia: it began to pay Ibn Saud for his oil in taxes rather than in royalties. The switch affected Aramco's tax relationship with the U.S. government as well. The tax payments gave the firm a dollar credit with Washington for every dollar paid to Saudi Arabia, while the previous royalty payments of a dollar had generated a tax credit of only 52 cents with the U.S. government. As a result of the change, Aramco would save approximately $80 million a year. Disturbed by this development, Democratic Senator Joseph O. Mahoney of Wyoming unsuccessfully suggested placing a tariff rather than a quota on petroleum shipments from Saudi Arabia. During the debate critics overlooked the fact that the increased Aramco payments to Saudi Arabia were an indirect way of extending financial assistance to that country. The U.S. government found this means "safer" than foreign aid, which Congress might reduce or terminate against the wishes of the Truman administration. The tax scheme also would discourage a nationalist takeover of the Aramco concession and counteract a possible British and Dutch challenge.

Ibn Saud was delighted with the new agreement. *Time* magazine opined that the American firm had made him the most generous offer in the history of Middle Eastern oil, since under the agreement the Saudi government was to receive 50 percent of the net income of Aramco. In return, Saudi Arabia promised to accept payment from Aramco in those currencies which the latter was paid for its oil, rather than in dollars and sterling exclusively. It was hoped that such an action would enable the American firm to do more business in countries like France and Italy. The major loser in this agreement was Great Britain, as the Anglo-Iranian Oil Company had been attempting to persuade the Iranian government to accept much lower royalties, and had been in the process of concluding a new contract with the Iraqi government, since it was one of the owners of the Iraq Petroleum Company.

Many observers have questioned whether Ibn Saud wanted the extra oil revenues to help uplift his nation socially, or wished to bolster his own luxurious life style and that of his relatives and friends. It was not, however, up to Aramco to dictate how the government of Saudi Arabia should spend its additional income; U.S. Ambassador J. Rives Childs was convinced of the futility of attempting to persuade the Saudis to implement much needed financial reforms. In any event, there is no doubt that there was a proliferation of new palaces, swimming pools, and Cadillacs in Riyadh during the post-World War II period. One thus could argue that Sheikh Salman al-Khalifah of Bahrain and his British advisor, Sir Charles Belgrave, spent that nation's oil revenues more wisely than Ibn Saud.

Ibn Saud, though, did decree some improvements for the benefit of his people. Once the railroad had been built from Dhahran to Riyadh at a cost of $50 million, the Saudi ruler proclaimed that a longer line connecting Riyadh and Mecca was to be constructed within five years. Measured against the cost of the railroad, the $400,000 that the U.S. government was extending yearly to Saudi Arabia under the Point IV program seems modest indeed; this sum was to be used to pay the salaries of experts on finance, agriculture, water, and public health.

Because Saudi Arabia was under authoritarian rule, there were those who were unhappy about conditions in the country, some of whom even dared to challenge the established order. In August 1953 nine employees of Aramco demanded "justice" for the 15,000 native Aramco workers. Saudi Arabia at this time had no labor law of any kind. When the U.S. company tried to settle the dispute the agitators, perhaps with Communist backing, escalated their demands.Even when the government of Saudi Arabia took charge of the negotiations, the agitators continued to adhere to a hard-line position toward Aramco and even stood up to the Ibn Saud regime itself. When the labor agitators were sent to jail, 13,000 Aramco workers walked off their jobs; during a riot in front of a police station, foreign laborers were attacked and a number of vehicles were stoned. Crown Prince Saud finally put an end to the twelve-day strike by threatening to send the workers back to their villages, where they had once labored for 7.5 cents a day. Confronted with this possibility, the laborers elected to remain members of a rootless industrial proletariat.

With the assistance of Aramco the Ibn Saud regime again employed the "iron fist" approach in the Buraimi Oasis boundary dispute. In 1952 the Arab Research Division of Aramco had published a book entitled *Oman and the Southern Shore of the Gulf* which was based on an Aramco study of the eastern part of the Hasa province. In August the Ibn Saud regime occupied the Buraimi Oasis by force, which led to a British blockade and renewed petroleum operations by the British in the disputed area.

Early in 1954 Great Britain agreed to submit the Buraimi Oasis dispute to international arbitration. The Americans looked upon this as a step in the right direction, and even the Saudis appeared to be favorably disposed at first. But by this time Aramco had officially adopted a position at variance with that set forth in private by one of its officers; throughout March, April, and May it continuously pressured the State Department into withholding its endorsement of arbitration. The State Department began to assume a more neutral position, as Great Britain prepared to go to arbitration, having withdrawn its prior condition that it be allowed to negotiate for concessionary rights in the disputed area. Meanwhile, in May Saudi Arabia had requested Aramco to drill for petroleum in the Buraimi Oasis, which the company was obliged to do under its contract. The British then threatened to shoot or arrest the members of the Aramco expedition, but after Secretary of State John Foster Dulles had protested their belligerent attitude, they contented themselves with dropping leaflets on this expedition.

At this point, the British made an agreement with the Sultan of Muscat, under which he would finance an army to safeguard Oman and reform the Trucial Oman Scouts which was led by British officers. In September 1955 the British delegate walked out of the boundary arbitration talks in Geneva, and in October Great Britain occupied both Buraimi and Oman with the locally raised and British directed forces. As a result, King Saud complained of "aggression in what rightfully is our territory," 30,000 Saudi tribal horseman demanded holy war, and Saudi Arabia entered into a defensive alliance with Egypt and Syria.

Secretary of State John Foster Dulles' Christmas Eve appeal to Great Britain, urging that that nation either withdraw from Buraimi and resume arbitration talks, or face the disintegration of the Anglo-American alliance, was to no avail.

One high-ranking British diplomat observed: "This is the last bit of empire we have. We must hang onto this 'thin red line' or we are finished."[56]

In the fall of 1957 Saudi Arabia and Kuwait entered into an agreement with the Japanese Oil Export Company concerning the offshore waters of the Neutral Zone. The Japanese-Saudi deal, which involved the creation of the Arabian Oil Company with a capitalization of $10 million, hardly pleased Aramco or the other foreign oil firms. Not only did the Japanese offer the Saudis 56 percent of the profits from the oil they pumped, but they also were willing to allow this formula to apply all the way across the board, encompassing even filling stations in Japan. In addition, one-third of the directors and employees of the firm were to be Saudis, while the Saudis were to have equal representation on the committee established to oversee the operation. This latter innovation apparently influenced Aramco, which in May of 1959 placed Sheikh Abdullah Tariki and one other Saudi representative on its board of directors.

During the same year there was a major change in the relationship among the four firms that comprised the Arabian-American Oil Company, that is, between the Texas Company and California Standard, and Jersey Standard and New York Standard. Throughout the 1950s both the Texas Company and California Standard had received "preferential dividends" of 10 percent on each barrel of oil they extracted in Saudi Arabia, which had amounted to $150 million apiece. This extra income was now to come to an end for the original two Aramco partners.

It was also in 1959 that Frank Hendryx, an American legal advisor to the petroleum directorate of Saudi Arabia, set forth the view that Saudi Arabia or any other sovereign state might legally alter, or even nullify, part or all of a concession contract. This opinion, first expressed in a paper given at the First Arab Petroleum Congress in Cairo, was highly disturbing to American and other foreign oil interests. Hendryx based his conclusions on seven earlier cases: one British, and six American.

Not surprisingly, this view encountered widespread opposition in the United States. Speaking in Dallas in 1960, Aramco General Counsel George W. Ray, Jr. noted that throughout their long relationship Saudi Arabia and Aramco had agreed on the sanctity of contracts; Hendryx's opinion, he felt, had no basis either in law or in morals. Moreover, eight years earlier, on December 21, 1952, the General Assembly of the United Nations had passed a resolution recommending that sovereign states give due regard "to the need of maintaining the flow of capital in conditions of security, mutual confidence, and economic cooperation among nations."[57] Six years after the UN resolution, in July 1958, the International Bar Association's Committee on Protection of Foreign Property in Time of Peace stated that the "... taking of private property in violation of a specific State contract is contrary to international law."[58] It should also be noted that the concept *pacta sunt servanda*, or respect for contracts, is a key provision of Islamic law.

During the final years of King Saud's kingship, as we have already pointed out, Prince Faisal was the "power behind the throne." In 1962 the Kennedy administration held talks with him in Saudi Arabia with the objective of encouraging reform within the country. On November 6 Faisal issued a ten-point program, calling for political, social, and economic change. A revolt in Yemen at this time saw the Saudis supporting the royalist cause; in 1963 when the Yemeni War threatened to spread across the border, the U.S. Government sent jet fighters to Saudi Arabia, and American military aid was to continue after Faisal assumed the throne in 1964.

Earlier, in 1962, the government of Saudi Arabia had set up the General Petroleum and Mining Corporation, or Petromin. Initially Petromin was to assume the internal petroleum marketing operations that had been handled by Aramco, later branching into other areas of oil activity. A year later, in 1963, Aramco agreed to the reduction of its concession to 125,000 square miles, and in 1967 the American firm sold its distribution facilities to Petromin, which was then building an oil refinery in which it owned 75 percent of the stock.

At the outbreak of the third Arab-Israeli war in June of 1967, Saudi Arabia declared a temporary embargo against the United States, Great Britain, and West Germany. But petroleum output in Saudi Arabia remained high. In 1968 Saudi oil production surpassed the 1 billion barrels mark annually for the first time, with Aramco and Saudi Arabia sharing in the enormous profits.

In June of 1968, however, the Saudi oil minister, Sheikh Ahmed Zaki Yamani, spoke at the American University at Beirut. Yamani, who had replaced the increasingly radical Abdullah Tariki, admitted that Petromin had been attempting to obtain a share in Aramco for some time. Yamani had favored participation privately for a number of years, but this marked the first time that he had endorsed the concept in public. Although he stressed in his Beirut speech that the more radical Arab countries were moving to take over foreign oil operations too rapidly, he nevertheless emphasized that take over was the eventual objective of the more conservative Arab nations as well:

> ...control over all oil operations is our objective. When this will be achieved is a big question, to which I'd hesitate to give a definite answer. But if our plans proceed without interruption ...I think the major part of our long-term plan will be achieved within 10 years from now. However, it may take up to 25 years to attain full control.[59]

It was not until March 1972 at a meeting of OPEC that Aramco agreed in principle to sell a 20 percent interest to the Saudi government. At the end of the year, after the major international oil companies had come to terms with five Persian Gulf states, Saudi Arabia obtained a 25 percent interest in the oil and gas producing operations of Aramco for approximately $500 million, with the option of increasing its participation to 51 percent by 1982. The purchase price was based on Aramco's net expenditures, adjusted for inflation, since 1932.

As these negotiations with Aramco were proceeding to their conclusion, the government of Saudi Arabia informed its American counterpart that it would guarantee an uninterrupted flow of quota-free, duty-free oil to the United States, if it were permitted to make heavy investments in the American petroleum industry "right up to the gasoline pumps." It was again Sheikh Yamani who made his surprisingly revolutionary proposal at the Middle East Institute in Washington. Among the more enthusiastic supporters of the scheme was James E. Akins, the director of the Office of Fuels and Energy of the Department of State. The departments of the Interior and Commerce also expressed approval of downstream investments by exporting nations in consuming countries.

Six months later, Saud al-Sowayal, the economic advisor to the Saudi ambassador, noting that some petroleum company executives were suspicious of the Saudis, also observed in Houston that the Saudis did not necessarily want a majority interest in the American firms in which they might invest. "They think," he lamented,"we want to make a hostage of the U.S. economy."[60] Yet growing dissatisfaction with American Middle Eastern policy led the Saudi government in the months ahead to threaten to curtail its petroleum shipments to the United States, a development that obviously threatened the proposed Saudi investment deal.

Despite its growing displeasure with the U.S. policy toward Israel, Saudi Arabia could hardly claim that it was not benefiting from Aramco's operations. Between 1962 and 1972 direct oil revenues to the Saudi government—most of them from the American firm—had increased from $410 to $2,779 million, as the nation's annual oil production approached the 2 billion barrel mark annually. Saudi Arabia was now the leading petroleum exporting nation in the world.

Summarizing Aramco's activities through the early 1970s, Joseph Walt has written:

> During its forty years in Saudi Arabia the oil company has provided the Saudi government technical assistance, agricultural aid, medical services, educational programs, trachoma and malaria research, and telecommunications service. Scrupulously avoiding involvement in domestic politics, it has carefully refrained from judgmental attitudes or negative criticism, even during those early years when it watched the Saudi monarchs squander millions on air-conditioned palaces and fleets of Cadillacs.[61]

Unfortunately, in 1973 Saudi-American relations worsened as King Faisal became aware that Egypt was going to war with Israel. In April of 1973 Sheikh Yamani visited Washington where he met with such high governmental officials as Henry Kissinger, George Shultz, William Simon, William Rogers, and Joseph Sisco. Rogers and Sisco attempted to interest Yamani in an alliance against Arab radicals and terrorists, but Yamani instead issued a public warning to the United States in the Washington *Post* about its Middle Eastern policy. He promised that his nation would increase oil production, if the United States cooperated with it.

As the fourth Arab-Israeli war approached, the Saudi government stepped up its campaign to pressure the U.S. petroleum firms into interceding with the American government, with the object of altering the latter's favorable attitude towards Israel. In going through the files of California Standard during the 1970s investigators for the Senate Foreign Relations Subcommittee on Multinational Corporations, who were following up leads furnished to them by newspaper columnist Jack Anderson, uncovered a document headed: "Confidential—Meeting with King Faisal," at which the Saudi monarch had suggested the need for a shift in U.S. foreign policy. The meeting apparently took place in May; within a week Aramco officials were lobbying at the White House and the State and Defense Departments on behalf of King Faisal's views. Then in June Mobil Oil placed a large advertisement in the New York *Times* calling for closer ties with the Arabs. An Exxon official also delivered a pro-Arab speech in New York City, while a former Socal chairman sent a pro-Arab letter to that firm's stockholders.

That fall, in a most unusual step, the chief executives of the four companies who were partners in Aramco dispatched a memorandum critical of Israel to President Richard Nixon on October 12, with a copy to Secretary of State Henry Kissinger. Even with these warnings, Secretary of the Treasury George Shultz talked of Arab "swaggering" at this late date; Shultz was only one of many government officials who believed that King Faisal was only bluffing in his hard-line stance.

The fourth Arab-Israeli war erupted in October. After the U.S. Congress adopted a resolution endorsing American aid to Israel, Saudi Arabia announced a 10 percent cut in its oil production and threatened to shut down its petroleum fields if the U.S. government continued to support Israel. Shortly thereafter the Saudis made good on this threat. Fortunately, however, the fourth Arab-Israeli war soon ended.

During the following year the movement toward the nationalization of Aramco progressed. The U.S. firm concluded an interim agreement with the Saudi government, under which it granted a 60 percent participation to the Saudi government, retroactive to January 1, 1974. The assassination of King Faisal in March 1975 did not halt the move toward nationalization. The main concern of the United States during the late 1970s was not whether or not the Saudis would obtain a 100 percent control over Aramco, but whether or not as one of the more conservative and pro-American states of the Arab world it would make its oil available to the United States, especially in time of crisis.

## Bahrain, Kuwait, and the Neutral Zone

Although the kingdom of Saudi Arabia occupies the bulk of the Arabian Peninsula, there are other smaller countries there as well that have been thrust into international prominence in recent years because they possess oil. One of these nations is Bahrain, a group of islands located just off the coasts of Saudi Arabia and Qatar in the Persian Gulf; the largest of these islands bears the name of the country. At one time Bahrain had a thriving pearl industry, but this has been on the decline since 1929, several years before the discovery of oil.

British influence has been strong in Bahrain since 1820. By the time of the start of World War I, in 1914, the sheikh then occupying the throne wrote to the

British political resident, "If there is any prospect of obtaining kerosene oil in my territory of Bahrain, I will not embark on the exploitation of that myself and will not entertain overtures from any quarter regarding that without consulting the Political Agent in Bahrain and with the approval of the High Government."[62] Eleven years later, in 1925, the Eastern and General Syndicate received a 100,000 acre concession there. This London-based firm was headed by Frank Holmes, a New Zealander, who then made unsuccessful attempts to sell his holdings to Anglo-Persian and Royal Dutch-Shell.

On November 30, 1927, the Gulf Oil Company obtained two option contracts from the Eastern and General Syndicate. Because of the Red Line Agreement, though, Gulf was unable to explore in Bahrain, so it transferred its holdings to California Standard in late 1928. Unfortunately, U.S. petroleum entrepreneurs still confronted a major obstacle. According to a ruling of the British Colonial Office that was handed down in October 1928, an American company (or any other foreign firm) could not control 51 percent or more of a Bahrain concession.

Then in March of 1929 Secretary of State Frank Kellogg had the American embassy in London ask the British government for a general policy statement on petroleum concessions in the Arabian Peninsula. Kellogg wished the embassy to emphasize that current U.S. legislation on concessions was extremely liberal. Two months later the British Foreign Office noted its willingness to admit in principle the participation of American oil interests in Bahrain but failed to issue a general statement of policy. Later in that year California Standard set up a British corporation in Canada, known as the Bahrain Petroleum Company, or Bapco.

The first representatives of California Standard arrived in Bahrain in May 1930, three months before the British Colonial Office approved the transfer of oil prospecting rights to it. By 1932 the first petroleum well had been brought in; the American role in the development of Bahrain's oil resources doubtless had a favorable impact on Saudi negotiations. In 1934 the first petroleum from Bahrain entered the world market, and by 1935 there were sixteen producing wells in Bahrain.

On March 23, 1934, the Minister to Iran, William H. Hornibrook, informed Secretary of State Cordell Hull that the Iranian government was maintaining that Bahrain belonged to Iran, and that under these circumstances the British and American oil interests desiring a concession in Bahrain had to obtain it from the Iranian government. These representations and protests were repeated in December. The State Department did not even reply but instead conferred with the British government over the matter; the Shah of Iran had protested to Great Britain at an earlier date, with an equal lack of success. The League of Nations also took no action, and Saudi Arabia and the other Middle Eastern countries refrained from recognizing Iran's claim to sovereignty over Bahrain.

With the threat of a possible Iranian takeover of Bahrain now on the wane, in 1935 California Standard sold a half interest in Bapco to the Texas Company, which had a large marketing complex in the Eastern Hemisphere. Bapco then began constructing a 10,000 barrel refinery. In 1940 the sheikh so extended the Bapco concession that it encompassed all of Bahrain. During that year Italian bombers attacked both the island and the Dhahran area of Saudi Arabia, but oil production continued.

After World War II, in 1952, Bapco entered into an equal profit-sharing agreement with the sheikh of Bahrain. Four years later, at the time of the second Arab-Israeli war, the British government sent troops to Bahrain because of the inflammatory activities of Arab nationalists in the country, who were threatening to destroy the refineries and the pipelines.

Like Saudi Arabia, Bahrain declared a total embargo on oil shipments to the United States during the fourth Arab-Israeli war of 1973. During this time Bahrain announced the termination of the agreement granting home port facilities to the U.S. Navy. By the middle of the 1970s even the smaller nations of the Persian Gulf no longer felt obliged to support the deteriorating American position in the Middle East.

To the north of Bahrain, sandwiched between Saudi Arabia and Iraq at the northern end of the Persian Gulf, lies Kuwait. In 1899, during the reign of Sheikh Mubarak the Great, Kuwait was placed under British protection to ensure its safety from German imperialists. At the end of World War I Ibn Saud also made an attempt to seize Kuwait, but he too was repelled.

Despite its relatively small size, Kuwait is incredibly rich in oil. In 1913 Sheikh Mubarak sent the British Political Resident a letter, in which he promised that "we shall never give a concession in this matter to anyone except a person appointed from the British government."[63] In 1925 Major Frank Holmes, who had received a concession for his Eastern and General Syndicate from Bahrain in the same year, obtained a Kuwait concession; then in a move that paralleled developments in Bahrain, after unsuccessfully approaching Anglo-Persian, Holmes sold his concession to Gulf Oil in 1927. The British government insisted that only British subjects or firms could obtain concessions in Kuwait. But Kuwait, unlike Bahrain, did not come under the scope of the Red Line Agreement of 1928.

Three years afterwards, in 1931, Gulf Oil sought the assistance of the State Department on behalf of its interest in Kuwait, after the State Department had interceded with the government of Bahrain two years earlier on behalf of California Standard. The State Department again insisted on an Open Door policy, and during 1931, 1932, and 1933 it carried on an extensive correspondence with the British government. In 1931 the British Colonial Office granted Anglo-Persian, which had withdrawn from Kuwait in 1926, another concession in Kuwait; finally in December of 1933 Anglo-Persian and Gulf Oil entered into an agreement under which they would jointly develop the petroleum of the sheikhdom. By this time Great Britain was softening its stance against foreign operations there.

A year later, in December 1934, these two firms obtained a concession from Kuwait and operated jointly as the Kuwait Oil Company. Gulf agreed not to market any of the petroleum produced by the Kuwait Oil Company in those places where it would compete with Anglo-Persian, particularly east of Suez. The joint undertaking involving American and British oil firms which failed to materialize in Iran during the early 1920s, became a reality in Kuwait during the mid-1930s. Backed by an exclusive seventy-five year concession, the Kuwait Oil Company began exploring and drilling for petroleum shortly thereafter. In 1938 it struck oil in what was the largest petroleum field in the world at that time, the Burgan field.

By 1942 there were nine producing wells in Kuwait. At this point the British army suspended oil operations in Kuwait for the remainder of World War II. Thus large-scale exporting of petroleum did not begin until 1946, and it was not until after the war that the United States opened a consulate in the country.

Lying between Kuwait and Saudi Arabia is that area known as the Neutral Zone, which is controlled by both governments. In 1948 Sheikh Ahmad of Kuwait awarded a sixty-year concession covering Kuwait's oil rights in the Neutral Zone to Ralph K. Davies and his American Independent Oil Company (Aminoil) for over $7 million. Aminoil consisted of a group of firms led by Phillips Petroleum and Signal Oil. A year later, King Ibn Saud of Saudi Arabia granted a concession covering Saudi oil rights in the Neutral Zone for sixty years to J. Paul Getty and the Pacific Western Oil Corporation. Getty received the concessionary right that Aramco had recently relinquished in return for an off-shore concession. Davies and Getty then coordinated many of their efforts, since their companies were to share equally in any oil extracted from the area. In 1953 Getty struck petroleum, and exports began the following year.

To return to Kuwait proper, in 1951 the government signed a new agreement with the Kuwait Oil Company, calling for an equal sharing of the profits. By 1955 production exceeded 1 million barrels daily, and the royalty payments to the sheikh amounted to approximately $250 million. Half of this money went into foreign securities, half into various types of improvements throughout Kuwait.

The early 1960s witnessed important political and economic changes in Kuwait. In 1961 the British voluntarily surrendered their extraterritorial rights there, and Kuwait became an independent, sovereign state. It then adopted a new constitution in 1962, and joined the United Nations in 1963, the same year that elections took place to the first National Assembly. As for oil developments, in 1960 Kuwait had set up a quasi-governmental agency known as the Kuwait National Petroleum Company. The government of Kuwait was to hold 60 percent of the Kuwait National Petroleum Company stock, and the private citizens of Kuwait the other 40 percent; it became the exclusive petroleum marketer in Kuwait during 1961 and built its own oil refinery for exporting purposes. In 1962 the Kuwait Oil Company surrendered half of its concessionary acreage to the state.

Between 1955 and 1965 Kuwait, despite its limited geographical size, led the Persian Gulf in oil production. The government of the country announced a total embargo on petroleum shipments to the United States in 1973 at the time of the fourth Arab-Israeli war, but it was not a major American supplier. A year later, in 1974, Kuwait endorsed a participation agreement with the Kuwait Oil Company, under which Kuwait would take over 60 percent of the operations and properties of the oil company. Then in 1975 the government of Kuwait announced that it was assuming 100 percent ownership of the British-American firm, but it did not approve the final agreement until 1976.

## Iran: From the D'Arcy Concession to World War II

In contrast with the nations of the Arabian Peninsula, the country of Iran, which was long known as Persia, has a historical tradition that dates back to before the birth of Christ. As in Turkey, there has been a major attempt since World War I to modernize Iranian life. Yet, despite a series of attempts to stimulate industrialization, Iran still remains basically an agricultural nation. Mineral deposits, aside from petroleum, are either of limited value or found in remote areas.

Both Great Britain and Russia attempted to expand their influence in Iran throughout the nineteenth century. As in the Ottoman Empire, there was competition for concessions; foreign entrepreneurs were interested especially in Iranian railroads, banks, tobacco monopolies, and public works. Under the Anglo-Russian settlement of 1907, the British obtained hegemony over the southern half of Iran and the Russians over the northern half.

During the early part of the twentieth century both Great Britain and Russia manuevered for Iranian petroleum. In 1901 a British adventurer, William K. D'Arcy, secured an extensive concession for oil and gas in the southern part of the country; at an earlier date, D'Arcy had made his fortune in the Australian goldfields. Seven years later D'Arcy struck petroleum in commercially significant quantities, and a year later, in 1909, D'Arcy formed the Anglo-Persian Oil Company in collaboration with the Burmah Oil Company. Iran was to receive 16 percent of the profits obtained from the concession. Under the leadership of Winston Churchill, however, in 1914 the British House of Commons engineered the purchase of a controlling interest in Anglo-Persian by the British Admiralty.

The Anglo-Persian Oil Company was less successful in its attempt to obtain a concession in the northern part of Iran, since the concession instead went to an independent Russian operator in March of 1916. Competing with Anglo-Persian for the concession were both the Sinclair and Standard interests. The former director of Iranian finances, Morgan Shuster, not only functioned as the agent of the American-Russian industrial syndicate, which represented Sinclair, but also as the fiscal agent of the Persian government in the United States. Standard attempted to win the goodwill of officials in Iran by advancing the financially shaky government a loan of $1 million, and after a long and bitter struggle that extended into the post-World War I period, eventually emerged as the victor over Sinclair. To William Appleman Williams, a "New Leftist" historian, this series of events remained "a bitter memory in the minds of Soviet leaders," who looked upon the affair as "proof of encircling capitalist imperialism."[64]

With World War I ended, Great Britain and Iran signed a treaty on August 9, 1919, that increased British control over Iranian finances, public services, and the army. In return Iran received a loan of £2 million, but the British were to enjoy control over Iranian tariffs and customs. The *Majlis* or Assembly, though, balked at ratifying the treaty, and Great Britain removed its military forces. Soviet troops then moved into northern Iran after the Communists had denounced earlier Tsarist treaties with Iran. After unsuccessfully protesting to the League of Nations, Iran signed a treaty of friendship with Russia in 1921 nullifying Iranian debts to Russia and Russian concessions in Iran.

That same year an uneducated soldier, Reza Khan, seized power in Iran, and restored central authority with the aid of the army. Reform had been fomenting there ever since the new constitution of 1906. Like his contemporary Mustafa Kemal in Turkey, Reza attempted to Westernize his country, but he was opposed by the traditional and conservative elements within the country. In 1922 Reza turned to the one-time economic advisor to the U.S. Secretary of State, Arthur C. Millspaugh, for aid in straightening out Iranian finances. Working under a five-year contract, Millspaugh reorganized the tax structure, made the collection of taxes more efficient, and balanced the budget. But finding the establishment of an Iranian republic impossible, Reza proclaimed himself Shah in 1925, thus ushering in the Pahlavi dynasty which was to last until 1979.

It is against this domestic and foreign background that the competition for petroleum concessions in northern Iran unfolded during the early 1920s. In 1916 a Russian entrepreneur named Khostaria obtained a concession in the northern provinces. Unfortunately for Khostaria, according to the Iranian Prime Minister, he had received it "under coercion by the Czarist Government and without the consent of the Persian Parliament."[65] Rather than attempt to operate in this unfavorable situation, Khostaria sold his rights to the Anglo-Persian Oil Company on May 8, 1920.

A year later, in 1921, Iran signed the previously mentioned treaty with the Soviet Union, which declared null and void not only the Khostaria concession, but also "all other concessions forced from the Government of Persia by the late Czarist Government for itself and its subjects."[66] Under the treaty Iran pledged to obtain Russian permission before transferring the Khostaria concession to a third state or its citizens. But earlier, in 1920, State Department oil expert Arthur Millspaugh had spoken with the Persian Minister, who told him that he personally favored the granting of a petroleum concession in the north to American interests. Noting that Iran was financially dependent upon Great Britain, the Persian Minister added that a private U.S. loan to his government would guarantee the authorization of such a concession, and that he had already approached American bankers on this subject. Millspaugh expressed his support for such a loan, which the Department of Commerce had suggested to the Department of State at an earlier date.

An increased American economic presence in Iran led the British Minister to Iran, Henry Norman, to recommend to his government that it promote Anglo-

American cooperation. The British Foreign Office, however, rejected his recommendation; the Foreign Minister of Great Britain, Lord Curzon, took the position that his government would allow only a symbolic American participation in the development of Iranian petroleum. This attitude did not persist for very long as the British soon came to prefer the possibility of an American presence in Iran to the activities there of Russian expansionists and left-wing nationalists. In the words of Sir Eyre Crowe, "Better Americans than Bolsheviks," while to another Foreign Office official the Russians were "the sole real peril."[67]

During 1921 the Iranian government entered into negotiations with Standard Oil for an oil concession in the northern provinces. On November 22 the American firm obtained a fifty-year concession under which Iran was to receive 10 percent of the gross value of all petroleum produced. The understanding greatly displeased Anglo-Persian, which then in conjunction with the British government complained to Standard Oil that it already held a concession there. Shortly thereafter Standard Oil and Anglo-Persian reached an agreement that they would develop the northern Iranian concession as equal partners, and on February 28, 1922, the State Department endorsed this understanding.

The loser in this arrangement was the Sinclair Oil Company, which according to its vice-president, A. C. Veatch, had been informed by Morgan Shuster, the fiscal agent of the Iranian government, that the reports of Standard Oil receiving a concession in the north were not true. Having learned that the rumors were true, Veatch wrote to Secretary of Commerce Herbert Hoover on December 23, 1921, that Sinclair had no intention of competing with another U.S. company in northern Iran.

Although the State Department was officially neutral in the rivalry between Jersey Standard and Sinclair for Iranian petroleum, its strong defense of the Jersey Standard/Anglo-Persian agreement showed an obvious partiality for Jersey Standard. In fact the State Department kept postponing an investigation into the validity of the Khostaria concession, which formed the basis of the Jersey Standard/Anglo-Persian agreement. When a preliminary report from the solicitor's office indicated that the Khostaria concession was not valid, Secretary of State Charles Evans Hughes not only failed to authorize further investigations, but he even withheld the preliminary findings from Sinclair.

In reaching the agreement with Anglo-Persian, Jersey Standard obviously felt that a conciliatory attitude toward the British in Iran might well lead to reciprocal benefits in Iraq. Furthermore, Jersey Standard did not have the transportation facilities necessary to ship crude oil from northern Iran to the Gulf, while it needed permission to transport petroleum through the Anglo-Persian concessionary area in southern Iran. Finally, Jersey Standard believed that an alliance with Anglo-Persian would help to provide security for a large U.S. investment in the Iranian petroleum fields.

Unfortunately, the Iranian government opposed this joint endeavor, preferring instead out-and-out American development of the concession. It took this position, however, only after the two foreign companies had made a joint advance of $1 million to the Iranian government in March 1922 against future royalties. Even before learning that a condition of the loan's acceptance was Iranian approval of the joint undertaking, the government had spent the money. The Iranian Assembly showed its displeasure at the Anglo-Persian/Jersey Standard alliance by amending the concession in June 1922, so as to allow the Iranian government to negotiate with Sinclair; the Persian Minister to Washington then began to encourage the latter firm. Although during this period Jersey Standard continued to discuss a possible loan with the Iranian government, nothing ever came of it.

Throughout subsequent months Sinclair continued to complain that it had not received proper support from the State Department. As the Soviet Union feared additional British economic penetration in Iran as a result of the proposed Anglo-Persian/Jersey Standard joint endeavor, it threw its weight against Jersey Standard and behind Sinclair, which had already obtained a petroleum concession from the Russians on Sakhalin Island in the Far East.

Sinclair won a temporary victory in June 1923, when the Iranian Assembly passed a law authorizing a petroleum concession in the north in return for a $10 million loan. In December Iran signed an agreement with a Sinclair representative that required the approval of the Iranian Assembly. The Tehran journal *Naserelmelleh*, though, strongly opposed the agreement, and charged on January 15, 1924, that Sinclair had given a $275,000 bribe to the Iranian Prime Minister.

Shortly thereafter Jersey Standard made it known that it would take the proper steps to protect its rights and to develop its northern concession. On February 21, 1924, the Persian Minister, Hussein Alai, wrote to Secretary of State Charles Evans Hughes, questioning the sincerity of Jersey Standard's claims, and lambasting it for allying itself with Anglo-Persian. Jersey Standard, Alai pointed out, had been "repeatedly warned by Mr. Shuster and myself of the strong feeling of suspicion inevitably entertained in Tehran, in view of past experiences, as to British motives and aims."[68]

By this time, Sinclair, too, had experienced problems with the Iranian government, after the American vice consul, Major Robert Imbrie, had been murdered by a Turkish mob in the summer of 1923. Although the official explanation for his murder was that Imbrie had been photographing a holy place and had thus enraged the local inhabitants, Harold Spencer, a long-time British secret service agent, offered a quite different reason. According to a New York *Herald Tribune* dispatch from Paris dated September 27, 1924:

> [Major Imbrie] was assassinated by a mob organized by financiers in the United States and England, who thought his influence might swing control of the Persian oil fields from the Shell group to an American syndicate in which the Sinclair group has the major interest.[69]

Shortly after Imbrie's death the Sinclair representative, Ralph H. Soper, left Iran as the Imbrie murder had sharply reduced the chances of floating an Iranian loan on the American market. Although the Iranian Prime Minister now tended to believe that Sinclair had lost interest in the northern concession, on October 14 Sinclair cabled the Iranian government that it would accept the concession, assuming that the loan was not a condition of the agreement. Sinclair, having obtained transit rights from the Russians, apparently expected to ship the oil through the Soviet Union, provided that the American government diplomatically recognized the Moscow regime. But this event in fact did not take place until 1933.

Back in the United States, the Teapot Dome scandal of 1924 tarnished Sinclair's reputation. Consequently, a year later Sinclair decided to halt its Iranian petroleum activities. In 1927 Jersey Standard, unable to reach an agreement with the Iranian government, also withdrew from the field. As a result, the Soviet Union became the leading foreign power in the north.

The British, however, remained in control of their highly profitable oil concession in southern Iran, but by 1932 the Iranian government for a number of reasons had become disenchanted with Anglo-Persian. It did not approve of the company's method of computing profits, or of the fluctuating royalties; in addition, during 1931 the government of Iraq had signed a new agreement with the Iraq Petroleum Company that was highly favorable to Iraq. Thus in November of 1932 the Iranian government decided to cancel the Anglo-Persian concession. The British sent warships to the Persian Gulf, and took their case to the council of the League of Nations and the International Court of Justice, but without result. According to Mira Wilkins, this was the first major clash between a Middle Eastern government and a petroleum firm.

By the spring of 1933 Iran and Anglo-Persian had agreed upon a new concession that included terms more favorable to the Iranian government. The geographical area of the concession was cut in half, and was to shrink to 100,000 square miles after 1938; Anglo-Persian lost its monopoly over the construction and operation of oil pipelines and was forbidden to transfer its concession to another party without the consent of Iran. Finally, under the agreement all of its Iranian holdings were to become the property of the Iranian government in 1993.

In the following year Anglo-Persian again became involved in a dispute with the Iranian government, which claimed that Bahrain was a part of Iran, and that British and American oil firms interested in Bahrain's petroleum should approach the Iranian government. Shah Reza won satisfaction from neither Great Britain nor the United States, but the British did change the name of Anglo-Persian to Anglo-Iranian in the following year in accordance with the wishes of the Shah, who wanted his country to be referred to as Iran by the outside world, as it had long been inside the land.

After 1927 the United States played only a minor role in Iranian petroleum. In 1932 the U.S. legation in Tehran sent a dispatch to the State Department, in which it bluntly declared that: "there is every just ground for armed British intervention in south Persia to protect and maintain protection of this, Britain's most important interest in Persia. . . ."[70] Although in 1937 the Amiranian Oil

Company obtained a drilling permit covering 200,000 square miles in northeastern Iran, this firm suspended operations a year later even though it held a sixty-year concession. The Seaboard Oil subsidiary was plagued by falling crude oil prices. According to State Department Economic Advisor Herbert Feis, Amiranian "never sank a drill. The supporters of the venture faltered over the economic and political hazards."[71] Despite this abortive undertaking, the American desire for Iranian petroleum escalated to a fever pitch during World War II.

## Wartime Oil Rivalry Among Three Allies

The outbreak of World War II found Shah Reza pro-German, anti-Russian, and anti-British. On August 25, 1941, shortly after Germany had declared war on Russia, Great Britain and the Soviet Union occupied Iran, forcing Shah Reza to abdicate in favor of his son. During 1942 Iran, Great Britain, and Russia signed a tripartite treaty of alliance; during the same year the United States began to make Lend Lease available to Iran. In September of 1943 Iran declared war on Germany, several months before Franklin Roosevelt, Winston Churchill, and Joseph Stalin met in a summit conference at Tehran.

At one point during World War II, there were as many as 30,000 U.S. troops in Iran. Arthur Millspaugh returned to straighten out Iranian finances, Colonel Norman Schwarzkopf reorganized the police, and an American military mission advised the army. As for the Russians, pro-Soviet elements in Iran organized the Tudeh party in 1941, and recruited many supporters in Azerbaijan and Kurdistan. Thus at the same time that the Iranian government was helping the Russians by permitting the Allies to ship an enormous amount of goods to them through Iran, it was worrying about Russian activities within the country.

Despite the attempt by Richard Cottam in his *Nationalism in Iran* to minimize the importance of oil in Iranian-American relations during this era, most authorities have emphasized its significance. During the winter of 1941-42, for example, the British requested Lend Lease funds for the construction of several pipelines across Iran. Without assurance that these pipelines would be made available to American oil firms after the war, though, the United States was hesitant to give its approval. Financial advisor Arthur Millspaugh also displayed a great interest in the petroleum of Iran. According to Justus Doenecke, "Millspaugh became the main subject of Iranian politics, for his oil activities and personal arrogance seemed even to American embassy officials to encourage the extreme nationalists."[72] Millspaugh also aroused the suspicions of both the Russians and the British.

Aside from the interest the U.S. government displayed in Iran, secret oil talks were taking place between the Standard-Vacuum Company and the Iranian government. Colonel John H. Leavell had concluded from aerial surveys that the prospects for successful oil exploitation in the north were excellent; consequently, in February 1943 the Iranian commercial attaché in Washington asked Standard-Vacuum if it would like to acquire a concession. Standard-Vacuum was interested in the proposal, but decided to wait until fall before responding fully. On

November 15 the State Department concluded that there was no reason why Standard-Vacuum could not go ahead with its plans for an oil concession in northern Iran.

The Iranian petroleum controversy reached its wartime climax in the following year. Both Standard-Vacuum and Sinclair sent representatives to Tehran early in 1944; in addition, the Shah appointed A. A. Curtice and Herbert Hoover, Jr. as his advisors on oil. At this point, though, the Soviet Embassy advised the press that Russia had prior rights to the petroleum of Iran. In September a Soviet delegation arrived in the country, demanding an exclusive, five-year concession for oil exploitation in the north.

Various interpretations were placed on this attempted push into northern Iran, and many American, British, and Iranian officials were highly suspicious of Soviet motives. Yet, according to George F. Kennan, at that time a counsellor attached to the American Embassy in Moscow, "The basic motive of recent Soviet action in northern Iran is probably not the need for oil itself but apprehension of potential foreign penetration in that area..."[73] Kennan's interpretation was later challenged by revisionist historian Denna Flemming, who in a two-volume study of the origins of the Cold War published in 1961 singled out petroleum as a major factor in Russian demands on Iran during World War II.

While oil negotiations were going on, the State Department took the position that there would no joint Anglo-American concession. A Sinclair representative felt that Iran would never grant a concession to Standard-Vacuum because it was too closely identified with the Anglo-Iranian interests. Although the State Department emphasized that it was assuming an impartial stance toward the Sinclair and Standard-Vacuum negotiators, the Chargé in Iran, Richard Ford, feared that the entrance of Sinclair into the negotiations would weaken the position of both petroleum companies.

Faced with unrelenting pressure from three sides, Shah Mohammed Reza Pahlavi and Prime Minister Mohammed Said informed the U.S. Ambassador, Leland Morris, on October 9 that they were suspending oil negotiations for the remainder of the war. While the British government adopted a cooperative attitude, the Russians attacked the Iranian government in their press and were partially responsible for the fall of the Said regime a month later. In sharp contrast, the United States opposed any action that might constitute interference in the internal affairs of Iran, and upheld the Tehran Declaration of December 1943 which guaranteed Iranian independence and sovereignty.

Despite Soviet belligerency, on December 5 the Iranian *Majlis* enacted a new law that made it criminal offense for a cabinet minister to enter into petroleum concession negotiations with foreign representatives without the prior approval of the Assembly. This law tied not only the hands of the present cabinet, but that of future cabinets as well. The Russian government found the law both objectionable and obstructive and demanded its revision. On December 28 the Soviet Union turned its wrath against the United States; Soviet Ambassador Andrei Gromyko wrote Secretary of State Edward Stetinnius a forceful letter setting forth the Russian position on Iran. Gromyko asserted that a Soviet petroleum concession would not jeopardize Iranian sovereignty, since the British long had held a concession there that did not threaten Iranian independence. A Soviet concession, he claimed, would be a means of lending economic assistance to Iran and would not violate the Tehran Declaration; therefore, there was no reason to postpone the granting of a concession until the postwar period.

On the other hand, the State Department's "Memorandum Concerning Iran," dated January 1945, took the position that a Russian concession would violate the three-power declaration. It went on to state, however, that the American government should not be responsible for the Iranian government rejecting the Russian request. It asserted that the United States was not playing the role of a "buffer" in Iran and was not preparing to defend that country by armed force. Instead, according to the memorandum, the American government should make a special attempt to bring Russian and British negotiators on Iran together for the purpose of formulating a joint Iranian policy.

At the Yalta Conference on February 8 Foreign Minister Anthony Eden stated that Great Britain was not trying to block the Soviet Union from obtaining petroleum from northern Iran, or from securing an oil concession should the Iranians be prepared to negotiate. Foreign Minister V. M. Molotov, though, remarked that the Iranian petroleum question was not that urgent. Equally important to the Americans and the British was the question of troop withdrawals from Iran, especially the Russian ones. With Stetinnius' support, Eden stated that troop withdrawals should take place before there was any additional discussion of oil concessions. But neither Roosevelt nor Churchill nor Stalin discussed Iranian affairs at Yalta, despite the proximity of Iran to the Soviet Union.

In the fall of 1945, after Japan's capitulation, the Seaboard Oil Company prepared to send a representative to Iran to negotiate an agreement for the development of the country's petroleum resources. Seaboard Oil wished to discuss a managerial relationship rather than a concession, although the managerial relationship would probably assume lines similar to those of a concession, since substantial foreign capital would be required. Far from greeting this proposal with enthusiasm, the Ambassador to Iran, Wallace Murray, suggested that its implementation might retard Soviet troop withdrawals and encourage a Russian demand for a Soviet petroleum concession. Standard-Vacuum and Sinclair, moreover, were likely to become involved. As a result, George V. Allen, the Deputy Director of the Office of Near Eastern Affairs in the State Department, wrote a discouraging letter on October 12 to the President of the Seaboard Oil Company, John M. Lovejoy, pointing out that the deadline for the withdrawal of foreign troops from Iran did not occur until March of 1946. On November 28, 1945, the Chief of the Petroleum Division, John A. Loftus, sent a similarly negative communication to Standard-Vacuum's President, Philo W. Parker, thus putting an end to any new participation by American oil firms in the development of Iranian petroleum at this time.

## Mohammed Mossadegh and Nationalizing Anglo-Iranian

During the immediate postwar years, Greece and Turkey were of greater importance to the United States than Iran. Although the American government did make a grant to the Iranians for military surplus purchases in 1947, the *Majlis* delayed action on this and no arms reached Iran until 1949. During that year the Shah visited the United States, and President Harry Truman promised Iran Point IV aid; the American government eventually made $500,000 available to Iran under this. In 1950 the Export-Import Bank authorized a $25 million loan for Iran, as the Iranian government had begun to implement its first seven-year plan of domestic reforms.

The United States, however, did not make the granting of an oil concession an integral part of its post-World War II relationship with Iran. The Soviet Union did. The Russians refused to remove their troops from northern Iran after World War II until the Iranian government signed a petroleum agreement with the Soviet Union. In April 1946 the Iranians reached an understanding with the Russians; under this a joint Soviet-Iranian oil company was created to prospect for petroleum in northern Iran. The agreement displeased the British, who were concerned about a provision that gave the Russians 51 percent of the stock in the firm for twenty-five years, and then 50 percent for an additional twenty-five years. The Iranians did not give their final consent to this understanding immediately, but instead entered into oil negotiations with the Americans. On July 30 the Shah told Ambassador George V. Allen that it would be necessary to balance a Russian petroleum concession in the north with an American one in the south; Allen, though, replied that while it was feasible to plan, internal and external political considerations forced the United States to wait on an actual concession.

At the end of 1946 Anglo-Iranian agreed to sell substantial quantities of Iranian crude oil to Jersey Standard and Socony-Vacuum at moderate prices over a twenty year period. As a part of the agreement, none of the petroleum was to be distributed east of Suez. In return, the two U.S. firms agreed to help finance a new pipeline that would extend from the Persian Gulf to the Mediterranean Sea. Although the deal was obviously beneficial to both parties, in London the *Economist* observed that "other countries have some excuse for nervousness," since it raised the threat of an "Anglo-Saxon corner on the world's oil."[74]

Iranian officials debated the merits of the April 1946 oil pact with the Soviet Union during the next year. Under Secretary of State Dean Acheson felt that if the *Majlis* rejected it, Iran should not grant a petroleum concession to anyone in its northern provinces. On October 22, 1947, the *Majlis* voted 102 to 2 to void the agreement, but this rebuff did not terminate Soviet interest in Iranian oil, as events during the years ahead would demonstrate.

Throughout 1948, as an anti-inflation move, the British Chancellor of the Exchequer, Sir Stafford Cripps, asked British firms to keep dividends to a minimum. Unfortunately, if Anglo-Iranian was to do so, it would result in a sharp reduction in petroleum royalties to Iran. Anglo-Iranian instead offered to revise its 1933 agreement with the Iranians; in July 1949, Anglo-Iranian reached a compromise that would double Iranian oil royalties. Far from being satisfied with the compromise, members of the *Majlis* interpreted it as an admission that the old royalty rate had in effect "cheated" Iran. The new reform government of General Ali Razmara therefore withdrew the Anglo-Iranian proposal from consideration by the *Majlis* at the end of 1950, without a vote on it.

Assessing the issues at stake in the Anglo-Iranian negotiations, Ambassador John C. Wiley wrote Secretary of State Dean Acheson on April 28, 1949:

> British Ambassador states AIOC production is at peak level and Iran Government is devising every possible means to cash in. It has made some 20 to 25 specific demands, including Iranization and modification of arbitration clause. Moreover, Iran Government desires 50% gross profits. British Ambassador states that so many elements enter into gross profits that this presents an impossible condition. Moreover, he says, Board of Directors AIOC almost entirely Scotch and will, under no circumstances, accept conditions which would deprive them of control.[75]

It became only a matter of time until the deteriorating relationship between the British oil firm and Iran terminated in an explosion. The final rupture occurred early in 1951, after a religious fanatic assassinated Prime Minister Ali Razmara, who had favored the *Majlis* decreeing a 50-50 formula, on March 7. Since Aramco had announced a 50-50 deal with King Ibn Saud of Saudi Arabia in December 1950, Anglo-Iranian had decided to follow suit in Iran. But on March 8 the *Majlis* oil commission adopted a resolution advocating the expropriation of Anglo-Iranian, then worth approximately $750 million. According to Anthony Sampson, an Iranian official told the Anglo-Iranian general manager at Abadan that the seizure would not have occurred if Winston Churchill still had been prime minister. Having expropriated Anglo-Iranian, Iran then set up a state firm, the National Iranian Oil Company.

The official American position toward the expropriation was a "hands off" attitude. Secretary of State Acheson later wrote of the British oil firm officials that "their own folly had brought them to their present fix, which Aramco avoided by (in Burke's phrase) graciously granting what it no longer had the power to withhold."[76] The U.S. failure to support the British more strongly on this issue disturbed Great Britain, which already was incensed by the critical remarks of the U.S. Ambassador to Iran, Henry P. Grady. Officials in Washington nevertheless felt that the Americans and British had many similar interests in Iran that might be undermined if they followed an identical policy, thus undercutting whatever influence the United States might have with the Iranian government. The State Department suggested settling the dispute through negotiation rather than by intimidation or threats of unilateral action; from the practical standpoint, the act of seizure might lead customers to seek oil elsewhere. Voice of America broadcasts at this time stressed the expropriation of petroleum properties by Iran without adequate compensation, instead of the act of seizure itself.

Rather than supporting Anglo-Iranian, a number of U.S. oil firms contemplated offering their technical assistance to Iran, which would undermine Great Britain's bargaining point: the lack of Iranian technical know-how to operate the petroleum properties. Although Assistant Secretary of State George McGhee asked the major American oil companies to adopt a hands-off position and not bid for managerial contracts from Iran, the U.S. petroleum firms refused to cooperate. Were they not to compete for these contracts, they claimed, then some non-American companies would inevitably obtain them to the detriment of the U.S. oil industry. Many American firms, too, were more disturbed by the precedent Iran had established in the expropriation of Anglo-Iranian than by the plight of Anglo-Iranian.

The American government had to consider the availability of an adequate supply of petroleum not only for the United States but for the free world as well. Under the leadership of Secretary of the Interior Oscar Chapman, who also was the Petroleum Administrator for Defense, and Director of the Office of Defense Mobilization Charles E. Wilson, the Foreign Petroleum Supply Committee was created on June 25, 1951. Through a joint voluntary agreement, nineteen U.S. oil companies that were active overseas served as members of the committee. On August 2 the committee obtained authorization to proceed with a four-point program, which it hoped would alleviate the looming oil shortage by increasing crude oil production in at least eleven nations and stepping up the manufacture of refined petroleum products in at least twenty-seven countries.

Earlier, on June 11, Prime Minister Mohammed Mossadegh of Iran had written a letter to President Harry Truman, attempting to justify the seizure of Anglo-Iranian. Among other things, Mossadegh accused the British oil firm of keeping inaccurate books. Speaking of the Iranian people, he complained that "the sole owners of the oil. . . have suffered these events for a good many years, with the result that they are now in the clutches of terrible poverty and acute distress."[77] Mossadegh went on to charge Anglo-Iranian with hiring secret agents, who had used economic pressure to undermine reform movements in Iran. At the same time the prime minister expressed friendship for the United States. Three days after the creation of the Foreign Petroleum Supply Committee, on June 28, Mossadegh again wrote Truman. Here he observed, concerning the Anglo-Iranian seizure, that: "the measures for the enforcement of the law were taken in a very gradual manner and with extreme care and caution...."[78]

During the following month, in July, President Truman sent Averell Harriman to Tehran as his personal representative. Harriman persuaded the Iranian Prime Minister to at least talk with the British, but when Great Britain offered an eight-point program for Mossadegh's consideration which Harriman described as a "reasonable basis for negotiations," Mossadegh totally rejected the plan. As Harlan Cleveland observed in the *Reporter*, the Prime Minister obviously "would rather see the oil flow into the sea than have the British remain on Iranian soil."[79] Without petroleum shipments there would be no funds with which to compensate Anglo-Iranian for its properties; under these conditions the takeover by Iran would amount to open confiscation. Yet Truman's suggestion on July 9 to allow the International Court of Justice to find a *modus vivendi* was also unacceptable to the Iranians.

In lieu of oil royalties, Mossadegh sought massive economic aid from the United States. At the time of the assassination of General Razmara a large "Point Four" program for Iran had been pending, but Mossadegh later sought ten times as much. If the United States granted Iran this huge sum, it would strain the Anglo-American alliance and postpone the settlement of the petroleum dispute; without U.S. financial aid Mossadegh would remain in a precarious position, which might encourage the Communists to attempt a coup. (Despite this threat, he was not seeking military assistance from the U.S.) Rather than adopt either alternative, the American government decided to offer the Iranians a compromise package valued at $23.5 million. In addition, it made an unsuccessful joint proposal in August with Great Britain for the settlement of the petroleum dispute, which featured an offer of $10 million in U.S. aid pending the resumption of oil revenues.

Late in 1951 Mossadegh visited the United States in search of a loan, but received no more than an understanding that the American government would treat any Iranian request with "sympathetic consideration." By this time the State Department had come to believe that Iran should sign a contract with another petroleum giant such as Royal Dutch-Shell to operate the Abadan refinery. While under the plan the Iranian title to the oil properties would be recognized, Iran would sell most of its petroleum to Anglo-Iranian for a number of years at the world price. Mossadegh remained adamant.

Prospects for a settlement dimmed even further in March of 1952, when the efforts on behalf of the International Bank to develop an agreement acceptable to both Great Britain and Iran collapsed. Four months earlier, the International Court of Justice had claimed that it had no jurisdiction over the dispute. The following month Mossadegh had rejected an Anglo-American offer to determine the amount of compensation due to Anglo-Iranian as a result of the expropriation.

That fall the newly appointed Iranian Ambassador to the United States, Allayhar Saleh, presented his credentials to President Truman on September 24. According to John F. Simmons, the Chief of Protocol, Saleh reported a hardening of Iranian public opinion against America:

> He described the sentiments of his people as gradually turning away from the United States, due to the association, in their minds, of our recent policies as being parallel to, and bound up with, the unpopular policies of Great Britain. He said that his people had a distinct resentment against past encroachments both of Russia and Great Britain, but that there was a danger now that the encroachments of Great Britain would become increasingly important in their minds. He said, more specifically, that it was very distasteful to Iranians to hear about recent oil policies in which the names of President Truman and Winston Churchill were associated. The implication of what he said was that the Americans, through this type of association, were rapidly becoming as much disliked as the British had always been.[80]

Republican Dwight Eisenhower won the presidency in the American general election held that November. A month later, on December 6, the State Department issued a press release that weakened the Anglo-American front. This document, which apparently was issued with the acquiescence of the incoming Eisenhower administration, stated:

> Under present circumstances, this Government believes that the decision whether or not such purchases of oil from Iran should be made must be left to such individuals or firms as may be considering them, and to be determined upon their own judgment. The legal risks involved are matters to be resolved by the individuals or firms concerned.[81]

By the middle of January 1953, a new round of talks between the British and Mossadegh was underway. At this point Mossadegh agreed to all the major points proposed and announced to his supporters in the Iranian *Majlis* that there might be a settlement of the petroleum dispute within a few days. In the search for the basis of the settlement, he even showed interest in the terms of the 1948 agreement between Great Britain and Mexico. He was reluctant, however, to agree to the British demand that Anglo-Iranian compensation should include damages for the loss of future business under the 1933 contract, which had been cancelled forty-two years before its scheduled expiration date. Unfortunately, Mossadegh's political support in Iran had begun to disintegrate seriously. As a result, by March 20 Mossadegh had returned to his original hard-line stance; he would only compensate Anglo-Iranian for its physical assets.

In a radio address delivered on this date the Prime Minister complained of "foreign machinations" and "greedy foreigners," describing the Anglo-Iranian operation as "a form of plunder for which there is no precedent anywhere in the world."[82] Yet this dramatic speech did not consolidate Iranian public opinion behind him. More and more Mossadegh became dependent on the Communists, whose mobs the police were unable to control. Desperate for a way out, the Prime Minister wrote to Eisenhower on May 28, requesting that he either solve the oil problem by having the United States purchase Iranian petroleum or extend additional aid to Iran. After withholding his response for a month, Eisenhower flatly rejected Mossadegh's request, taking a major gamble in doing so. At this point Eisenhower recommended referring the question of compensation to some neutral international body.

On August 13, the Iranian political scene exploded when the Shah fired Mossadegh after Mossadegh had usurped his prerogative in deciding to dissolve the *Majlis*. Mossadegh was to be replaced by General Fanzullah Zahedi, but he ignored his dismissal, and the Shah departed for Baghdad and Rome, and Zahedi for the provinces. The possibility of a Marxist takeover in Iran loomed larger than it ever had before. When mobs tore down the statues of the Shah's father, though, the royalist elements of the Iranian army formed a countermob; shortly afterwards the forces of the Shah and Zahedi won a minor tank battle. Zahedi also enjoyed the support of Kermit Roosevelt, a CIA operative who was on the scene in Tehran. After the Shah had returned to Iran on August 22, Mossadegh was placed under arrest, and Zahedi took office with a new cabinet.

Although Zahedi had been suspected of pro-Nazi sympathies during World War II, the West believed he was more likely to reach an understanding with the oil companies than was Mossadegh. Eisenhower displayed his goodwill toward the new regime by granting Iran $45 million on an emergency basis. In contrast, Democratic Senator Thomas Hennings of Missouri demanded that a congressional investigation be held to determine whether the oil cartel "undermines the integrity of the ideals professed by Americans in their relations with friendly nations."[83] It was quite apparent to everyone that the major American oil companies had not been buying Iranian petroleum, since exports from Iran had dropped precipitously from $400 million in 1950 to $2 million for the two-year period between July 1951 and August 1953.

Once Zahedi had taken office, petroleum negotiations over the seizure of Anglo-Iranian resumed. A leading figure in these talks was Herbert Hoover, Jr., a special oil advisor to President Eisenhower. Early in 1954 the National Security Council concluded that the American petroleum firms should enter into an international consortium to market Iranian oil, and in doing so should not become subject to antitrust prosecution. Armed with this decision, Hoover went ahead with working out the details of a petroleum settlement. Many of the big firms and the independent companies in the United States, though, were opposed to the resumption of Iranian oil production on an unlimited basis, since it would glut the world market.

Finally, in the summer of 1954 an agreement was reached, through the efforts of Hoover and of Loy Henderson, the American Ambassador to Iran. Although both the petroleum fields and the Abadan refinery were to remain under the control of the Iranians, a consortium of the seven major oil companies and the Compagnie Française des Pétroles (CFP) was to operate them. Technically there was to be a holding company and a subsidiary operating company, which were to be incorporated under Dutch law. The foreign petroleum interests obtained a 50-50 profit sharing formula, as well as the privilege of running both the oil fields and the Abadan refinery as they saw fit. Once again, profits were to be in sterling, while the bulk of the sales was to be in the Eastern Hemisphere.

Under the terms of the consortium, Anglo-Iranian (soon to change its name to British Petroleum) was to hold 40 percent of the stock, five U.S. companies (Jersey Standard, California Standard, the Texas Company, Socony-Vacuum, and Gulf) another 40 percent, Royal Dutch-Shell 14 percent, and Compagnie Française des Pétroles 6 percent. A business historian, Mira Wilkins, has concluded that the decision of Jersey Standard to invest in Iran at this time was political; as one of its directors noted at a later date, "We would have made more money if we had done added drilling in Saudi Arabia; we had plenty of oil there; we were pushed into the consortium by the United States government."[84] Other American oil men, including Ralph Davies of the American Independent Oil Company, which was operating in the Neutral Zone of Saudi Arabia and Kuwait, were displeased with the establishment of the consortium. American Independent demanded of the State Department that it be allowed to take part in the consortium, although, as Herbert Hoover, Jr. noted its role if any was the responsibility of the consortium, not the U.S. Government.

It might appear on first glance that the consortium agreement was in direct violation of antitrust legislation. But, according to Attorney General Herbert Brownell, this was not the case. Brownell wrote Eisenhower on September 15, 1954, that:

> it is my opinion that these agreements, in their present form and if they remain unaltered, in view of the facts and circumstances which now characterize the production and refining of Iranian oil and the determination by the National Security Council that the security interests of the United States require that United States oil companies be invited to participate in an international consortium to contract with the Government of Iran, for the production, refining and acquisition of petroleum and petroleum products from within the area of Iran defined in these documents, would not in themselves constitute a violation of the antitrust laws, nor create a violation of antitrust laws not already exisiting. . . .[85]

In 1977, however, historian Burton I. Kaufman claimed that the Justice Department granted immunity reluctantly, and only because President Eisenhower and the National Security Council placed great pressure on it to do so. His claim is substantiated by Senator Frank Church's Subcommittee on Multinational Corporations, which found that the Antitrust Division indeed had raised objections.

But, as the Idaho Democrat complained two decades later, the Eisenhower administration did not consult Congress. During 1956 Democratic Representative Emanuel Celler of New York, then the chairman of a House antitrust subcommittee, asked Secretary of State John Foster Dulles for various documents relating to the consortium plan. Although Dulles did give to Celler three legal opinions that Attorney General Brownell had supplied concerning the joint purchase of Iranian petroleum, he resisted releasing any more documents. In the opinion of Dulles, their publication would have an adverse effect on the conduct of American foreign policy.

The consortium arrangement was probably the best deal that Iran could obtain under the circumstances. The Iranians, though, were not told that the eight participating companies had devised an "aggregate programmed quantity" formula, which was designed to restrict future production and thus prevent an oil glut. This was not made public for a number of years. It is hardly surprising that former Prime Minister Mossadegh concluded that this deal meant "the enslavement of my country for 40 years;"[86] other Iranian nationalists expressed similar views. But General Zahedi, the new prime minister, believed that the consortium arrangement was "the key to the rehabilitation of our economy."[87] Apparently most members of the *Majlis* felt the same way, for that body ratified the pact in October.

During 1955 another American group joined the consortium, and the five U.S.

companies that were already members gave up 1 percent of their holdings. This new coalition, Iricon, was led by the Atlantic Richfield Company (Arco), and included the American Independent Oil Company (Aminoil). The overall American share remained at 40 percent, with the original U.S. participants now holding 7 percent of the shares apiece. There were no other fundamental changes in the operation of the consortium until May 1973, when under a new agreement the National Iranian Oil Company took over the consortium's reserves, facilities, and operations.

## Iran: From the Consortium to the Embargo

A year after the establishment of the petroleum consortium, Iran, in a major diplomatic move, joined the Baghdad Pact which Iraq, Turkey, and Great Britain had signed. Iraq was to withdraw from the pact in 1959. The Baghdad Pact greatly displeased the Russians, as well as some of the Arab nations. Although the United States did not ratify it, the American government set up a military mission in Iran and supplied military equipment to the Shah. Between 1945 and 1961 the United States gave Iran $1.1 billion in foreign aid, most of it for economic assistance. The Iranians, however, did not turn their backs entirely on the Russians, and in fact relations between the two countries improved following the conclusion of a non-aggression pact; during the 1960s Iran received $840 million in economic aid from the Soviet bloc. On becoming President, Richard Nixon countered the Soviet assistance by increasing arms shipments to Iran in an attempt to stabilize the Persian Gulf area.

With respect to domestic developments, in 1956 the Shah launched his second seven-year program of domestic reforms, which was to be financed by oil revenues and outside grants. Six years later he implemented still another seven-year plan. Although the Shah visualized his "white revolution" as an alternative to a Marxist "red revolution, " his reforms displeased Communists, large landowners, and religious conservatives. By 1979, even though there had been an increase in Iranian oil production and revenues during the 1970's, the opposition from these groups had reached such a point that the Shah was forced to leave the country.

But petroleum activities in Iran since 1954 were by no means confined to the consortium. In 1957 Enrico Mattei's Ente Nazionale Idrocarburi entered into an agreement with the Iranian government that seemingly threatened the 50-50 formula the American, British, French, and Iranian oil negotiators had ratified in 1954. Under the 1957 agreement, the Italian state corporation was to share petroleum income with the National Iranian Oil Company and then hand over to the government of Iran half of its 50 percent share. During this year Iran adopted new petroleum legislation that was designed to open the whole nation to oil exploration.

Then in 1958 Indiana Standard entered into a partnership with the National Iranian Oil Company to explore an offshore area in the Persian Gulf of 4 million acres. Although under the petroleum legislation adopted in 1957 the NIOC had the right to operate as an oil company on its own, it also had the privilege of entering into new petroleum agreements with foreign firms. The Indiana Standard agreement marked the first foreign exploration contract signed anywhere, that combined a cash bonus (for unproved offshore acreage) with government participation. According to Iranian prime minister Manucher Eqbal, the agreement was "unbelievably profitable to Persia and unprecedented in meeting her aspirations."[88]

During the years following the agreement several key Iranian government officials visited the United States, including Prime Minister Eqbal, who met with President Eisenhower in Washington in October of 1959. Eqbal noted that Iranian oil exports had increased since the establishment of the consortium but that prices had recently fallen, thus causing the government of Iran to face current and future budget deficits. Eisenhower suggested that the Iranians market their petroleum in India, Southeast Asia, and the Philippines, only to have Eqbal counter with the observation that Iran controlled a mere 12 percent of the oil the consortium produced. In June of 1964 the Shah met with President Lyndon Johnson, and in an atmosphere of great warmth and cordiality expressed his fear that the oil companies might give preferential treatment to Arab producers, since OPEC was "an instrument of Arab imperialism."[89] The Shah added that Russian Premier Nikita Khrushchev, eager to deny petroleum to the West, was backing Arab designs on Khuzistan.

By the mid-1960's, however, the so-called Seven Sisters were becoming increasingly alarmed by the Shah's demands for additional oil production, which would increase the petroleum revenues he needed, not only for military expenditures but also to implement social and economic changes. The Shah complained to British Petroleum, which in turn complained to the State Department; in October 1966 the five American oil firms received assurances from the State Department that the U.S. Ambassador to Iran would do everything possible to restrain the Shah. A month later, though, international petroleum consultant Walter Levy warned Under Secretary of State Eugene Rostow of the consortium's secret agreement limiting oil production, only to have the State Department release a memorandum stressing the need for the American government to restrict its involvement in Iran's petroleum affairs.

Apparently, one of the partners to the secret agreement, probably the French member of the consortium (CFP) which was unhappy at the restrictions on Iranian oil output, leaked its terms shortly thereafter. It became necessary for the companies to relinquish a quarter of the consortium's concession, and surrender some of their petroleum to the National Iranian Oil Company. In return the NIOC agreed to sell its petroleum only in exchange for goods from behind the Iron Curtain, so to avoid upsetting the free world market.

When Israel invaded Egypt in June 1967, the Arab states, led by Iraq, agreed to a petroleum boycott of the Western powers. The Shah, while maintaining a neutral stance towards Israel, refrained from joining the boycott. By November he was pressuring the State Department for an annual increase in oil production of 20 percent, rather than 12 percent, as a reward for taking "grave petroleum risks." Although Eugene Rostow had warned that conditions were deteriorating in the Middle East, the spokesmen for both the "Seven Sisters" and the independent oil companies, who met with Rostow at the State Department in March 1968, countered his warning with a proclamation that Iran should receive equal treatment with Saudi Arabia. James Akins, Deputy Director of the Office of Fuels and Energy, made one of the more outspoken assessments of the Iranian situation. On March 18 he wrote to Robert L. Dowell, Jr., the Petroleum Officer at the American Embassy in Tehran, that: "The Shah is an oriental despot and the oil executives are dinosaurs, if they come to blows it could be the battle of the century."[90]

By this time the Iranian government had begun to enter into agreements with other private and state oil firms that were not members of the 1954 consortium. In 1966 the National Iranian Oil Company signed an exploration and production service contract with the Enterprise de Recherches at d'Activités Pétrolières (ERAP) of France; in 1969 the NIOC reached similar agreements with both an all-European consortium led by ERAP and Continental Oil. The Continental Oil pact, which was the first one that an American company had ever entered into individually in this area, focused on a 5,000 square mile tract in southern Iran. Two years later, in the summer of 1971, a Japanese firm agreed to enter into a joint venture for at least nine years with the NIOC to develop the Lurestan tract, west of Kermanshah. At the same time, Mobil, formerly Standard-Vacuum. also signed an agreement with the NIOC to exploit 1,500 square miles of the Strait of Hormuz, and then obtained a one-third interest in the Japanese Lurestan holdings in the fall.

As for the consortium, during the spring of 1970 it agreed to a 15 percent increase in royalties and tax payments, thus raising the Iranian receipts from oil for that year to over $1 billion. Although technically the consortium agreement was scheduled to run through 1979 with three optional extensions, in March of 1972 the Shah announced that the time was rapidly approaching for Iran to obtain complete control over petroleum operations. The Shah offered the members of the consortium two alternatives: either allow the contract to run out and face a loss of preferential status or even expropriation, or agree to an immediate Iranian takeover with guaranteed supplies of oil for a defined period at special prices. The petroleum firms chose the latter approach, and beginning in May 1973 the National Iranian Oil Company inaugurated a program under which the consortium, now known as the Oil Service Company of Iran, provided technical expertise and service operations for the NIOC.

In the summer of 1973 the National Iranian Oil Company entered into a contract with Ashland Oil under which it obtained a 50 percent interest in that firm's refining and marketing operations in New York State, including a refinery in Buffalo and 180 service stations. Meeting with President Nixon in Washington, the Shah announced that Iran would supply Ashland with 100,000 to 200,000 barrels of crude oil daily for twenty years. It was Iran, then, rather than Saudi

Arabia, that became the first Persian Gulf state to enter into such an understanding with American petroleum interests. That fall Iran, again remaining neutral, refrained from joining the oil embargo various Arab states proclaimed in conjunction with the fourth Arab-Israeli war.

## The Movement Towards State Control in Libya and Algeria

Unlike the Middle East, the commercial exploitation of the petroleum resources of North Africa has been a post-World War II phenomenon, primarily taking place in Algeria and Libya. After a long and bloody struggle, the former French colony of Algeria obtained its independence in 1962. Ahmed Ben Bella then became the first premier (later president) of the new nation, but he was deposed and imprisoned by Colonel Houari Boumedienne following a bloodless coup in the summer of 1965. Boumedienne, however, continued the socialistic policies of his predecessor.

Although the country is rich in mineral deposits of which phosphates, iron, lead, zinc, and coal are the most important, Algeria remains primarily an agricultural nation. Petroleum production began on a limited scale as early as 1914, but large-scale oil exporting did not take place until the final days of French rule. By this time various U.S. petroleum firms had begun to bid for and receive concessions.

In 1967 a drastic change occurred in Algeria's oil policy, when it expropriated the refining and marketing operations of Jersey Standard and Mobil, and seized the facilities of Sinclair, Phillips, and Tidewater. President Boumedienne declared that the U.S. petroleum firms would be reimbursed for their properties, but did not quote a specific amount; according to one estimate, Jersey Standard and Mobil would lose respectively $21 million and $6 million as a result of the expropriations. In the opinion of the American oil companies, these acts of seizure were a gesture of defiance in response to the conciliatory attitude that was emanating from the heads-of-state conference then meeting at Khartoum.

President Boumedienne made a general pledge to eliminate both U.S. and British economic interests from Algeria at the time of the expropriations. All foreign oil companies were required to export their crude production, a stipulation directed more against Royal Dutch-Shell than the U.S. firms. Algeria then turned to another socialistic nation, Iraq, and reached an agreement in 1967 to cooperate in the development of their oil fields. During this year, too, the state-owned Egyptian Petroleum Company began to purchase crude oil from the Société Nationale Pour le Transport et la Commercialisation des Hydrocarbones, the Algerian state-owned oil firm. Set up in 1963, SONATRACH at first dealt only with transportation and marketing, but the Boumedienne regime expanded its functions to cover various other aspects of the petroleum industry.

The expropriation campaign progressed even further during 1969. In October Algeria obtained 51 percent of the assets of the Getty Petroleum Company. Then in 1970 it seized the holdings of the remaining American oil producing firms then active in the country, as well as those of Royal Dutch-Shell, West Germany's Elwerath, and Italy's Ausonia, all of which had been unwilling to turn over to

Algeria a majority share in their holdings. In February 1971 the Algerian government next took over 51 percent of the interest in the operations of the French petroleum firms holding Algerian concessions; after a year long French boycott failed, there was a settlement largely on Algerian terms.

Despite these developments, in March of 1973 the Sun Oil Company and SONATRACH signed a joint venture exploration pact, under which they would investigate the petroleum possibilities of 2.3 million acres of central Algeria for three years and eastern Algeria for four. Sun was to spend $32.5 million on this undertaking, and it was to have a 49 percent interest in the new project, thus leaving SONATRACH a 51 percent interest. The fact that it had not been involved in the earlier Algerian seizures undoubtedly made it easier for this firm to sign an agreement with the Algerian government.

It was during 1973 that Algeria also became the first member of OPEC to demand "contributions" in the form of capital investments from its customers to counteract the depletion of Algerian petroleum reserves. If an oil firm chose to be an active participant, it was to assume a 49 percent share in a joint venture exploration pact with SONATRACH: if it decided to be a passive participant, it was to contribute to a SONATRACH exploration program in proportion to the numbers of barrels of crude oil it extracted.

In a related development Algeria became the site of the last conference of the non-aligned, less-developed nations to be held before the imposition of the Arab oil embargo in late 1973. Convening at Algiers between September 5 and 10, this gathering made four major recommendations: the encouragement of the creation of producers' associations; the declaration that every nation should be allowed to control its own natural resources; the assertion that acts of seizure should not be subjected to international adjudication; and the endorsement of any moves designed to expel Israel from those territories it had seized in 1967.

Once the fourth Arab-Israeli war had broken out, Algeria reduced its petroleum output by 10 percent. This was not as drastic an action as the total oil embargo such Persian Gulf states as Saudi Arabia, Kuwait, Bahrain, and Dubai imposed on shipments to the United States. Algeria's North African neighbor, Libya, though, did resort to a complete petroleum embargo against the Americans. Admittedly the Libyans embraced some of the same reform ideals the Boumedienne regime advocated, but the government led by Colonel Muammar el Qaddafi had earned a reputation for radical ideas and bizarre actions that set it apart from Algeria.

The postwar history of independent Libya is divided into two periods that are as different from each other as night is from day: the highly conservative regime established by King Idris I, and the volatile socialist administration set up by Colonel Qaddafi and a group of army officers after a successful coup on September 1, 1969. Once an Italian colony, Libya remained under British military rule from 1942 until 1951, because of the inability of the Allies and the United Nations to determine its future status; it finally became independent in the latter year. During the next decade Libya was one of the leading African recipients of U.S. grant aid and Point IV assistance, and a site of American military activity.

Aside from oil, Libya has no important mineral resources. It was only as recently as 1955 that the modern Libyan petroleum industry had its birth, when Libya approved a new oil law which was to prove both far-sighted and constructive. Although nine petroleum firms had received permits to engage in geological reconnaissance, the initial concessions date from November 1955. The first U.S. oil firms to operate within the country were Esso International (Jersey Standard), Texaco, California Standard, and Gulf. Large exploration blocks were granted, and drilling was both early and rapid; by 1958 significant petroleum discoveries had been made. The oil, having a low sulphur content, enjoyed widespread acceptance on the Atlantic Coast of the United States because its users could reduce pollution. Within three years Libya was producing petroleum for export, and selling the oil at a price somewhat cheaper than its Middle Eastern counterparts.

During the early 1960s, the Libyan government dealt with foreign petroleum firms on the basis of a 50-50 split of profits and a one-eighth royalty. By the end of 1963 the major oil companies had spent $700 million in their quest for Libyan petroleum. After Occidental took over the Mobil concession, it discovered oil almost immediately. Oxy and its maverick president, Armand Hammer, added a new twist to the scramble for concessions by offering to invest 5 percent of its pretax profits in Libyan agricultural development. The firm then generated further goodwill by striking a large underground deposit of water near the ancestral home of the king.

Before the Occidental strikes, in 1965, Libya had been producing 1.2 million barrels of oil a day. At that time Oasis and Jersey Standard were monopolizing production, but by 1968 no fewer than thirty-seven companies (seventeen of them producing) held concessions there. As the decade neared its end, Libya surpassed Kuwait in oil production and was even challenging Saudi Arabia.

It was at this point that King Idris, who had kept his nation neutral during the third Arab-Israeli war, was overthrown by a group led by Colonel Qaddafi. During 1970 the new regime pressured Great Britain and the United States into removing their military personnel from the country; the focal point of this activity was the multi-million dollar installation at Wheelus Field. Also in 1970 the new revolutionary Libyan government demanded that the U.S. petroleum firms either launch a multi-million dollar oil exploration program or lose their holdings. Qaddafi, realizing that Oxy was the most vulnerable American petroleum firm then operating there, cut back its production in retaliation for Oxy's reduction in exploration expenditures since the 1969 coup.

The key American figure in the 1970 confrontation between the Libyan government and Occidental was Armand Hammer. Hammer first approached President Gamal Nasser of Egypt, one of Qaddafi's heroes, and asked him to try to restrain the Libyan leader. Since Nasser feared that an oil shutdown in Libya would undermine Libyan subsidies to the Egyptian army, he advised Qaddafi to go easy. It was rumored that Occidental was attempting to enlist the aid of the

CIA in toppling the Qaddafi regime, but Hammer vigorously denied the rumors. Yet Hammer did approach the chairman of Exxon, Kenneth Jamieson, in July with a request that Exxon supply Oxy with cheap oil, in return for which Hammer would stand firm on the industry price. Jamieson refused the proposal.

In the same month Libya seized the four petroleum distributing companies active in that country without expropriating the foreign investments in oil production. Late in August the Occidental manager in Libya warned Hammer that he had better settle with Qaddafi before the Libyan government expropriated the properties of his firm. Hammer thus left for Tripoli to meet a team of negotiators led by Deputy Minister Abdul Salaam Jalloud. The talks were hardly routine; according to Daniel Yergin, "To show his displeasure during one negotiation, he [Jalloud] rolled up a Western company's proposal into a paper ball and tossed it back into the faces of the company's representatives."[91] By September 4, the day after Hammer left Libya, his negotiators had consented to increase the royalties and taxes due Libya by 20 percent.

Within a few weeks Exxon, Mobil, and British Petroleum agreed to increase the posted price for crude oil by 30 cents a barrel retroactive to September 1, with annual 2 percent increases through 1975. They also consented to raise Libya's half share of the profits by 5 percent. This agreement by the American and British oil firms resulted in what many authorities regard as the most significant change in petroleum prices in a generation. The Texas Company and California Standard especially wished to avoid a showdown with Qaddafi, since they had a joint Libyan concession that was in danger. It is highly questionable, however, whether this revolutionary nation was the place and the time for the oil companies to take a defiant stand. By the end of the year a Libyan National Oil Company (LINOCO) had been set up to explore for oil, and to market petroleum products domestically.

Relations between the foreign oil firms and the Libyan government underwent their most explosive series of climaxes during the six months immediately prior to the 1973 Arab oil embargo. On May 24 of that year the Qaddafi regime denied the right to export petroleum from Libya to Nelson Bunker Hunt, whom the Libyan leader personally detested. In early June the Libyan government expropriated the 50 percent interest Hunt held in the Sarir oil field, which was valued at approximately $140 million. It had already seized the British Petroleum share of this; Hunt had then refused to market illegally the BP oil. In a speech celebrating the third anniversary of the expulsion of U. S. forces from Wheelus Field, Qaddafi announced the expropriation of Hunt, and charged that the American oil companies in general had been tools of the U.S. policy of domination in the Middle East.

Again the Libyan government promised to compensate Hunt for the seized holdings, but it refrained from quoting a sum. At this time British Petroleum had yet to be paid for its share in the Sarir oil fields. While promising Hunt compensation, the Qaddafi regime nevertheless claimed that the concession (which had been granted in 1955) was illegal, since individuals were not allowed to operate in Libya on the same basis as companies.

Although President Anwar Sadat praised the expropriation of Hunt lavishly, Egypt did not implement similar policies. In the United States, the American government concluded that the Hunt seizure was invalid under international law and informed the Libyan government to this effect. Yet the U. S. government, while claiming that the expropriation was an act of political reprisal with "no public purpose behind it," had admitted on June 11 that Libya did have the right to seize foreign controlled industries, provided that it offered prompt, adequate, and effective compensation.

The uproar over the Hunt expropriation did not moderate Libyan policy. In August it seized 51 percent of the assets of Occidental in return for an immediate cash payment of $135 million. Hammer stated that his firm had acquiesced in what he described as a "transfer," but what Libya plainly labelled as a nationalization. At the same time that it expropriated Occidental the Qaddafi regime instructed Amoseas, the Libyan subsidiary of California Standard and the Texas Company, to cut its oil production by more than one-half. It then proclaimed the seizure of 51 percent of the interests of Continental Oil, Amerada Hess, and Marathon Oil and stated that compensation would be paid over the next two years on the basis of their net book value. Having disposed of the independents, the Libyan government announced in September that it was expropriating 51 percent of the interests of major firms it had not previously seized, including Exxon and Mobil.

At the time of the fourth Arab-Israeli war, Libya imposed a total embargo on oil shipments to the United States. The American position on Israel gave Qaddafi an additional reason to undermine the position of the U.S. and other foreign petroleum concessionaires in Libya. By May 1974 the Libyan government had expropriated the holdings of Royal Dutch-Shell, which offered the most resistance, Amoseas, and Atlantic Richfield. Libya then enjoyed control over approximately two-thirds of its petroleum output, with the National Oil Company (established in 1970) supervising production and marketing. Many observers, however, wondered whether the unpredictable Qaddafi would be able to run the Libyan petroleum industry effectively as a business enterprise, or whether his mercurial temperament would impinge upon the operations of the National Oil Company, with possibly disastrous economic consequences for Libya.

# 6. The Far East and South Asia

## The United States and Chinese Oil, 1883-1945

While China has ranked high in coal production, even in recent times, its output of iron has been of less consequence, and its petroleum output has been quite restricted. It has been such non-industrial products as tea which have long played an important role in its foreign trade. The first U.S. ships visited China as early as 1784; by the middle third of the nineteenth century American commerce with China had proliferated. As time went on, there gradually arose the myth of the Great China Market, although U.S. trade with China has never constituted more than 2 percent of American foreign commerce. Nevertheless, the United States became so concerned over equal trading opportunities in China, that in 1899 and 1900 Secretary of State John Hay announced the Open Door policy to safeguard these opportunities from the encroaching forces of European imperialism.

American merchants had begun shipping petroleum products to China as early as 1863. Twenty years later, in November 1883, the American legation in Peking notified Secretary of State Frederick Frelinghuysen that the Shanghai Consul General had reported "acts of interference" with the sale of U.S. oil there and in Ningpo. The Chinese foreign office, however, claimed that this was not the result of official government policy, but rather due to the ignorance of minor officials. As a result, John Russell Young of the American legation expressed the optimistic view that "petroleum is working its way into the confidence of the people."[1]

Far from being ended, the friction over oil products continued. In February 1887 the American legation protested the levying of a new kerosene tax of 90 cents per case in Canton. Although many Chinese were disturbed about the explosive nature of kerosene, the governor of Canton explained that the new tax had been levied because a series of robberies had forced the people to pay additional taxes for police. Yet in the long run, these controversies did not impede American shipments of kerosene to China; by 1899 these shipments were totalling 15 million gallons annually.

For the next twenty-five years almost all of the petroleum products imported into China came from the United States. Around 1905, though, a number of Chinese began to boycott American goods in China, Japan, the Philippines, Hawaii, and the United States itself to demonstrate their resentment at the growing mistreatment of the Chinese by certain bigoted Americans. While a New York Standard Oil representative informed the State Department that the boycott "would be a grave disaster to the petroleum industry of the United States,"[2] the Chinese pictured American businessmen in China as acting like wild Indians, and threatened Standard properties in Canton and Amoy in retaliation.

Despite the furor this episode aroused, Standard's total sales and its share of the Chinese market actually increased between 1905 and 1910 from 80 million gallons (52 percent) to 96 million gallons (60 percent). In 1911 Standard and the recently merged Royal Dutch-Shell agreed to divide the Chinese and Japanese market between themselves. Unfortunately, rumors that Standard had supported the overthrow of the Díaz regime in Mexico and had backed the revolutionary Sun Yat-sen in return for a promise of oil concessions did little to enhance its position with the Chinese government. In the fall of 1911, however, a revolution drove the hopelessly corrupt Manchu dynasty from power, and on January 1, 1912, China became a republic.

The new Chinese government, not wishing to turn to an international consortium of bankers, approached Standard Oil of New York for a loan in 1913. In return for a $15 million loan Premier Hsiung Hsi-ling offered oil monopoly rights in Shensi Province; Standard Oil, however, wanted a petroleum concession in China without having a loan to the Chinese government as an integral part of the agreement. When Standard's representatives in northern China tried to proceed on this basis, China attempted to play the American firm against Japan by forcing it to bid against Japanese offers. At that point, the U.S. company balked at further negotiations.

In January 1914 the new American Minister to China, Paul S. Reinsch, helped to formulate a tentative agreement under which the Chinese government would obtain a loan, as well as a one-third share in a Sino-American development company. Once again the agreement's loan requirement proved unacceptable to Standard, which withdrew its representative from Peking to the great displeasure of the Chinese officials. Minister Reinsch, though, persevered and was able to persuade the Chinese that Standard, unlike the Japanese, did not pose a political threat to the country. At the same time he convinced the American petroleum officials that they should be more generous to the Chinese.

As a result, China and Standard signed an agreement on February 10 under which Standard Oil of New York was required only to help China obtain a loan in the United States. In return, the petroleum firm received what amounted to a sixty-year monopoly over the oil of several northern Chinese provinces, provided that petroleum was found in them. If oil was located within Shensi or Chihli districts within a year, a Sino-American development company would be set up in the United States; Standard would own 55 percent of the company's stock and the Chinese government 37.5 percent, with a two-year option to purchase the other 7.5 percent.

It was perhaps inevitable that such a lucrative contract would disturb the Japanese and the British, who demanded similar rights elsewhere. In China itself the notables of Shensi, who had not been informed about the negotiations, feared the growing power of the Peking regime. The central government made further enemies by seizing the oil deposits in the provinces of Shensi and Chihli.

Within a month after the signing of the agreement, though, the Chinese government issued a revised set of mining regulations that placed a road block in the way of Standard Oil of New York. Under these, foreigners could hold no more than half of the total shares in a mining enterprise, and had to submit any dispute to arbitration by the director of mining supervision. When Standard sent its operatives into Shensi after having received discouraging reports about Chihli, a local bandit, one White Wolf, prevented the petroleum firm from operating there; Premier Hsiung Hsi-ling even had the company's carts seized, on the grounds that the Chinese government needed them for its army. These obstructionist tactics lessened with the outbreak of World War I, but the Chinese government continued to refuse to substitute another province for Chihli.

There was still another major obstacle to the successful execution of this enterprise. While the Chinese government envisioned a $100 million capitalization for the development company, which would form the backbone of a national petroleum industry, producing and marketing both crude and refined oil, Standard had in mind a far more limited operation, capitalized initially at $1 million. Such a restricted concept was intolerable to the Chinese, who postponed a decision on the level of capitalization until geologists could determine just how much petroleum there was to exploit.

By the summer of 1915 Standard Oil of New York officials had concluded—with complete justification—that it would be bad business to invest millions of dollars in such a shaky undertaking. On August 9 William Edward Bemis, the Vice-president of Standard, informed Premier Hsiung Hsi-ling that he was suspending further talks until his firm had determined the oil potential of the Shensi fields, which now seemed less promising than they had originally. Minister Paul S. Reinsch concluded that Standard had decided to limit its operations in China to marketing. This assessment proved correct, for in February 1917 Standard notified the Chinese government that it was not interested in the development company. Under the final settlement, Standard received $543,703 in partial reimbursement for its development expenditures. In assessing the events, Noel H. Pugach has concluded that "American economic expansion seemed to have suffered another major defeat in the Far East," and noted that the events of this period well demonstrate the problems involved when a foreign business attempts to operate in China.[3]

Although American oil firms were active in China during the interwar decades, their operations did not generate the huge quantity of diplomatic correspondence that their dealings in Japan, Sakhalin, and Manchuria did. Since most market-seeking U.S. manufacturers usually make their initial investments in the more industrialized nations, it is noteworthy that Standard Oil of New York had a large interest in Chinese oil marketing during the 1920s. Unfortunately, however, Standard suffered enormous losses through looting there, and the State Department was unable to obtain compensation for it. Even as late as 1935 vandalism was still a problem in China.

Late in 1925 T. V. Soong, Minister of Finance and head of the Cantonese government's Oil Monopoly Bureau, approached the Standard manager in southern China, Cameron, about a grant of an oil monopoly to the U.S. company. Observing that the price of petroleum was high, Soong suggested that Standard might supply most, if not all, of the oil and gasoline that was consumed locally, as well as distributing petroleum products under government supervision. The company's manager in Canton, though, opposed the plan, since the British and the other American firms could legitimately object that it excluded them from their markets. On January 7, 1926, the Minister to China, John V. A. MacMurray, informed Secretary of State Frank Kellogg that he had advised Standard the U.S. legation could not support any understanding that restrained trade in violation of treaty provisions and the antimonopoly policy of the American government. In June the Cantonese government renounced the proposal.

Nine years later, in 1935, the Benedum Trees Company informed Nelson Johnson, the Minister to China, that it had suggested to the Chinese Minister of Industry that it might help erect a petroleum refinery in China. Under the proposed partnership, Benedum Trees was to enjoy access to China's oil supplies in the future and the exclusive right to refine this petroleum. After Johnson suggested to the State Department that the plan appeared to create a state monopoly in China, the Acting Secretary of State, William Phillips, informed him on March 1 that the proposal was in obvious violation of the Nine-Power Treaty of 1922.

In 1937, when Japan attacked China, the Standard-Vacuum Company (the joint operating enterprise of Jersey Standard and Standard Vacuum in the Orient since 1933) and the Texas Company complained that large quantities of kerosene and gasoline were being smuggled into the Tientsin area. Apparently the smugglers were importing both Japanese and Asiatic Petroleum Company (Royal Dutch-Shell) products, as well as those of the two American firms. Although Standard-Vacuum and the Texas Company requested that customs authorities be asked to set up a preventive office at the Tientsin East Station, Secretary of State Cordell Hull advised Nelson Johnson, now the Ambassador to China, on July 1 that the embassy should refrain from making the specific suggestion, but that it should not object to the U.S. oil companies doing so.

Later in the month, during a congressional debate on the American policy toward China, isolationist Republican Representative Hamilton Fish of New York asked whether or not there was a single good reason for keeping American troops on the Chinese mainland. During the ensuing debate it was brought out that American gunboats were patrolling the Yangtze Kiang River to protect the investments of U.S. petroleum interests there. Fish made the suggestion that both U.S. boats and men should be withdrawn from China. Democratic Senator J. Hamilton Lewis of Illinois made a similar proposal to the Senate on the following day.

As the Japanese gained an ever increasing hold on northern China, in the summer of 1938 reports began to circulate that a new firm, the North China Petrol Company, would begin operations under their auspices with an initial capitalization of $20 million. By this time Japanese oil was dominating the market in northern China, and the American and British petroleum firms were on their way out. After Ambassador Johnson had complained to Secretary of State Hull about these developments, Hull wrote to Salisbury, the First Secretary of the embassy at Peking, on July 8, observing that:

> we apprehend in these reported developments a definite move on the part of the Japanese to establish in Chinese territory under their control monopolies similar to those in Manchuria and as such inimical to American trade interests; and we consequently feel that a protest should be made in the most emphatic terms against the organization of the proposed companies.[4]

Two months later, on September 14, the Ambassador to Japan, Joseph Grew, advised Secretary Hull that Seihin Ikeda, the Minister of Finance as well as the Minister of Commerce and Industry, had told him there would be no Japanese petroleum monopoly in northern China. Despite severe handicaps in the Yangtze Valley, both Standard-Vacuum and Royal Dutch-Shell managed to conduct a substantial trade with China for three more years. Beginning in January 1939, however, the American Embassy in Peking started to make representations to the Japanese Embassy concerning episodes of interference with U.S. petroleum commerce, chiefly kerosene and candles; during the next eighteen months there were no less than twenty incidents involving cities or areas in northern China and Inner Mongolia. This count, moveover, does not include those cases of interference that occurred in such southern China cities as Canton, Amoy, and Shanghai.

The various types of harassment included price fixing, the levying of taxes, the establishment of monopolies, and currency restrictions. Although some of the obstacles were removed in response to U.S. representations, most of them remained in effect; the Japanese justified the interferences on military grounds. As the Japanese Minister for Foreign Affairs, Yosuke Matsuoka, verbally pointed out to Ambassador Grew on December 17, the Japanese wished to prevent petroleum products from reaching Chiang Kai-shek.

By March of 1941 the Japanese not only had prevented U.S. oil shipments from Canton to other Chinese ports, they also in effect had brought about a monopoly on behalf of Japanese petroleum dealers. The stocks of Standard-Vacuum and the Texas Company at Chefoo were placed under the control of the Japanese special military mission there on July 28; although the U.S. firms regained freedom of access to their business premises on the following day, police guards remained on hand. On the same day, moreover, the Japanese had also occupied Standard-Vacuum and Texas Company properties in Tsingtao and forbidden officials of the firms to remove their stocks and property from the premises. This sort of harassment was to continue until the Japanese attack on Pearl Harbor, after which Japan seized outright all U.S. properties in those parts of China under its military rule.

## The American Quest for NEI Concessions

Unlike China and Japan, which were never under European rule, the East Indies (Indonesia) remained a colony of the Netherlands until the end of World War II. Indonesia, which has significant deposits of tin and has served as an important source of rubber, has long been the leading oil producer of the Far East. Less known is the fact that American merchants had begun shipping petroleum products to what was then the Netherlands East Indies in 1865, before the first petroleum well was ever drilled there. When the initial attempt to drill for oil was made in 1871, it was done with a rig from the United States.

The first discovery of petroleum in commercial quantities did not take place until 1885 in northern Sumatra. Five years later, in 1890, The Royal Dutch Company for the Working of Petroleum Wells in the Dutch East Indies was incorporated at The Hague. But by 1896 a rival had arrived on the scene, for during that year Marcus Samuel bought oil properties in Borneo, just prior to establishing the Shell Transport and Trading Company, Ltd. in 1897. These firms were to join forces in 1906.

In 1894, even before Samuel's arrival, the managers and agents of Standard Oil had begun to assume an active role at Batavia and a number of other key cities scattered throughout the Orient. But the negotiations between Standard Oil and the Moeara Enim Petroleum Company for exploring a concession in the interior broke down in 1898, after the Dutch Colonial Minister, J. T. Cremer, had applied pressure. According to John F. Fertig of Standard Oil, "The principal opposition comes from the Brokers and Speculators, who see in the entrance of the S. O. Co. an end to speculation, and a beginning of sound business."[5] Moeara Enim then obtained a mining permit on its own and contracted with Shell, which earlier had refused to sell out to Standard, to market its products. Standard's position in the Netherlands East Indies suffered an additional blow in 1899, when an ordinance was passed that allowed only Dutch citizens to serve as directors of oil companies operating in the colony.

Although eighteen petroleum firms had been active in the Netherlands East Indies in 1900, a mere twelve years later there was only Royal Dutch-Shell, which at that time took over its last competitor, the Dordtsche Petroleum Company. By that year, 1912, the Netherlands East Indies had already produced approximately 100 million barrels of oil.

Standard Oil, having already been forced out, then formed the NKPM (Nederlandsche Koloniale Petroleum Maatschappij) through its marketing affiliate in the Netherlands, the American Petroleum Company. The new Dutch concern with a Dutch board of directors sought a concession at Djambi in central Sumatra, only to have Royal Dutch-Shell challenge it by making a higher bid. Aware that NKPM was Standard Oil, the government of the Netherlands suspended the issuance of any new concessions in 1913. Four years later, a new mining law designed to appease Dutch Socialists went into effect. Under the new law the government would enjoy at least half ownership in every future petroleum concession that it granted. On July 20, 1918, still another mining law became operative, which stated that the discovery of petroleum, coal, or iodine did not necessarily grant the discoverer the right to obtain a concession for exploitation; after discovery, the Netherlands would either exploit the concession, or contract for a corporation to do so.

A. M. Snouck Hurgronje, an official of the Ministry for Foreign Affairs at The Hague, offered the typical Dutch defense against accusations that the laws and policies constituted a program of discrimination against foreign capital. He wrote to the American Legation in the Netherlands on December 2, 1920:

> ...before the modification of the regime of 1913 no distinction was made between foreign and Dutch capital for the exploitation of petroleum concessions; that between 1913 and 1918 no concession was granted, and that under the new regime of 1918 the possibility existed as much for companies with foreign interests as for those with Dutch interests to obtain a contract for exploitation.[6]

Hurgronje went on to point out that U.S. capital already had invested in numerous agricultural and mining projects in the Netherlands East Indies, and that, much to the satisfaction of the Dutch government, a number of American citizens were involved in these enterprises. According to Hurgronje, the Netherlands had been reluctant to grant exclusive oil mining concessions in the East Indies to a single American company, because it had been disturbed by "the lack of interest on the part of American capital to obtain and exploit petroleum concessions in the Indies."[7]

The official American perspective was best indicated by Consul Charles F. Jewell's dispatch of January 20, 1920, to the State Department, entitled "Mineral Oil Concessions in Netherlands India." While admitting that the Dutch laws technically did not exclude U.S. interests, Jewell noted that the government of the Netherlands was able to discriminate against these interests through its interpretation and application of the measures. The State Department apparently agreed with Jewell, since on May 17 it made a report to President Woodrow Wilson in which it complained that for many years U.S. companies had been unsuccessful in their quest for petroleum rights in the Netherlands East Indies. On July 20 Secretary of State Bainbridge Colby indicated his dissatisfaction with Dutch policy in a letter to the American Minister to the Netherlands, William Phillips.

Events took a new turn on September 4, when NKPM, the Nederlandsche Koloniale Petroleum Maatschappij, obtained some mining concessions in the

Netherlands East Indies. Several of these were for oil, several others for coal, but it apparently was not made clear to the State Department at first that NKPM had obtained these licenses from third parties, rather than from the government of the Netherlands. Henry Fletcher of the Department of State eventually advised the Dutch legation in Washington on March 14, 1921, that: "The company takes the position, based upon the above considerations, that the [NKPM] has not increased its holdings in the Netherlands East Indies by reason of the 10 concessions recently granted."[8]

A month before the September NKPM transaction, Sinclair had approached Minister to the Netherlands William Phillips in search of U.S. diplomatic support for its application for a part of the Djambi concession, which Sinclair considered to be the only significant oil field in the Netherlands East Indies. On September 15 a Sinclair representative met with the Dutch Minister for Foreign Affairs, J. H. van Karnebeek. At this time van Karnebeek informed him that the government was about to lay before the Second Chamber of the Estates General a bill, originally introduced in 1907, granting to the Bataafsche Petroleum Mattschappij, the principal Royal Dutch-Shell subsidiary in the Netherlands East Indies, the right to develop the Djambi concession. Harry Sinclair then met personally with BPM officials in London in December to discuss the possibility of a subcontract, but by this time he had begun to lose interest in the Djambi concession. Further alienated by their vague answers, Sinclair admitted that his basic concern merely was to obtain petroleum for American ships operating in the area.

Back in Washington, the State Department had concluded that BPM was not going to enter into a subcontract with an American oil firm, and consequently was moving toward a harder stance toward the Netherlands. On April 21, 1921, Secretary of State Charles Evans Hughes instructed Minister Phillips to throw his weight against the Djambi bill, unless the concession was divided and American participation guaranteed. When Phillips then complained to Foreign Minister van Karnebeek, van Karnebeek replied that it would not be advisable for his government to halt the Estates General debate on the bill just to make the United States happy. The Netherlands, moreover, had already made a verbal commitment to the BPM to permit it to develop Djambi, which in van Karnebeek's opinion was not the only valuable underdeveloped area in the Netherlands East Indies.

A week later, on April 28, the Socialist member of the lower house who had succeeded in defeating the first Djambi concession bill proposed an amendment that would divide the proposed concession between BPM and NKPM. The amendment was defeated, and the Second Chamber approved the forty-year Djambi concession bill by a vote of 49 to 30; after a brief debate, the First Chamber passed the measure, and the Queen now affixed her signature to the bill. Although according to Peter Mellish Reed these events "in many ways represented a defeat for American diplomacy and for Jersey Standard,"[9] van Karnebeek then advised Phillips on April 29 that his government hoped that American capital still would participate in the development of the Netherlands East Indies.

Assessing the circumstances surrounding the passage of the Djambi concession bill, the American Commercial Attaché at The Hague, Caldwell Johnston, wrote the Bureau of Foreign and Domestic Commerce on May 25, 1921, that:

> A very remarkable situation exists and could hardly be paralleled elsewhere. It is no doubt that even in Russia in the old days such methods could not have been carried out. In the first place, in the Second Chamber a measure is passed because vital information is withheld or suppressed at the critical time, and then jammed through the Senate under threat of resignation of the Cabinet unless the Bill was passed. [This] exceeds anything short of absolute autocratic disregard of popular will.[10]

Yet for several reasons a change of policy was in the air. The Dutch were becoming increasingly disturbed by the discriminatory actions of the American government against Royal Dutch-Shell, which was seeking petroleum leases in the United States, and the Japanese had begun to show an interest in the mineral resources of the Netherlands East Indies. Early in 1923 the Dai Nippon Oil Company made an offer of 2 million guilders for an oil concession on Celebes. By this time, too, the feeling was growing that there were important petroleum fields in the Netherlands East Indies outside of the Djambi concession. This feeling was reinforced by a concession at Talang Akar, an "unpromising site" over 100 miles from Djambi in southern Sumatra, that Royal Dutch-Shell sold to NKPM, only to have it spout petroleum in 1922.

The American press was angered by Standard's defeat in the Estates General. To cite two examples, the St. Louis *Post Dispatch* observed that "Holland and Great Britain appear to have arranged an oil alliance," while the Mobile *Register* complained that "the Dutch Government apparently has played fast and loose with our Minister, and it should now clear up its ambiguous attitude toward the admission of American capital." Only the New York *Journal of Commerce* reminded its readers that the American government had also sinned.[11]

Even more important, on the other side of the Atlantic a number of Dutch politicians and newspapers also believed that the Dutch Colonial Ministry had misled Minister Phillips, since they felt the petroleum fields outside of Djambi were of doubtful value. One Dutch Senator, Embden, was quoted as complaining that "the treatment of the United States was very offhand and unsatisfactory and could only cause irritation and that it was, moreover, bad commercial policy."[12] But this support was balanced by a number of anti-American utterances; the *Rotterdammer* opined that "the interference in our affairs by the Government of the United States must be stamped as entirely out of place."[13]

On March 8, 1923, the flagship of the U. S. Asiatic fleet arrived in Batavia, accompanied by a squadron of six destroyers. It also stopped at Surabaja and Makassar. Although the American government did not openly state the purpose of the visit, it seems likely that there was a direct relationship between it and the efforts of the State Department on the behalf of U.S. petroleum entrepreneurs in the Netherlands East Indies. Naval diplomacy, of course, had been invented long before this visit, and had been practiced by the American government, as well as by the governments of various other nations, on a number of occasions.

In the next year California Standard undertook a geological survey of Java, Borneo, Madura, and Sumatra. During the months that followed the survey, there was intermittent talk of new U.S. oil concessions in the Netherlands East Indies, but no real progress occurred until January 14, 1925, when Colonial Minister de Graaff called in Jersey Standard and NKPM officials for a series of discussions. De Graaff proposed a new concession that would not be purchased from other oil companies active in the Netherlands East Indies but instead would be obtained from the Dutch government. On June 29 the contract negotiations were concluded. Of the 240,000 hectares involved, there were to be two sites in the Talang Akar area comprising 75 percent of the total concession, as well as smaller plots on Java and Madura; in return, NKPM was to pay to the Dutch government 20 percent of the profits, and a 4 percent land tax. The U.S. firm also was required to employ Dutch nationals in a ratio of three-to-one.

The plan suffered a major setback on July 28, 1926, when the Volksraad of the Netherlands East Indies apparently bowed before Royal Dutch-Shell pressure and rejected the contract by a vote of 15 to 19. The Dutch government blamed the defeat on the Socialists, who were displeased because the bill did not provide for governmental participation in the management of the concession. Especially disturbed by this rejection were the Dutch Ministry of Colonies, and the pro-American Governor-General of the Netherlands East Indies, also named de Graaff, whose explosive reaction led a leading Indonesian newspaper to suggest that he should curb his temper.

Ignoring the action of the Volksraad, in April 1927 the Dutch government presented the NKPM contract to the Second Chamber of the Estates General. On October 27, 1927, that body passed the bill, and on February 8, 1928, the First Chamber followed suit; then on July 17 the Minister of Colonies signed the contract, promising that NKPM would receive another NEI oil concession in the immediate future. That December Jersey Standard obtained an additional 360,000 hectares in the Atjeh, Indragiri, and Palembang districts of Sumatra, and on northern Celebes. Nevertheless, Secretary of State Frank Kellogg did not issue a statement that the Netherlands was no longer categorized as a nonreciprocating nation under the terms of the Public Land Leasing Act of 1920, until the Minister of Colonies had signed the NKPM contract on July 17.

According to Peter Mellish Reed, the U.S. petroleum companies would not have obtained a larger share of the petroleum concessions of the Netherlands East Indies if it had not been for the active support of the American government. Reed observes:

> From records of State Department correspondence, one is led to believe that Standard's activities in the Dutch colony were of little success until the State Department lent a hand by encouraging a revision of Dutch petroleum policy so that American interests would be able to receive equitable treatment with Dutch-owned companies.[14]

Yet Standard did spend over $3 million on social serviccs in the Netherlands East Indies between the two world wars. While much of this was for housing, there were expenditures for schools, recreation, sanitation, and hospitals. Between 1929 and 1938 Standard also sponsored an antimalaria campaign on its concessions.

In 1924 Jersey Standard had accounted for only 5 percent of the total petroleum output of the Netherlands East Indies, while BPM produced 95 percent. By the outbreak of World War II fifteen years later, California Standard and the Texas Company also had become active in the NEI, and the American oil interests were now producing 28 percent of the total Indonesian petroleum output. Although the Gulf Oil Company had been interested in the Netherlands East Indies from 1928, when it began its geological surveys, to 1932, when it withdrew from there, the refusal of the Volksraad to approve its formal request for a contract killed its interest. Despite the activities of Jersey Standard, California Standard, and Texaco, as of 1940 only 7 percent of the capital invested in the Netherlands East Indies came from the United States. While Dutch investments comprised no less than two-thirds of the total, there were also British and Chinese investments that were greater than those from the United States.

## Japan and Sakhalin after World War I

The post-World War I era found Japan in a strong position towards the Far East. As an ally of the United States during World War I, it left the war as one of the victors rather than one of the vanquished. Unfortunately, between the two world wars the climate of opinion in Japan gradually shifted from internationalism and democracy to nationalism and totalitarianism, and towards an armed conflict with America.

In 1931 Japan occupied Manchuria, converting it into the puppet state of Manchukuo in the following year. Japan withdrew from the League of Nations in 1933, and in 1934 it abrogated the Washington treaties of 1922. Less than a year before attacking China, in November 1936, Japan signed the anti-Comintern Pact with Nazi Germany. The Japanese movement toward militarism accelerated between 1938 and 1940 with the enactment of the National Mobilization Law and the call for the establishment of a Greater East Asia Co-Prosperity Sphere that was to be dominated by Japan. In the spring of 1941 the Japanese signed a neutrality pact with the Russians, which the Soviets were to break only after the United States had dropped an atomic bomb on Hiroshima during the summer of 1945.

Japan adopted this increasingly bellicose position even though it is a country lacking in natural resources. Its population, moreover, increased from 43.8 million in 1900 to 73.1 million in 1940. Yet as the nation became more and more industrialized it became necessary for it to import larger and larger quantities of raw materials to make up for its lack of natural resources. The desire to guarantee access to raw materials was a major consideration in Japan's decision to implement an imperialistic program in the Far East.

When one thinks of U.S. relations with Japan with respect to petroleum, his or her attention focuses on the period just prior to Pearl Harbor, and on the strong desire on the part of many Americans to deny to the Japanese strategic materials vital to the prosecution of a major war in the Far East. Actually, though, oil was a key issue in American-Japanese economic relations throughout the entire interwar period, extending back even to the immediate post-World War I years. Involved in this persisting controversy was not only Japan proper but also Sakhalin and Manchuria.

Under the peace treaty ending the Russo-Japanese War in 1905, Japan strengthened its position in both Korea and Manchuria, but failed to obtain the entire island of Sakhalin from Russia. As World War I drew to a close, Japan invaded Siberia in April 1918. More than one historian has suggested that President Woodrow Wilson sent American troops to Siberia during this period, not only to counteract the Communist regime which had recently come to power in Russia, but also to check Japanese territorial imperialism.

Although in 1919 Sinclair became the first American petroleum firm to seek entry into Sakhalin, in the following year Sinclair's agent in the Far East warned the home office that Japan wanted the Russian, or northern, half of the island. This prophecy proved quite accurate, since Japan indeed did seize northern Sakhalin in July 1920. It had decided to act at this time for a number of reasons: U.S. troops had been withdrawn from Siberia; civil government in the Russian Far East had proven unstable; and finally, China was in no position to offer effective opposition.

The State Department's reaction toward this act of territorial aggression was one of disapproval. Secretary of State Charles Evans Hughes told the Japanese on May 31, 1921, that the United States could "neither now nor hereafter recognize as valid any claims and titles arising out of the present occupation and control."[15] Rather than compromise on this, Hughes reiterated this stand at the Washington Disarmament Conference a year later. Baron Kijuro Shidehara, however, had stated that the military occupation of northern Sakhalin was only a temporary measure. According to him, it was "fixed and settled" policy for Japan to respect the territorial integrity of Russia, not to interfere with her internal affairs, and to support the principle of equal commercial and industrial opportunity for all nations in the Soviet Union. The American Secretary of State in turn declared that he expected the Japanese to live up to these principles.

Secretary Hughes, in fact, had warned Sinclair as early as September 27, 1921, not to close an agreement over the petroleum reserves of northern Sakhalin. Hughes pointed out that conditions there were not stable, and that the U.S. government had not recognized the Russian regime; later attempts by Sinclair to persuade the State Department to reconsider its position proved fruitless. Another State Department advocate of the hard-line policy toward Sinclair's Sakhalin plans was De Witt Poole, the chief of the Division of Russian Affairs. At the same time that Hughes wrote Sinclair, Poole wrote to Hughes: "If we encourage the Sinclair people to conclude a definite contract, we create, of course, a concrete issue with Japan."[16] Apparently both Hughes and Poole felt that State Department support for the Sinclair project might have a deleterious effect on negotiations at the upcoming Washington Disarmament Conference.

Sinclair, however, was determined to obtain petroleum from northern Sakhalin. On January 7, 1922, it signed a contract and supplementary agreement with the Sakhalin government at Chita, Siberia. The contract, which ran for thirty-six years, granted Sinclair the right to explore for petroleum throughout Russian Sakhalin during the initial year; by the end of the fifth year Sinclair was required to restrict its choice of lands to an area totalling 10,000 square *versts*. But the royalties and the taxes were not especially heavy, nor was the land rental fee.

Equally important was the supplementary agreement, which indicated that support from Washington was expected. The Russians realized in this pre-Teapot Dome era that Harry Sinclair had close ties with key officials of the Harding administration, and therefore thought that he could manipulate support for the agreement. The most critical stipulation was that the Russians had the right to cancel all provisions of the agreement, if during the first year of the contract they became convinced by "any acts or declarations of the United States Government that the latter will not give support to the company in its execution of said concessional agreement."[17] As the negotiators were well aware that Japan might block Sinclair from exploring in northern Sakhalin, Article 35 of the contract invoked the concept of *force majeure*, granting to Sinclair an extension of time to fulfill its contractual obligations in the event that interference took place.

Although Secretary of State Hughes obtained from the Japanese a statement of intent respecting the withdrawal of all troops from Sakhalin at the Washington Disarmament Conference, no specific date was set. The Japanese delegation, however, did promise to support equality of economic opportunity throughout the Far East, including northern Sakhalin.

In November of 1922 northern Sakhalin formally became a part of the Russian Soviet Federated Socialist Republics, and two months later the RSFSR ratified the Sinclair contract. Yet far from helping Sinclair to obtain support from Washington, the signing probably hurt its cause, since the State Department's attitude toward the Bolshevik regime in Moscow was far more negative than it had been toward the short-lived Far Eastern Republic of which northern Sakhalin had been a part.

As the Japanese were in effective control of northern Sakhalin and were to remain there until 1925, the Russians proposed a tripartite agreement among Russia, Japan, and Sinclair. But when the Russians laid this proposal before the Japanese with the blessing of Sinclair, Japan asked instead for a monopoly concession extending for a ninety-nine-year period with the provision that Japan would give Russia 5 percent of the oil produced.

Faced with lessening chances of success, Sinclair wrote the State Department in February 1923 that it would be necessary to begin early exploration in northern Sakhalin. It asked the State Department to inform the Japanese that the Sinclair geological party was on the way. The Japanese, though, were not about to allow the Sinclair interests to explore for petroleum in northern Sakhalin even with official U. S. support, and the Japanese Foreign Minister, Count Chika Uchida, reminded the Japanese Embassy in Washington that this was the case.

De Witt Poole's fellow opponent at the State Department of Sinclair operations in northern Sakhalin was the chief of the Far Eastern Division, John V. A. MacMurray. Poole and MacMurray went so far as to recommend that the State Department unofficially notify Japan that it would not assist Sinclair. In contrast, Stanley Hornbeck of the Office of the Economic Adviser was more sympathetic to the American oil firm's request; he suggested that the issue at stake was whether or not the U.S. government would support American business enterprises in Russia. Hornbeck, though, emerged as the loser in this confrontation. Although Hughes did not advise the Japanese that the State Department would not assist Sinclair, he did tell Sinclair that he would not inform Japan the Sinclair party was en route.

At the same time, the State Department rejected a similar request from California Standard. California Standard's oil rights were not very extensive and had been granted by the Tsarist government. The company pointed out that it obviously could not explore in northern Sakhalin without U.S. government support, but Hughes replied that the State Department could not grant California Standard a favor that it had denied to Sinclair.

In the fall of 1923 Sinclair sent its expedition to Sakhalin, consisting of two company officials and an interpreter. Far from being greeted with open arms, the American party spent its first night on wintry Sakhalin in a room that housed sixteen other men and a chicken coop. Two weeks after their arrival, the Americans were dispatched from the port of Alexandrovsk without exploring the oil concession. However, since its representatives had been denied entry into northern Sakhalin, Sinclair could now invoke with justification the *force majeure* provision of its contract, thus extending the contract.

The State Department remained silent during this fiasco, thus forcing the Russian government to try to work out an understanding with Japan over the petroleum reserves of northern Sakhalin. Accordingly, the Soviet regime notified Sinclair that it had not lived up to its contractual obligations. In May 1924 it further informed Sinclair that it either had to commence operations in northern Sakhalin within six months, or face the cancellation of its contract. At this time there was little likelihood that Sinclair would begin exploring in the immediate future.

But more was involved here than the attitudes of the governments involved. Commercial, fishing, and lumbering interests in Japan now were calling for an end to the occupation of northern Sakhalin, since it never would become a permanent part of Japan. The shift in public opinion led to the start of negotiations between Russia and Japan; the Japanese agreed in principle to evacuate the Russian half of the island once it was determined what amount the Soviets would pay to Japan in return. On January 20, 1925, the two governments signed an agreement, under which the Japanese were to develop 50 percent of the petroleum lands of Russian Sakhalin and Russia was to retain the remaining 50 percent. Should Russia decide at some future date to lease part or all of its petroleum lands to foreign interests, Japanese nationals were to enjoy an equal opportunity to acquire them.

The understanding undercut the position of Sinclair in northern Sakhalin, which never had been very secure in the first place. The Russians initiated legal action against Sinclair as soon as the Russo-Japanese agreement was signed, and a court hearing was set for March 11, 1925. At the hearing Russia claimed that Sinclair had failed to meet its assigned time limit with respect to exploration, and that a time extension under the *force majeure* clause was not justified, because this clause only covered new or unexpected developments. The Japanese, it will be recalled, had occupied northern Sakhalin militarily prior to the negotiation of the contract. Sinclair lawyers took the position that the Japanese ban on oil exploration had rendered the *force majeure* clause operative, and that Article 35 had been so written as to cover interference by the Japanese army.

Having been placed in an increasingly hopeless position, Sinclair hired the services of former Secretary of State Robert Lansing, who they hoped could influence the State Department. Reminding the new Secretary of State, Frank Kellogg, of Japanese pledges at the Washington Disarmament Conference, Lansing wrote on March 30, 1925, that "unless action is taken by the United States Japan stands to benefit by her own wrong and to extinguish forever those American rights."[18] When Japan recognized the new Soviet regime diplomatically, Lansing argued, it thereby recognized all prior acts of this regime, including the Sinclair contract.

Kellogg, though, submitted a memorandum to a cabinet meeting during March questioning whether or not the new Japanese concession in northern Sakhalin violated the earlier pledges made at the Washington Disarmament Conference. According to Kellogg, the Russo-Japanese agreement of 1925 did not violate "Open Door" policy as long as other nations enjoyed access to the remaining 50 percent of the oil properties of northern Sakhalin. President Calvin Coolidge supported Kellogg on this issue, and Kellogg's stand proved to be a popular one in the United States.

Two months later, in May, the Soviet Supreme Court nullified the Sinclair contract, after Sinclair had rejected a Russian offer in March to return the 200,000 rubles posted as a guarantee, and to negotiate a new contract for an unnamed concession. Although the State Department had concluded that Sakhalin was of questionable value to Japan, by 1928 the petroleum concession was producing 150,000 tons of crude oil annually, and the Japanese were purchasing most of the Russian oil production on the island.

While it may be argued that the disputed ownership of northern Sakhalin militated against the American government assisting Sinclair, the government did support those American petroleum companies seeking oil in Mosul when Turkey and Iraq were disputing its ownership. Although the U.S. government did not officially recognize the Soviet Union until 1933, it would seem that it could have exerted unofficial, or indirect diplomatic pressure on Russia to help Sinclair obtain the concession on northern Sakhalin. In summarizing the episode, Floyd J. Fithian has concluded that "the treatment afforded Sinclair in his quest for Sakhalin oil was a curious deviation from the general policy of the United States on this issue."[19]

**The Japanese Petroleum Control Bill and the Manchukuo Oil Monopoly**

Although the American public was not overly concerned about the Japanese invasion of Siberia and northern Sakhalin during and after World War I, it began to turn against Japan in 1931 when Japan seized South Manchuria by force and set up the puppet state of Manchukuo. Henry Stimson, then Secretary of State, addressed a note to both Japan and China setting forth the so-called Stimson Doctrine. In this he affirmed that the United States did not "intend to recognize any treaty or agreement...which may impair...the sovereignty, the independence, or the territorial and administrative integrity of China...or the Open Door Policy."[20] The Stimson Doctrine, though, did not include economic sanctions, only moral condemnation, and the Japanese remained in control of Manchuria throughout World War II.

As Japan moved into Manchuria in 1931, it attempted to integrate the economies of that nation with its own, thus generating a major controversy over Manchurian petroleum. In 1932 American oil firms provided 55 percent of the total imports of Manchukuo, and Japan only 10 percent. But despite the pro-U.S. orientation of its trade, in 1934 the puppet government of Manchukuo set up a petroleum monopoly, much to the displeasure of the American oil concerns.

U.S. petroleum firms also were displeased with oil developments in Japan itself. During 1931, 1932, and 1933, after Mitsuibishi Oil had built a refinery and other firms followed suit, there was a decline in Japanese imports of refined petroleum from the United States. New York Standard, which merged with the Vacuum Oil Company in 1931, until then had furnished approximately half of the imported refined petroleum, while Royal Dutch-Shell had provided the remainder. (Other U.S. firms had also helped to meet Japan's requirements for crude oil.) In 1929, however, the Great Depression had led to a disastrous price competition.

Since cartels in Japan enjoyed the right to regulate major companies involved in basic economic pursuits, in 1932 the petroleum cartel obtained the right to fix oil prices. The Japanese firms controlling the cartel then set such low rates that Royal Dutch-Shell and New York Standard had to sell their gasoline at a loss. New import quotas also were established, and the newly created Japanese import firms obtained oil from the Soviet Union, which enabled them to undersell cartel members.

That the Japanese petroleum industry was in a state of flux during the early 1930s made it almost inevitable that someone in Japan would propose decisive governmental action. It should not be surprising then that Keizaburo Hashimoto, the president of the Nippon Oil Company and the Manchurian Oil Company, suggested to the Diet—where he had considerable political influence—that a petroleum monopoly should be established under governmental auspices. Hashimoto advanced the plan as a national defense measure, although several Japanese naval officers testified before the Diet that the navy did not need additional oil reserves at that time.

On March 27, 1934, the Diet passed a Petroleum Control Bill, which imposed price controls on oil, fixed importing and refining quotas, and required all importers and refiners to keep six months of petroleum stocks on hand in Japan. Under the new bill, importers and refiners were to disclose upon request the complete details of their business operations; they also were to provide storage facilities for petroleum at their own expense, and to sell to the Japanese government on demand as much oil as it desired. Obviously Hashimoto's objective was to so increase the operating costs of foreign petroleum firms that they could not compete in Japan. His only defeat came when the Diet rejected a governmental subsidy for Nippon Oil that would have covered the cost of building the storage tanks.

To make matters worse for the foreign oil companies, the Bureau of Mines, which was nationalistic, anti-Western, and pro-military in orientation, was to administer the law, and regulations for its enforcement were to take effect on July 1. Fortunately for the foreign petroleum firms, however, Saburo Kurusu, the Chief of the Commercial Affairs Bureau in the Japanese Foreign Office, was handling the diplomatic protests at this time; Kurusu had an American wife, and was more friendly toward Western representations. The conflicting attitudes of these two bureaus, which continued to manifest themselves in the months and years ahead, were only one aspect of the great rivalry between the liberal civilians of the country and the nationalist militarists, which was to culminate in the eventual triumph of the latter.

That July both the American and British ambassadors received instructions from their governments to complain to Japanese officials that the establishment of a petroleum monopoly in Manchukuo would violate not only the Nine-Power Treaty of 1922, protecting the sovereignty and territorial integrity of China, but also the "Open Door" principle. The Japanese government replied that it could neither block the investment of Japanese capital in the Manchurian Oil Company, nor force the government of Manchukuo to abandon its plans for controlling the petroleum industry there. It was the position of the Japanese government that the Nine-Power Treaty did not apply to Manchukuo.

Ambassador to Japan Joseph Grew, after discussing the matter with his British colleague, Sir Robert Clive, concluded that it was futile for the United States to protest the oil monopoly in Manchukuo to the Japanese government. It was at this time that Grew suggested that the American and British governments consider the imposition of a partial or complete embargo on crude oil exports to Japan and Manchukuo, a plan that Standard-Vacuum officials had proposed to him on August 20.

Two days later Standard-Vacuum and Royal Dutch-Shell officials met in

Washington with Under Secretary of State William Phillips, and on the following day talked with Secretary of the Interior Harold Ickes. Ickes, however, became suspicious of the petroleum firms' true motives for the embargo, and in his conversations with Walter Teagle of Jersey Standard and Sir Henri Deterding of Royal Dutch-Shell he learned that Royal Dutch-Shell had the greatest stake in the Japanese oil market. Ickes then raised the question of the lukewarm British attitude toward economic sanctions against Japan at the time of the Manchurian crisis of 1931. Rather than strike the first blow itself, the State Department took the position that Great Britain should make the initial protest to Japan, although Teagle claimed the British had already done so.

Standard-Vacuum, Royal Dutch-Shell, and the Texas Company, acting independently of the State Department, then agreed on August 31 not to supply petroleum to the new Manchukuo refinery. Unfortunately, this proved to be a futile gesture, since California Standard and Union Oil continued their sales; to assure a united front, it would be necessary for the American government to prevent the noncooperative U.S. companies from shipping crude oil to Japan and Manchukuo, but there was no official support at the State Department for such a general embargo. Secretary Hull, who favored a general policy of relaxing rather than increasing trade restrictions, feared it would lead to the deterioration of relations between Washington and Tokyo.

On October 20 the Finance Minister of Manchukuo presented an oral statement, outlining the proposed scheme, to representatives of Standard-Vacuum and the Asia Petroleum Company. The Manchurian Oil Company was to receive a monopoly over the sale of both crude and refined petroleum (exclusive of lubricants) in Manchukuo. Since 40 percent of the stock in this firm was held by the South Manchuria Railway Company, half of whose stock was owned by the Japanese government, and another 40 percent was held by private Japanese investors, the Manchurian Oil Company for all practical purposes was a Japanese firm. The monopoly was to purchase crude oil and refined oil from abroad as needed to satisfy the wants of Manchukuo; petroleum was to be allocated among the foreign oil firms then operating there on a quota basis. On November 13 the Cabinet and Privy Council of Manchukuo approved the plan.

A week earlier, on November 5, Koki Hirota, the Japanese Foreign Minister, had told the American and British ambassadors that the Manchukuo government had informed him that the new petroleum monopoly did not discriminate against foreign entrepreneurs. At the same time he gave them a memorandum that boldly proclaimed: "The plan of the Government of Manchukuo for the control of the oil industry is a project of that Government itself and is not within the knowledge or concern of the Imperial [Japanese] Government, and the Imperial Government is not in a position to give any explanation with respect to it."[21] Secretary of State Hull noted that this attitude was highly dishonest, and on November 30 the United States informed Tokyo that it regarded Hirota's statement as unconvincing.

With respect to the Japanese Petroleum Control Bill enacted earlier that year, both Grew and President Hashimoto of the Nippon Oil Company agreed that it was basically a military measure designed to facilitate national defense, but Grew believed the forced storage of oil by U.S. firms violated the 1911 Treaty of Commerce and Navigation between America and Japan. The Japanese government, though, informed him on November 20 that it would not be swayed by American, British, or Dutch representations over any aspect of the petroleum law. Grew nevertheless concluded that the Japanese were becoming more reasonable; according to his analysis, they wished the oil talks to be transferred from diplomatic to private channels.

The American and British oil interests began conversations with a group of Japanese officials at the Tokyo Club on January 9, 1935. By April 13 they had reached an understanding that not only would reduce the stock holding requirements to three months, but also would sanction an increase in the foreign companies' oil quotas. This so-called "five-point plan" was informal in nature, and was neither guaranteed nor published; had it become common knowledge, ultra-patriotic groups might have bitterly attacked it.

That September the Japanese Foreign Office announced a postponement in the inauguration of the six months storage provision from October 1, 1935, to June 30, 1936. The announcement seemingly conflicted with the three months' stock holding requirement of the "five-point plan." On October 7 Secretary Hull suggested to Edwin L. Neville, the Chargé in Japan, that if the Japanese government refused to compensate Standard for any costs incurred in the storage of more than a three months' supply of petroleum, the American government should point out that the stock holding provision violated the 1911 treaty between Japan and the United States. A month later, on November 15, Neville offered the following assessment:

> The situation now appears to be that the Japanese Government prefers to have the foreign oil companies continue in business in Japan, but is prepared to do without them if they do not wish to operate under existing conditions. It is not believed that this attitude can be considered entirely as a "bluff" on the part of the authorities.[22]

Discussions took a new turn in December, when Standard-Vacuum and Royal Dutch-Shell offered to make available to Japan information on the hydrogenation process for making petroleum from coal. Both the State Department, which was aware of the offer, and the oil firms realized the data would be used by the Japanese for military purposes. Although it did not restrain Stanvac and Shell, the State Department did instruct the American Embassy in Japan to inform the Japanese government that the United States had nothing to do with the proposal. At this time the research on hydrogenation was not too far advanced, and the Japanese apparently felt that they were on the verge of discovering the process themselves.

Negotiating from what it regarded as a position of strength, on December 23 the Ministry of Commerce and Industry verbally notified the foreign oil companies that the "five-point plan" was not binding and that the offer to furnish data on the liquefaction of coal would have no effect on the mandatory stock keeping provisions of the petroleum law. The stage thus was set for the expulsion of foreign oil firms, but this drastic step never took place. In fact, refined petroleum imports from the United States continued to rise annually from 1933 to 1940, with the exception of 1938.

Oil negotiations between the Americans and the Japanese failed to lead to a final resolution of the controversy throughout 1936. Finally in January of 1937 a written understanding was reached between the American and British oil interests and the Japanese government. The foreign oil interests received a written assurance that they would not be the object of discrimination, and that the storage facilities needed to hold the reserve stocks would not be built at their expense. But the Japanese government retained its six months oil supply requirement and its general right to regulate the petroleum industry. Ambassador Grew concluded in a report dated April 2 that the Japanese government "prefers that the oil companies remain in the position of lawbreakers (not having stored six months' stocks as required by law), in order that they will have no claim to rights or privileges and that their quotas may be cut at any time to make room for the products of the projected hydrogenation plants."[23]

American oil rights in the Far East faced an additional threat later that year when Japan invaded China, as Stanvac and Texaco were selling refined products in China at this time. On December 12 the Japanese military caused a furor when it bombed an American gunboat, the *Panay*, on the Yangtze Kiang River, and at the same time attacked three Stanvac vessels. The Japanese government quickly compensated Stanvac for this outrage, thus avoiding another extended diplomatic interchange. The fact that the *Panay* had been guarding the plants of the American petroleum company later led perennial Socialist party presidential candidate Norman Thomas to demand in a speech at Madison Square Garden: "Don't send our sons to war for the petroleum industry."[24]

The hydrogenation issue again arose in late 1938, when the Japanese began to pressure U.S. oil companies for data. Doubtless the approach of World War II, which might be dated from the Japanese attack on China in the previous year, had made Japan more eager than ever to secure alternate sources of oil. On this occasion the State Department, rather than taking a neutral position, suggested that Stanvac stall, which it proceeded to do.

Earlier, in February 1935, officials in Manchukuo proposed that the three foreign oil companies operating there furnish the new petroleum monopoly with limited amounts of gasoline and kerosene. In return, these firms were to enjoy preferential rights at competitive prices. Standard-Vacuum, however, did not care for the plan. On March 22 Secretary Hull advised Ambassador Grew that it was up to the oil companies to decide whether or not they wished to sell petroleum to the Manchukuo monopoly.

The Manchukuo petroleum monopoly officially began operations on April 10, and within the next few days both Ambassador Grew and the British Ambassador, Sir Robert Clive, protested its establishment to the Japanese foreign minister. On April 13 Grew left a note boldly asserting that "the American Government is constrained to express its considered view that upon the Japanese government must rest the ultimate responsibility for injury to American interests resulting from the creation and operation of the petroleum monopoly in Manchuria."[25] Grew had received word on April 12 from Mukden that the Dairen Customs had denied the Texas Company the right to import a shipment of lubricating oil, although technically the product lay outside the jurisdiction of the Manchukuo petroleum monopoly. At the same time the Dairen Customs issued a notice prohibiting foreign oil companies from exporting their stocks without the permission of the Manchukuo government. Unlike Texaco, both Standard-Vacuum and Royal Dutch-Shell refused to declare their stocks under these circumstances.

Back in Washington, the State Department considered, and then rejected, a crude oil embargo against Manchukuo. The Chief of the Division of Far Eastern Affairs, Stanley Hornbeck, pointed out that such an embargo would not work unless it included Japan as well, since the Japanese simply would supply Manchukuo with its needed crude oil. Charging more for products going to Manchukuo than for products going to Japan would be an equally futile gesture for the same reason.

After consulting with Standard-Vacuum officials, Secretary of State Hull telegraphed Ambassador Grew on April 24 that under the circumstances the foreign oil firms should settle with the government of Manchukuo, and leave the question of possible claims and government actions for the future. Subsequent conferences between the American and British embassies and the petroleum companies led to the conclusion that the companies should sell their plants and equipment to the oil monopoly, but withdraw their stocks if possible. The Manchukuo Ministry of Finance, though, stated on May 11 that it would be impossible to refund the import duty on the reexported stocks under the existing customs law but that it would be glad to discuss with the companies those circumstances that made it necessary for them to reexport their stocks.

As negotiations continued throughout the summer, on August 23, the petroleum advisor of the Manchukuo government, Tsuge, complained that the petroleum companies' asking price for their properties was too high. The companies had taken the position that as going concerns the properties had considerable value. In the interim the Manchukuo oil monopoly gave the remaining agents of the foreign petroleum firms the right to continue selling their duty-paid stocks until they were exhausted, a process the oil companies assumed would require only three or four months. As the year drew to a close, the haggling over the purchase price for the petroleum firms' properties continued; officials of the Finance Department of the Manchukuo government told a Standard-Vacuum representative on November 13 that the government would not pay compensation for loss of business.

Evaluating the status of these rather ineffective negotiations as of December 1935, Ambassador Grew noted:

> ...it became evident that the foreign companies could not expect anything more favorable than a partial settlement based exclusively on the physical value of the properties and without consideration of the principle of compensation for loss of business. It was therefore believed that the companies should proceed to the filing of claims.[26]

After eighteen more months of stalled negotiations, in May 1937 the Chief of the General Affairs Bureau of Manchukuo raised the possibility of improved relations between that nation, the United States, and Great Britain, since the settlement of intangible items in the oil companies' claims would inevitably require diplomatic negotiations. While playing down the political aspects, Manchukuo began to show a growing willingness to pay compensation in addition to the assessed value of the physical properties, if the petroleum firms set a reasonable selling price. That August, however, a comparison of the figures quoted by Standard-Vacuum and the purchase offer of Manchukuo revealed a large discrepancy. Since by this time the Texas Company had disposed of its physical properties, the Manchukuo government proposed to pay it a lump sum for goodwill out of the settlement it had offered the other two companies. Oddly enough, despite this unresolved haggling, American exports of petroleum to Manchukuo increased from $782,000 in 1936 to $3,436,000 in 1937, and rose even higher in 1938.

Despite their various defenders, Secretary of State Cordell Hull and the State Department have not received unanimously favorable accolades from U.S. diplomatic historians for their support of the American oil companies against Japan in Manchukuo. For example, Charles Callan Tansill's revisionist monograph *Back Door to War: The Roosevelt Foreign Policy 1933-1941* offers this criticism of New Deal diplomacy:

> The friction between the United States and Japan over Japanese commercial policies in Manchukuo was entirely needless. Secretary Hull was determined to press for the continuance of a trade principle (Open Door), even when its partial abrogation meant an increased volume of American trade. He seemed to be unaware of the ominous fact that his notes were creating a backlog of ill will that might later burst into the flames of war.[27]

### Japan and the Netherlands East Indies before Pearl Harbor

American concern for Far Eastern petroleum began to intensify as soon as World War II broke out on September 1, 1939. The petroleum monopoly in Manchukuo no longer was the irritant to Japanese-American relations which it once had been; instead, the critical issue now became Japanese access to the extensive petroleum output of the Netherlands East Indies. The United States was faced with a serious dilemma: if it refused to supply Japan with oil, Japan might move against the sprawling Dutch colony to the south.

On December 19, 1939, the Universal Oil Products Company told representatives of the Japan Gasoline Company that it would not execute a contract it had signed with the Japanese firm on August 19, 1938, because of advice it had received from the State Department. Under the contract Universal Oil Products would have granted the Japan Gasoline Company a license covering its polymerization process for the production of iso-octanes in return for $1 million dollars. At that time the State Department had made no objection to this contract with a firm basically engaged in research, the invention of various refining processes, and the licensing of its patents.

As a result of this *volte face*, however, the Japanese Ambassador to the United States, Kensuko Horinouchi, advised Secretary of State Cordell Hull on January 6, 1940, that the contract-breaking was in violation of the Treaty of Commerce and Navigation between America and Japan. In his reply, dated January 27, 1941, Hull justified the State Department's intervention by pointing out that the contract had been signed prior to "...the continued widespread bombing and machine-gunning of civilian populations in certain areas...."[28] He explained that these practices had caused the U.S. government to adopt a policy of withholding the delivery of technical processes for the production of high quality aviation gasoline and added that the United States desired to conserve various commodities and technical processes for the sake of national defense.

Nevertheless, the flow of petroleum products from the United States to Japan continued. The total value of American petroleum exports to Japan in 1940 was $53,133,000; in this year alone the United States exported three times as much gasoline to Japan as it had in any of the three preceding years. In addition, during 1940 the Japanese were to import 556,703 barrels of aviation gas. Finally, large quantities of crude oil were making their way to Japan from America.

The Japanese also were obtaining huge amounts of petroleum from the Netherlands East Indies at this time. In the middle of July Japan notified the NEI government that it wished an oil agreement, and thus was preparing to send a large mission there. In its reply, dated July 26, the NEI government stated that it was "prepared to stimulate and facilitate exports to Japan with all the means at its disposal—bearing in mind Japan's minimum demands...."[29] While the Dutch Minister for the Colonies, who was residing in London, was anxious for the NEI petroleum firms to make a deal with the Japanese, the State Department was far from enthusiastic. It objected to the unusually large oil shipments to Japan and hoped that they would be of short duration.

In the last half of 1940, following the advent of the Prince Konoye cabinet in Japan, there was a heated debate inside the Roosevelt administration over petroleum shipments to Japan. Secretary of War Henry Stimson and Secretary of the Treasury Henry Morgenthau believed that an oil embargo would force the Japanese to adopt a position of greater restraint. On July 25, in the absence of Secretary Hull from Washington, they persuaded FDR to halt the exporting of steel and petroleum products and scrap iron in the name of national security. Under Secretary of State Sumner Welles protested that this policy would lead to war with Japan; as a result, on the next day the Roosevelt administration limited its embargo to aviation gasoline and scrap iron. But as the Director of the Division of Monetary Research of the Treasury Department, Harry Dexter White, pointed out in a brief memorandum to Morgenthau dated August 6, its actual effect was open to dispute:

> What effect the recent addition of (a) aviation gasoline, (b) aviation lubricating oil, and (c) tetraethyl lead to the list of commodities requiring an export license will have on Japan's military effectiveness seems to be unknown in Washington. If what some experts say is true, the restriction has considerable effectiveness, but if what others say is correct, it is almost completely ineffective. The numerous oil experts interviewed are not in agreement on important points...[30]

The American government may have decided to adopt a firmer stand toward Japan because of certain remarks which the British Ambassador to the United States, Lord Lothian, allegedly made following a dinner at the British Embassy in Washington on July 18. Secretary of War Henry Stimson had earlier criticized the British for "giving in" to Japan on the Burma Road, whereupon the agitated Lothian complained about U.S. shipments of aviation gas to Japan. Lothian reportedly promised: "If you will stop shipping aviation gasoline to Japan we will blow up the oil wells in [the] Dutch East Indies so that the Japanese can't come down and get that...".[31] Lothian, though, did admit that this was his own idea and not the policy of his government. Secretary of the Treasury Henry Morgenthau obviously was skeptical about his remarks, as he told Harry Dexter White that "this fellow Lothian, I am convinced, only thought of that because he got angry at Stimson...".[32]

Not every prominent American official held constant on the question of oil for Japan. A number vacillated or even changed their minds, including the Ambassador to Japan, Joseph Grew; Grew had long opposed sanctions, but he gradually became more hawkish. On September 12 he observed in a lengthy dispatch:

> If we conceive it to be in our interest to support the British Empire in this hour of her travail, and I most emphatically do so conceive it, we must strive by every means to preserve the status quo in the Pacific at least until the European war has been won or lost. In my opinion this cannot be done...by merely registering disapproval and keeping a careful record thereof...Until such time as there is a complete regeneration of thought in this country [Japan], a show of force, together with a determination to employ it if need be, can alone contribute effectively to the achievement of such an outcome and to our own future security.[33]

On August 3 the Japanese Embassy complained about the presidential embargo on the exporting of aviation gasoline, which, according to another decree issued on July 31, did not apply to the Western Hemisphere: "...Japan would bear the brunt of the virtual embargo," its note suggested; "the resultant impression would be that Japan had been singled out for and subjected to discriminatory treatment.[34] In October the Japanese Embassy informed the State Department that the Japan Gasoline Company had filed a damage suit against the Universal

Oil Products Company, which on the advice of the State Department had refused to execute the contract it had made with the Japanese firm. The Japanese government suggested as a compromise that the Department of State use its good offices to grant a license for the exportation of all of those processes that did not relate immediately to the production of high quality aviation gasoline.

After the July 1940 embargo began, the British government expressed its fear to officials in Washington that the embargo would lead to a Japanese invasion of the Netherlands East Indies. The British also were disturbed because some American producers had charged that Great Britain had pressured FDR into restricting exports so that the British and the Dutch could increase their sales of petroleum products to Japan.

Although the American government did not want the petroleum properties of the NEI to fall into Japanese hands, some U.S. officials were concerned that the Dutch government would destroy them as a defense measure. This, the War Department believed, inevitably would lead to war with Japan. In mid-August the Consul General at Batavia, Foote, reported that such a plan was in existence, and involved all of the refineries, wells, and petroleum of the Netherlands East Indies; of these properties, approximately half were American-owned, with a total value of around $300 million. But the destruction did not take place at this time. On November 13 the local oil companies and the Japanese oil importers signed an agreement at Batavia calling for the annual exportation of 1.8 million tons of petroleum products from the NEI to Japan.

During the winter of 1940-41 the British government again brought up the subject of restricting the petroleum trade. With World War II now a year old it was assuming a more hawkish stance. Not only did Great Britain want to effect a reduction in the oil flow from the NEI by having the companies fail to meet their quotas, it also desired to curtail that from the United States by removing petroleum tankers from the trade. The American response, as given by Stanley Hornbeck, pointed out that both the British and the Dutch in the past ". . . had begged us to go lightly and to think long and hard in relation to petroleum and other embargoes lest we produce repercussions in the Far East which would be unfortunate for all concerned, but particularly for them."[35] Hornbeck, it should be noted, was a hawk, not a dove.

U.S. relations with Japan itself continued to worsen. At the beginning of 1941 Stanley Hornbeck received authorization from Sumner Welles to ask the Standard-Vacuum Company to withhold from Japan full or prompt deliveries. The petroleum firm agreed. Shortly thereafter Secretary of State Cordell Hull gave his approval to a series of measures halting the export of metal drums, containers, storage tanks, and oil drilling equipment. Even these steps did not satisfy the proponents of vigorous action against the Japanese, since as of the spring of 1941 U.S. oil companies were supplying approximately 75 percent of Japan's petroleum needs. The May issue of *Amerasia* complained that ". . . by continuing to provide Japan with fuel for her bombers, battleships and industrial machinery, we have enabled her not only to strengthen her position in China, but also to advance into Indochina and Thailand, so that she is now in a better position than ever to launch an attack on the Dutch East Indies and Malaya."[36] As *Amerasia* correctly pointed out, Japan had plans to bring the NEI within the confines of its Greater East Asia Co-Prosperity Sphere.

That June Secretary of the Interior Harold Ickes, who was now in charge of American domestic petroleum policy, came out in favor of a U.S. oil embargo against Japan. Secretary of State Cordell Hull, however, opposed Ickes' plan. Caught in the middle of this debate was President Roosevelt, who was confronted with the dilemma of increasing U.S. naval strength in the Pacific, while still maintaining an adequate naval force in the Atlantic, should the American government decide to impose an oil embargo on the Japanese.

The petroleum controversy reached its climax a month later, after FDR had discussed various aspects of the question with Japanese Ambassador Nomura on July 24 in the presence of Under Secretary Welles and Admiral Stark. If Japan, the President noted, "attempted to seize the oil supplies by force in the Netherlands East Indies, the Dutch would, without the shadow of a doubt, resist; the British would immediately come to their assistance; war would result between Japan, the British, and the Dutch; and in view of our own policy of assisting Great Britain, an exceedingly serious situation would immediately result."[37] On the same day, apparently without knowing about this conference, 1940 Presidential nominee Wendell Willkie suggested a U.S. embargo of petroleum shipments to Japan. Two days after Japan went through with its invasion of southern Indochina, on July 26, the Roosevelt administration froze Japanese credits in the United States.

In an unexpected move, the Netherlands East Indies decided on July 28 to require special permits for all exports to Japan. At the time it also noted that should Japan not conduct itself properly in the future, it would face total economic ostracism on the part of the NEI. Despite Dutch inquiries, however, Under Secretary of State Sumner Welles would not guarantee that the United States would defend the Netherlands East Indies militarily, if the Japanese attacked the latter.

While no promise to the Netherlands was forthcoming, on August 1 the President did place a virtual embargo on the exporting of motor fuel and lubricating oil to Japan without technically referring to the country by name. At a meeting at the State Department three days later, Assistant Secretary of State Dean Acheson discouraged the Standard-Vacuum Company from applying to the NEI government for export licenses for contractual petroleum shipments to Japan. He pointed out that it was up to the Netherlands East Indies to decide whether or not to continue oil shipments there.

Some Japanese militants demanded an immediate response to the hardening American policy, but the Chief of the Naval General Staff, Admiral Osani Nagano, counselled caution. Advising Emperor Hirohito to avoid war if possible, Nagano suggested that with the oil embargo operative, Japan had at the most two years of petroleum reserves, thus making the outcome of a military challenge to the United States problematical at best. It should be noted that this heated debate occurred only four months before the Japanese attack on Pearl Harbor.

Assessing the petroleum refining and "cracking" capacity of Japan at that time, Clayton D. Carus and Charles L. McNichols have written that:

> In 1941 there were twenty-five petroleum refineries in Japan proper, with a total daily crude capacity of about 62,000 barrels, and an indefinite number of small primitive plants, most of them in the oil fields of west-central Honshu, whose combined intake was less than 1,500 barrels daily.
>
> The so-called "cracking" capacity for the production of gasoline hardly exceeded 15,000 barrels a day at the outbreak of the war. This was distributed among twelve plants, all of them built under the supervision of American engineers and using either the Cross or the Dubbs systems.[38]

Once the United Stated officially entered World War II, Japan failed to expand its petroleum refining facilities at home to meet the increasing demand. In explaining why, Carus and McNichols note:

> For one reason, the shortage in shipping that developed immediately after the beginning of the war made it more feasible to build refineries near the conquered oil fields in the East Indies. For another, the Japanese have few men skilled in constructing and installing modern, and therefore complicated, refinery equipment; even with the help of German technicians wartime construction must have been limited.[39]

Japan experienced serious problems in its attempt to secure a petroleum supply elsewhere throughout World War II. The Netherlands East Indies oil production of 62 million barrels in 1941 fell sharply to 20 million barrels in 1943, following the Japanese occupation of the N. E. I. in February 1942. The main reason for this decline was that military authorities had destroyed the Standard-Vacuum refinery at Palembang, Sumatra, despite the Japanese threat to kill all the European inhabitants if this occurred. Demolition experts also sealed the wells at Talang Akar with scrap iron and cement and destroyed the pipelines.

By July, however, the Japanese had managed to patch up the pipelines. They also had captured the Shell Pladjoe refinery outside Palembang almost intact, while they drilled new wells to restore production. As a result, Japan was able to produce an average of 20,000 barrels of oil daily in the Netherlands East Indies, and at one time the output reached 35,000 barrels. But Allied air raids during February 1944 and January 1945 again put an effective end to refinery operations there.

## The Indonesian and Vietnamese Experiences Since World War II

During the years that followed the close of World War II American interest in Far Eastern petroleum centered on Indonesia, formerly the Netherlands East Indies, and South Vietnam, once a part of French Indochina. Although prior to 1945, the United States had not been concerned about the oil potential of South

Vietnam, it had been vitally interested in the long-functioning petroleum fields of Indonesia. The scope of the Indonesian oil reserves is a known factor, but the extent of the Vietnamese ones became the subject of a bitter controversy in 1970, in part because of a misquoted figure that was widely circulated. The U.S. government officially intervened in an attempt to engineer a settlement between the Indonesian government and the American petroleum companies in 1963, but eight years later in the case of war-torn Vietnam it tried to minimize the importance of Vietnamese oil. For these reasons the developments pertaining to petroleum took quite dissimilar paths in these two Far Eastern countries.

Despite the termination of World War II in September 1945, the Japanese remained in control of the Standard-Vacuum facilities in South Sumatra until June 30, 1946, at which time the Indonesians took them over again. The Indonesians kept these facilities until July 1947, when the Dutch army arrived; at that time the withdrawing Indonesians damaged a number of wells and storage tanks. Thus it was not until October 1947—two years after the end of the war in the Pacific—that the returning American oil men were able to repair the facilities at Palembang enough to begin processing the crude oil left behind by the Japanese. Jersey Standard and Socony-Vacuum hoped to be operating Palembang by the summer of 1948 at its 45,000 barrels per day pre-World War II capacity. Caltex, however, did not reclaim its fields until January 8, 1949.

In that year the Netherlands officially recognized Indonesian sovereignty, although the Indonesians had declared themselves independent four years earlier. A year later Indonesia became a unitary republic. Quite naturally these developments affected the nation's petroleum relations with America. In September of 1954 Standard-Vacuum reached an agreement with the Indonesian government, under which it would expend between $70 and $80 million on a four-year modernization program that would train Indonesian administrators and technical workers in the process. The program was to include not only the expansion of its Palembang facilities, but also the construction of a ninety-mile pipeline. Since at this time Standard-Vacuum needed to supply the Orient and the Indian Ocean marketing areas with crude oil from its own fields, it negotiated a new pact with the Indonesian government.

Although there was a major revolt in Celebes during 1957, Caltex, the California Texas Oil Company, also was willing to risk its investment capital in Indonesia. During the early part of 1957 Caltex announced that it would increase the value of its Indonesian facilities from $50 to $75 million, and in the process raise its daily petroleum output there from 120,000 barrels to 180,000. Among Caltex's contemplated projects were additional exploration work in central Sumatra, a pipeline from there to the Strait of Malacca that would be paralleled by a forty-mile road, and a deep water port at Dumai.

It became apparent in the following year that both Stanvac and Caltex may have made a mistake in deciding to expand their activities in Indonesia, when both U.S. firms had to shut down their central Sumatra operations for two weeks while President Sukarno's troops expelled the insurgents from the jungles. The rebels had threatened to destroy the American oil installations if the Indonesian government launched an all-out attack on them. In addition, Communist labor leaders on Java recommended that "inspection teams" of workers infiltrate the petroleum centers. But despite these problems, as of 1958 Caltex and Stanvac under the 50-50 formula were contributing approximately $125 million annually to Indonesia; they also were employing 16,000 Indonesians, and were producing around two-thirds of the total Indonesian oil output.

One might cite other Stanvac and Caltex benefits for Indonesia as well. By the end of the decade Caltex had built 550 miles of road, Stanvac 775; Stanvac also had constructed three airports in Sumatra, while Caltex had repaired and drained the governmental airport at Pakanbaru. In addition, Stanvac had installed 125 miles of telephone wires, and Caltex had inaugurated a radio telephone system. Both firms had built hospitals and had undertaken free inoculation programs. Furthermore Caltex had constructed a technical and academic training center at Rumbai, Stanvac a technical high school at Pendopo.

Their records of performance, though, did not satisfy Sukarno. In 1960 the Indonesian government decreed that petroleum was "wealth which is controlled by the state"; it then altered the role of the oil firms to that of contractors. Sukarno also demanded that the 50-50 split be changed to a 60-40 spilt, under which Indonesia would receive the three-fifths. While the petroleum companies did not protest this change, they did object to basing the split on the posted price rather than the realized one. There was also the question of whether the taxes and duties would be paid before or after the split. Further problems arose over what price Indonesia would offer for the marketing and refining facilities, and the amount of hard currency it would need to operate them. Negotiations over the various aspects of the petroleum controversy dragged on until the final months of the Kennedy administration.

During the early 1960s the three major foreign petroleum firms that were operating in Indonesia were Caltex Pacific (California Standard and the Texas Company), which had an investment of $180 million; Stanvac Indonesia (Jersey Standard and Socony Mobil), with $280 million; and Shell Indonesia, with $650 million. By 1963 these three companies were doing an annual export business of $211 million, and they also dominated the Indonesian domestic market.

In April of 1963 the Sukarno regime accused these firms of stalling the petroleum negotiations, and ordered them either to reach an agreement by June 15 or face the shutdown of their Indonesian operations. On May 18 the U.S. Embassy in Djakarta cabled Washington, "U.S. - Indonesian relations have reached most serious turn since Indonesian independence. We are dangerously close to the end of the road."[40]

The pessimistic U. S. oil companies then approached Under Secretary of State Averell Harriman and Assistant Secretary of State for Far Eastern Affairs Roger Hillsman. Harriman recommended a meeting with Sukarno in Tokyo, where he would be freer from domestic Indonesian pressures. The American mediator was to be Democratic Lieutenant Governor Wilson Wyatt of Kentucky. U.S. Ambassador to Indonesia Howard Jones, who was regarded with a critical eye by the petroleum firms, and professional oil consultant Walter Levy, who had played a major role in settling the Iranian oil controversy in 1954, also were to be present. The key Indonesian representative aside from Sukarno was Chaerul Saleh, the Minister of Basic Industry and Mining.

After several days of continuous negotiating in Tokyo, the two sides reached a compromise settlement. The petroleum firms agreed to a 60-40 split of the oil profits in favor of Indonesia, and approved time payments by the Indonesian government for the oil companies' assets. Sukarno consented to a 60-40 split based on the actual rather than the posted price. Under the agreement the U.S. firms would not pay Indonesia taxes out of their two-fifths, and Indonesia would make a realistic assessment of the value of the petroleum facilities. The U.S. firms also obtained the right to explore for oil in Indonesia over twenty-five years, as well as the privilege of continuing both their domestic and export marketing operations.

While Jones praised the presentations of Wyatt and Levy, Wyatt characterized the settlement as a draw. The chief executive of Jersey Standard, Jack Rathbone, was more enthusiastic in his public statements. "Considering that we have been negotiating with a government that is both irascible and irrational," he observed, "I'm contented with the way this has worked out."[41] On the occasion of the actual signing of the document, moreover, both President John F. Kennedy and Sukarno issued messages of congratulation. Behind the scenes, Wyatt pointed out that strong support for the Clay report on foreign aid in the U.S. Congress had meant that there would be a cessation of foreign aid to Indonesia, if there had been no oil settlement. As the Indonesian economy had been deteriorating in recent years, the country was in no position to absorb the severe dislocation that would have resulted from the termination of exports. Probably the biggest loser as a result of the agreement was the Chinese Communists, who would have gladly negotiated for access to the rich Sumatran fields.

Two years later, in 1965, a wholesale slaughter of Indonesian Communists occurred, and the subsequent political shake-up witnessed the military gain in power and Sukarno lose. It was at this time that Lieutenant General Ibnu Sutowo obtained control over petroleum matters, and the Indonesian government moved into domestic marketing. On December 31, Royal Dutch-Shell sold its facilities to the state for $110 million. During 1967 General Sutowo set up an Indonesian petroleum office in New York City, and in 1968 established an integrated government oil company: Pertamina. By the end of the decade Indonesia had purchased the Stanvac refinery; Pertimina now became the sole domestic refiner of petroleum in Indonesia. Nevertheless, during the late 1960s Japanese private companies, as well as French and Italian state ones, began to play a role in developing Indonesian oil, and the Japanese were especially helpful in assisting the state petroleum company.

Beginning in 1966, prior to the creation of Pertimina, Indonesia began signing a new series of contracts with U.S. oil firms. The first was concluded in 1966 with the Independent Indonesian American Petroleum Company (IIAPC), which was headed by a group of Colorado and Montana oil men. In February 1968 Cities Service, Ashland Oil and Refining Company, and the Australian Drilling Company contracted with Pertamina to explore 56,000 square miles off northeastern Java. Like the IIAPC agreement, these contracts were to run for thirty years and provided for an initial investment of at least $7.5 million, an exploration period of at least six years, and a 65-35 split of the profits in favor of Pertimina. An earlier agreement which Indonesia had signed in late 1967 with the Continental Oil Company had contained a production sharing contract, and had set the division of profits at 67.5 to 32.5 percent. Among the other major firms that followed suit were Union Oil and Sinclair Oil.

In 1969 Phillips Petroleum and Superior Oil took over a part of the AGIP (the Italian state oil company) offshore concession off West Irian. By this time Gulf, Mobil, Frontier Petroleum Corporation, and Ray M. Huffington had already increased their oil acreage in Indonesia. Jersey Standard and Shell, however, had not signed new contracts with the Indonesian government. That summer White Shield, which admitted that it was copying the tactics of Armand Hammer's Occidental, obtained an offshore concession and defeated twenty-two other rivals in the process. Among other things, White Shield (which was in fact only the service arm of a drilling syndicate) promised to invest 20 percent of its profits in the country and to extend a 5 percent equity interest to Indonesians, if oil were discovered.

Then in 1971 the Indonesian government signed an agreement with Caltex—whose properties were to revert to the state upon the expiration of its work contract in 1983—that allowed it to operate for another eighteen years, in return for advance rent payments of $18 million in fifteen annual installments. At the same time Calasiatic/Topco obtained new thirty-year production-sharing contract acreage on Sumatra; Atlantic Richfield (Arco) had already won a new eight-year production-sharing contract in Kalimantan, or Indonesian Borneo. By 1972 sixteen of the twenty-four companies engaged in offshore drilling in Indonesian waters were American.

Summarizing the petroleum situation in Indonesia as of 1973, Louis Kraar observed in *Fortune*: "Thanks to Pertamina's success, Indonesia has emerged from a long twilight of mismanagement and near bankruptcy to regain its old position as a major supplier of petroleum to the world."[42] Today not only does Indonesia own its oil industry, but it also controls the management of it; American and other foreign companies merely execute "production-sharing contracts" in such areas as exploring, drilling, and extracting. Gulf Oil was even a joint partner in a scheme to package and market fertilizer for Java, and former Governor Edmund G. Brown, Sr. of California played the roles of both investor and operating executive in this arrangement.

Despite its successes, Pertamina has had its critics. Sutowo and the rest of the army leadership was accused of personally benefitting from its operations, and an Indonesian investigating commission charged that Sutowo was operating a "state within a state." There also were complaints about unpaid taxes and heavy debts, but these criticisms did not keep two dozen foreign corporations from investing more that $1 billion in Indonesia between 1966 and 1973.

Although the prolonged agony of Vietnam has been burned permanently into the consciousness of innumerable Americans, it is necessary to place the developments pertaining to Vietnamese petroleum in their proper historical and economic setting. Prior to 1954 Vietnam was under French rule as a part of French Indochina, but the Communist Viet Minh had been fighting a civil war there since 1946, and had consolidated their position most effectively in the northern part of the country. French Indochina was of such strategic importance in the eyes of the American government that the latter extended three billion dollars in military aid to the doomed French colonial forces, but it did not send in troops. The defeat of France at Dienbienphu in 1954, along with the inability of the French to obtain strong allied support, led to a nine-power agreement on the status of Vietnam at the Geneva Conference. Under this agreement there was to be a temporary partition of Vietnam, dividing that nation into a Communist North Vietnam and a South Vietnam, whose fate was to be decided in a general election scheduled for 1956, an election that was never to take place. There then ensued a bitter civil war that lasted for two decades.

After 1960 South Vietnam underwent an economic transformation. Although rubber rather than rice had earned most of the foreign exchange before the civil war, rice exports declined to the point where the nation actually had to import rice. The country also lacked the mineral deposits of North Vietnam; what minerals there were, moreover, were not exploited on a large scale. Thus it is not surprising that reports of possible important petroleum fields offshore South Vietnam in the 1960s attracted widespread attention both within and without the country. By this time, of course, the United States had become deeply involved there militarily.

At the end of 1968 the United Nations' Economic Commission for Asia and the Far East (ECAFE) began what turned out to be highly successful oil explorations off the southwestern coast of South Vietnam. In the spring of the following year some of the seismic survey data that had been collected was sold to a group of primarily American oil firms, including Esso, Phillips, Marathon, and Union. Around this time, too, the U.S. Navy concluded that the offshore area from Korea to Thailand "might contain potentially one of the most prolific oil reserves in the world;"[43] this petroleum would be especially valuable because of its low sulphur content. By June of 1970 the *Petroleum Weekly* was comparing that body of water centering around Vietnam with southern Louisiana and Texas from the standpoint of oil production.

Not surprisingly, news of these developments had spread far beyond Southeastern Asia and the United States by this time. On May 22 the Swiss weekly *Die Weltwoche*, which was widely circulated in Western Europe, published an article entitled "The Hushed Up Sea of Oil." The article raised the question: "Is the ideological conflict the only concern of the United States' engagement in Indo-

China?"[44] It also noted that Amoco and Gulf already had obtained concessions from Thailand and Cambodia for the entire width of the Gulf of Siam.

During that summer Saigon divided its offshore water into eighteen large blocks, and then rewrote its investment laws so as to encourage the entry of foreign capital. On December 1, South Vietnam proclaimed a new petroleum law with tax rates that were beneath the average global level, and embarked on a public relations campaign. When the South Vietnamese government announced at the end of the year that it would accept tenders for leases beginning at the end of February of 1971, approximately two dozen foreign oil firms, most of them American, descended on Saigon.

The Vietnamese Provisional Revolutionary Government, however, declared that North Vietnam regarded these petroleum concessions as invalid. The Russians, according to Agence France Presse, believed that Southeastern Asia oil discoveries had played a major role in shaping U.S. policy towards Indochina since early 1970. In March the French, who had been in Vietnam long before the Americans, offered to screen all bids and oversee operations for the Thieu regime then in power. This proposal was not acceptable to the petroleum firms, who were suspicious of the French motives.

Back in the United States, then cruelly torn by the indecisive and drawn-out Vietnam War, there were protests against the operations of U.S. oil companies in South Vietnam. Among the members of Congress to call for an official investigation were Democrat William R. Anderson of Tennessee, Democrat James Abourezk of South Dakota, and Democrat Bella Abzug of New York. In addition, more than 10,000 persons wrote letters of inquiry to the Senate Foreign Relations Committee over a period of several weeks.

As protest mounted, the State Department played down the significance of South Vietnamese oil and the official American role in developing it. On March 16, 1971, Secretary of State William Rogers stated at a press conference that the existence of petroleum reserves in South Vietnam had not affected U.S. foreign policy there in the slightest. The Assistant Secretary of State for Congressional Relations, David M. Abshire, wrote a letter to Senate Foreign Relations Committee Chairman J. William Fulbright on February 10, claiming that the American government had not provided technical assistance to South Vietnam in exploring for offshore oil. Abshire pointed out that the total dependence of South Vietnam on imported petroleum at that time amounted to $30 million annually.

In March Texaco informed the office of Representative James Abourezk that it had no specific interest in South Vietnam. It thus took a position diametrically opposite to Esso, which had approached the State Department on the subject. When the Economics Ministry in Saigon announced on June 10 that it would open bids for oil exploration on its continental shelf, which it hoped would yield the South Vietnamese treasury as much as $1 million daily, thirteen of the thirty firms to express an interest were American. British, French, and Italian companies also took part in the bidding, which was only to last for sixty days.

It should be noted, though, that the widespread interest in the oil of Southeastern Asia to a considerable degree was the result of a monumental error in a UPI dispatch released in February 1970. According to the dispatch, U.S. geologist James Gauntt in assessing the offshore oil potential of Thailand, Indonesia, Malaysia, and Vietnam had observed that within five years the average daily production of this area would exceed 400 million barrels, ten times the current world production. Actually, Gauntt had predicted a 4 million barrel daily output for the future, a mere 1 percent of 400 million. The mistaken figure appeared in print in such publications as *Die Weltwoche, Atlas, World Oil, Le Monde,* the *Nation*, and *The Village Voice*, before Leon Howell caught and exposed the error in the July/August 1971 issue of *Current*.

With the signing of the Vietnamese peace treaty during the Nixon administration and the fall of South Vietnam during the Ford administration, the emotional level of the debate over Vietnamese policy declined in intensity. Writing retrospectively in *Asian Survey*, David G. Brown concluded that petroleum considerations had only a slight impact on the American involvement; "if anything," Brown observed, "the U.S. found the possibility of the existence of such reserves a political embarrassment."[45] While a petroleum law had been passed and an oil board set up at the time when the Americans were playing a major role in South Vietnam, the United States sought to avoid political complications over petroleum at that time, and the South Vietnamese wished only to affirm its national sovereignty over their mineral resources, as many other nations were doing.

## The U. S. Oil Firms and the Indian Subcontinent, 1902-1973

Despite its huge population and widespread poverty, India is a nation so rich in natural resources that it perhaps ranks only behind North America, the Soviet Union, northwestern Europe, and Brazil. Its deposits of coal, iron, mica, bauxite, and manganese are especially noteworthy. The Indian reserves of petroleum, however, appear to be relatively small, although recent discoveries of oil have added to them.

Since India remained a part of the British Empire until after the end of World War II and Burma was not separated from India and given self-government until 1937, any discussion of American petroleum relations with Burma prior to 1937 necessarily involves India as well. On March 20, 1902, the Colonial Oil Company, a subsidiary of Jersey Standard, applied to the Burmese government for a petroleum prospecting license. Burma rejected the application because of the hostility of the Viceroy of India, Lord Curzon. When the Anglo-American Oil Company, a Standard Oil subsidiary chartered in 1888, made a similar attempt in June, it met a similar fate. Lord Curzon even denied the U.S. Consul General in India, General Patterson, an interview in this connection, having his secretary instead write a letter stating: "It is not desired by the Government of India to introduce any of the American Oil Companies, or their subsidiary companies, into Burma, and that an interview with the Viceroy would not be attended with any other result."[46] When William H. Libby of Jersey Standard talked with the Bengal Chamber of Commerce in Calcutta, the response was equally negative.

Three years later, in 1905, Standard Oil officials tried a new approach. They applied for permission to operate a refinery at Rangoon and to erect storage tanks for bulk distribution. This, too, fell upon deaf ears; Lord Curzon still held office in India. Further investigation, however, revealed that the departure of Curzon would by no means pave the way for Standard Oil's entry into Burma, since Queen Victoria apparently had issued an edict barring from the country any concern in which John D. Rockefeller or J. P. Morgan had an interest. To protect the position of the Burmah Oil company—which was one of the earliest commercial producers and exporters of petroleum—the British attempted to discourage foreign competition; Burmah Oil was not to enter into any agreement with any American trust. Standard Oil, though, was to find a sympathizer in none other than Marcus Samuel of Shell, who complained to a parliamentary committee about the exclusion policy in Burma that summer.

A dozen years later, in 1917, Standard Oil of New Jersey also failed in its efforts to lease the oil lands at Sylhet in the Indian state of Assam. Under the so-called permanent settlement, most of the land there was held through absolute title to both the surface and subsurface rights; the Standard agent thus attempted to lease rather than purchase the property in question. The response of the Indian government was to issue a new regulation in October, making it illegal for an owner to transfer his interest in a mine, including any mineral deposits.

On November 25, 1921, the British Ambassador to the United States, Sir Auckland Geddes, struck a more hopeful note for American petroleum interests when he branded two British documents dating from 1884 and 1885 as forgeries. The documents had barred U.S. oil companies from doing business in either India or Burma. On December 10, however, Secretary of State Charles Evans Hughes replied that it was not yet entirely clear to him whether the challenged information "was wholly erroneous or was substantially or in part correct." Hughes pointed out that earlier in 1921 the British government had issued an official memorandum affirming that in India "prospecting or mining leases have been, in practice, granted only to British subjects or to companies controlled by British subjects."[47]

For the next six years the pattern of discrimination against U.S. oil firms continued. American consular reports described the British policy on petroleum ownership and production in India as "one of entire exclusiveness"; New York Standard, which was marketing oil, told the State Department that it was not even permitted to buy a warehouse in Burma. In 1926 Royal Dutch-Shell asked New York Standard to join it in a boycott of the purchase of Russian crude, but New York Standard refused to do so, since this step would have undercut its position in India at the expense of Royal Dutch-Shell. Royal Dutch-Shell then retaliated by inaugurating a kerosene price war in the fall of 1927.

In March 1928 the British petroleum firms operating in India asked that government to impose a tariff, which would exclude U.S. oil companies. The Indian governmental inquiry board pointed out it was the British combine that had been guilty of dumping, not the American firms. At the same time Royal

Dutch-Shell attempted to persuade the government of Burma to pass a protective tariff against the Russian oil that New York Standard was selling. Fortunately for the U.S. petroleum companies operating in South Asia, however, Sir Henri Deterding reached an agreement with New York Standard later that year which in effect stopped the Indian oil war.

The long-standing discrimination against U.S. petroleum firms again came to the fore in the spring of 1941, when British and Indian representatives arrived in Washington to discuss a new treaty of commerce, navigation, and consular rights. The Chief of the Division of Near Eastern Affairs Wallace Murray protested the suggestion that Article 8 should be altered. According to Murray, "Such a deletion would result in our acquiescence in the continued enjoyment by British oil companies of exploratory and extractive privileges in India not accorded to American firms by virtue of legislation existing in India since 1885 forbidding corporations controlled by foreign interests to engage in the extraction of petroleum."[48] The High Commissioner of the Government of India, Sir Firoz Khan, stated that granting an American oil concession would have an unfortunate impact on public opinion, and that Indian corporations simply lacked the capital to invest in the United States. The state of Baluchistan was the focal point of the fruitless discussions that followed; Khan was reluctant to allow U.S. capital to enter it, and this was the only part of India in which the American petroleum firms were interested.

Two and a half years later the entry of U.S. oil companies into India still remained a touchy issue. On December 15, 1943, Secretary of State Cordell Hull in a letter to Admiral William D. Leahy pointed out the disadvantages inherent in the construction of an oil refinery in Bombay. Not only were there restrictions on the marketing of American-produced oil in India, but also such a refinery would be allowed to import foreign petroleum for processing. Furthermore, Bombay had certain economic liabilities as a location, and the materials that were needed to build a complete set of facilities were not available there. Above all, as Hull pointed out, the United States wished to give first priority to the American position in the Middle East.

On August 15, 1947, Great Britain ended its rule over India with the partition of that nation into the separate sovereign states of India (Hindu) and Pakistan (Moslem). These newly independent nations immediately clashed over the ownership of Kashmir, as they were to do again in 1965. Following the assassination of Mohandas Gandhi in January of 1948, Jawaharlal Nehru emerged as the leader of India and occupied that position until his death in 1964.

Post-partition developments relating to petroleum and the American role therein have been volatile. In 1949 both U. S. and British oil companies sent a technical mission to India at the request of the Indian government to look into the possible construction of a refinery. The petroleum firms offered to build two such refineries themselves, if they were allowed to market the products at a price 10 percent higher than the world parity price. The proposal was unacceptable to the Indian government, but the foreign oil companies began to display a more conciliatory spirit after the Iranian nationalization of 1951. Burmah-Shell, Standard-Vacuum, and Caltex set up three refineries during the 1950s, after the Indian government had decided to reduce its petroleum import bill. In 1954 oil imports had amounted to $200 million, or 15 percent of the total amount of Indian imports.

Once these privately owned refineries became operative K. D. Malaviya, the Minister for Mines and Fuel, declared that their prices were too high for refined products, and that they were processing too much crude oil. When the Indian government set up a study group, the Damle Committee, during the 1960s to investigate these accusations, this committee "recommended greater discounts to be obtained by the companies from their Associates and imposition of additional non-recoverable duties by Government and reductions in distribution and marketing charges and the profit margin of the companies."[49] After initially hesitating, the government of India agreed to accept these recommendations. The petroleum firms, though, would not consent to reducing their prices on refined products; they instead proposed—and the government failed to approve—a plan under which crude oil would be imported at reduced prices, and the petroleum products that were then refined in India would then be sold at the suggested Damle Committee price levels. The eventual outcome of the deadlock was an uneasy compromise.

The Indian government then decided to set up three refineries of its own, two in the east and one in the west, and by 1966 the combined capacity of these refineries was 51 million barrels a year. In 1965 the Indian Oil Corporation had obtained exclusive control over the import trade in oil products, and by 1967 because of transport priorities and filling station land preferences, it was supplying over one-third of that nation's market. During this period, moreover, India decided to reject an offer by the Tenneco Company to explore for petroleum offshore in the Bombay High region, although only U.S. firms at that time had the necessary deep-sea drilling technology. In 1970 the foreign petroleum companies operating in India agreed to submit to a demand for reduced imported crude oil prices.

By this time the Esso Standard Refining Company of India Ltd. had adopted a definite bargaining stance toward the Indian government. Esso's maneuverings were directed by Deputy General Manager T. E. Wallace, who previously had worked for Esso in Indonesia. The American firm, which wished to merge its refining with its marketing facilities, offered to set up a new concern in which the Indian government would hold 51 percent of the stock, or share majority control with Indian private capital. Esso never before had agreed to accept a minority position in a firm operating in a Third World country.

Rather than consent to this unique proposal, the Indian government ignored it. Two-and-a-half years later, toward the end of 1972, Esso further jolted Indian political and industrial leaders by offering either to sell India a 74 percent interest in its holdings there, or to transfer its entire operation to the Indian government in return for adequate compensation. The value of Esso's Indian properties at this time was approximately $55 million, it was faced with vanishing profits from its operations there, and was no longer able to market automobile accessories that had been manufactured by local small businessmen under the Esso brand name. Accordingly, it was willing to sell most, or all, of its holdings in India. The Indian government, however, felt the offer constituted a bid to establish an "adverse climate for the investment of private Western capital, particularly American, in India."[50] Furthermore, since its operating agreements with other foreign oil firms were to expire in only seven years, the government was hesitant to give its assent to the plan.

In 1973, the year of the Arab oil embargo, India was producing 7 million tons of crude petroleum. Unfortunately, this only furnished about one-third of its total need. It is, therefore, not surprising that India was one of the countries outside the United States and Western Europe that suffered the most when oil prices rose sharply. Despite the continued talk of Third World solidarity, it is apparent that the petroleum rich countries are actively engaged in the quest for profits, if only because they need these funds to finance expanding governmental operations. As a result, countries like India, which for decades were so concerned about the activities of American and other multinational oil firms, now find themselves at the mercy of OPEC.

# 7.

# Observations and Conclusions

Living in an age when the "Seven Sisters" have been cast in the same scapegoat role with respect to the world economy as the munition makers were in World War I by the Nye Committee of the mid-1930s, it is not easy to structure and analyze objectively the large quantity of information presented here on U.S. oil policy and diplomacy during the twentieth century. As the British economist Alfred Marshall warned many years ago, "general propositions in regard to either competition or monopoly are full of snares."[1] Since, as Allan Nevins pointed out in his biography of John D. Rockefeller, "some journalists and some politicians will utter sweeping and dogmatic statements upon an industrial aggregation like the Standard Oil. . . ,"[2] the author offers the following observations and conclusions in the same cautious spirit one uses in crossing an uncharted mine field.

According to one thesis, the multinational oil companies basically have played an exploitative role. Terence McCarthy has complained that:

> capitalist cartels and trusts shift gallon by gallon or ton by ton, bits of one country to other countries. And as they sell off one country bit by bit they use the proceeds to buy up leases on other countries mile by mile which they then sell off bit by bit until nothing is left behind but holes in the ground, some narrow and deep, some broad and shallow—until all the Third World becomes West Virginia.[3]

Critics in Third World nations frequently have drawn highly colorful parallels in attempting to describe what they regard as the negative impact of Standard Oil and other petroleum giants on their countries. In the narrative we have seen the Rockefeller firm compared to both an octopus and a hempen rope. During the World War I era one member of the Ecuadorian House of Representatives pictured Jersey Standard as a boa, "the hissing of which is heard on the shores of the Zambesi and the Lualaba, which swallows little birds like Nicaragua and chokes powerful antelopes like Mexico."[4] In 1918 the Venezuelan author José Pocaterra even drew a parallel between the U.S. oil men and the Spanish *conquistadores*. Terms such as "profiteers," "exploiters," and "plunderers" also have appeared in the press of non-Third World European nations such as Germany at the time that the Truman administration was considering criminal prosecution of the oil giants. The vocabulary usually associated with the "robber baron" interpretation of the late nineteenth-century's Gilded Age in the United States appears to have been used internationally for many years in critical assessments of the petroleum multinationals.

There are many who believe that the oil giants have exercised, and continue to exercise, considerable power over foreign nations, as well as American oil policy and diplomacy. Writing in 1935, Samuel Guy Inman claimed that between 1910 and 1927 the American oil companies played the leading role in shaping U.S. foreign policy toward Mexico. Halfway around the world, during the late 1960s, the Soviet Union charged that oil discoveries in southeastern Asia had influenced American policy in Vietnam and had encouraged a continuing U.S. presence there. But despite the hearings and reports during the 1970s of Senator Frank Church's Subcommittee on Multinational Corporations, which publicized widely the power and influence the "Seven Sisters" enjoy today, many scholars, probing the more distant past, have reached the conclusion that businessmen have had little real influence over U.S. government policies. The Chairman of the Board of Anglo-Persian noted in the 1920s that American firms found themselves involved in difficulties so frequently, because the American government did not provide them with steady diplomatic support.

Democratic Senator Estes Kefauver of Tennessee thus may have exaggerated the power of the oil giants when he observed during the 1950s that the State Department sneezed when they took snuff. Yet it is unarguable that the U.S. government has attempted to safeguard the position of American petroleum firms operating abroad on occasions too numerous to mention, not only by making diplomatic representations, but also by employing various diplomatic weapons and even displaying military shows of force. This has been true of not only Republican presidential administrations but also of Democratic ones.

As the *Foreign Relations* volumes published to date have only reached the Korean War, it is difficult to cxamine comprehensively the case studies focusing on petroleum between that event and the Arab oil embargo of 1973. There is, however, the necessary data to analyze episodes involving petroleum from the latter part of the nineteenth century through the middle of the twentieth. Perhaps the earliest of these episodes were the 1882 U.S. complaint when Turkey proposed raising its 8 percent duty on imported American oil, and the 1887 American protest when the government of Canton levied a new tax on U.S. kerosene entering China. These diplomatic representations were to become more and more common after World War I and eventually involved every major geographical region of the globe: Europe, Latin America, the Middle East, and the Orient.

Over the years the granting of petroleum concessions to American investors has been a major concern of foreign governments. There have been occasional attempts by the host nations to invalidate these at a later date, either because the original contract was defective or because the concession holder had failed to carry out his contractual obligations. President Arturo Illia of Argentina thus annulled various foreign oil contracts in 1963. An equally serious challenge to U.S. oil entrepreneurs came from the establishment of state petroleum monopolies in other countries, a phenomenon that has become more and more common in recent years. These can be found in such Latin American nations as Argentina (1922), Bolivia (1936), Mexico (1938), Colombia (1948), Brazil (1953), and Venezuela (1960), as well as in such Middle Eastern nations as Iran (1951), Kuwait (1960), Saudi Arabia (1962), Algeria (1963), Iraq (1964), and Libya (1970). State oil monopolies also have been set up elsewhere in such places as Spain (1927), Manchukuo (1934), and Indonesia (1968, a combination of three national companies). A number of these foreign governments, moreover, established state petroleum monopolies prior to expropriating the oil properties, not afterwards.

Even if they refrained from invalidating concessions, establishing state petroleum monopolies, and expropriating holdings, foreign governments could still harass American oil firms by imposing new taxes or customs duties, or by raising those already in existence, on U.S.-owned petroleum concessions or on crude oil and petroleum products imported from America. Back taxes, too, could be claimed by the host government, even to such a point that they became confiscatory. Furthermore, foreign governments could reduce the maximum price the U.S. firms were allowed to charge for their crude oil or petroleum products, or they could reduce the size of the quotas the American firms were allowed to import, sell, extract, or refine.

There were various other actions that the U.S. petroleum firms found equally disturbing. These included favoring of foreign (that is, nonnative) oil companies over American ones; stipulating that the proceeds from petroleum sales remain in the host country for a certain period of time; forcing the storage of large quantities of oil; refusing permission for geologists to investigate claims; mandating the employment of nationals in the petroleum industry; and demanding oil payments in certain currencies.

Unfortunately, there is no simple answer to the question of the effectiveness of U.S. diplomatic representations made over the years in an attempt to minimize the impact of such negative actions on the American firms. The results have been mixed indeed. A careful investigation of the case studies in this volume reveals that official U.S. protests were sometimes successful and sometimes unsuccessful; in some cases they led to a compromise settlement, and in some cases the outcome is not clear from the available information. At various times the American government has refrained from making diplomatic representations because it felt that either the U.S. firms had a weak case, or the issue at stake was none of its business. A classic case of diplomatic nonsupport by the American government was Sinclair's abortive petroleum concession in Sakhalin in the early 1920s.

That the American government has not been more effective in its diplomatic representations on behalf of U.S. oil companies is perhaps surprising in view of the wide variety of diplomatic weapons available to it. These include refraining from extending official recognition to a regime; holding up senatorial approval of a treaty with the nation involved; cutting off or reducing foreign aid; demanding the payment of an existing loan or discouraging the granting of new loans; blocking the purchase or sale of other products; and warning U.S. investors to avoid that particular country. At various times and in various places the mere threat to follow one of these courses has been enough to discourage a foreign government from discriminating against U.S. oil interests.

Still another weapon is the petroleum embargo, a device the United States employed long before the Arab world applied it. In July 1940, for example, the United States halted the exporting of aviation gasoline and scrap iron to Japan; in August of 1941 it placed a virtual embargo on the shipment of motor fuel and lubricating oil to that nation. During World War II, moreover, the United States suspended petroleum exports to the neutral Franco regime in Spain when that country did not embargo wolfram shipments to Germany. More recently, in 1965 the American government imposed an embargo on oil shipments to Rhodesia. One also could cite the Roosevelt administration's moral embargo on petroleum exports to Italy at the time of the Italo-Ethiopian war, even though this no more stopped Mussolini than the embargo of oil to Rhodesia three decades later frustrated Rhodesian independence.

Since the 1973 Arab oil embargo intermittent talk about possible American military intervention in the Middle East to safeguard U.S. access to the enormous petroleum reserves there has created a veritable furor. Yet U.S. military activity of a more limited sort on behalf of American oil interests occurred on occasion during the earlier decades of the twentieth century. In 1909, for example, the U.S. Navy stopped off at Smyrna on its way home from its world cruise, and took aboard a number of Turkish naval officers in an apparent attempt to generate some goodwill on behalf of Rear Admiral Chester. Similar episodes involving Mexico and the Netherlands East Indies occurred during the early 1920s.

The onset of World War II witnessed more extensive activity by the American military. In January 1942 the U.S. Air Force was assigned the task of helping to protect Curaçao and Aruba, the sites of important oil refineries; as the war progressed, the American military came to play a larger and larger role in safeguarding the production, refining, and transportation of petroleum elsewhere in the world. Although Josephus Daniels once remarked that the purchase of an oil well did not automatically grant an American warship the right to control the government of a host nation, there have been occasions when foreign governments actually have welcomed the U.S. military presence to help them safeguard petroleum properties.

Ironically, despite the large arsenal of diplomatic weapons which long has been available to the United States as a great power, including the imposition of embargoes and even military intervention, the governments of foreign nations have nevertheless continued to discriminate against American petroleum interests abroad, even resorting to such a drastic step as expropriation at times. Examples of the seizure of U.S. oil properties throughout the twentieth century might be cited from all over the world: Russia, Hungary, Rumania, Spain, Mexico, Bolivia, Peru, Algeria, and Libya. Although the Soviet Union and its Eastern European satellites are members of the Communist bloc, the other nations cited here fall somewhere on the economic spectrum between capitalism at one extreme and Marxism at the other.

Not only have a number of those nations bent on expropriation not been deterred by the fact that they were risking a confrontation with a great power, they have made the decision to seize American petroleum holdings even though in the short run this step might prove economically disadvantageous. Because the host country frequently is not prepared to find, extract, refine, or distribute oil, state petroleum monopolies set up in the aftermath of expropriation sometimes have experienced severe difficulties in their early years of operation, for example Pemex. Quite obviously many foreign governments have allowed political, social, and cultural factors to take precedence over economic ones when they made the decision to expropriate holdings.

Nevertheless, the American government has not always been consistent in its response to the seizure of U.S. oil properties abroad. After Fidel Castro's 1960 expropriation of several American petroleum firms, the Eisenhower administration cancelled the $150 million bonus the United States had been paying to Cuba for sugar imports. In contrast, during the aftermath of Peru's 1968 seizure of the International Petroleum Company, President Lyndon Johnson's Secretary of Agriculture granted that nation a 5 percent increase in its sugar exports to America.

Of course, the United States has not been the only country whose nationals have played the role of oil entrepreneurs abroad. There have been numerous occasions when the American government has accused foreign governments of favoring non-American petroleum interests. Since Royal Dutch-Shell and Anglo-Persian (later Anglo-Iranian and finally British Petroleum) are two of the "Seven Sisters" that are either wholly or partly British owned, it is not surprising that the greatest challenge to American petroleum companies operating abroad has come from British interests. Even before the end of the nineteenth century U.S. oil entrepreneurs experienced discrimination at the hands of the British in both India and Burma. The Anglo-American petroleum rivalry was particularly bitter in Latin America just before, during, and after World War I in such nations as Colombia, Mexico, and Costa Rica. U.S. oil interests did not even gain entry into the Middle East until the time of the Red Line Agreement of 1928, and despite that understanding and the Achnacarry Agreement of the same year, an undercurrent of Anglo-American hostility persisted into World War II.

Considering the rivalry that existed between Great Britain and the United States, both capitalistic nations, it is hardly surprising that American oil relations with the Soviet Union at times have been characterized by an even greater degree of friction. Even before the Bolshevik takeover late in 1917, Russian oil was competing with American oil in places like the Ottoman Empire. While U.S.-Soviet petroleum relations did become more harmonious during the 1930s than they had been during the 1920s, following World War II the Soviet establishment of a sphere of interest over Eastern Europe, which featured among other things the expropriation of foreign petroleum holdings, led to a renewal of friction. In addition, it should be remembered that Russia played a major role in the Cuban seizure of U.S. oil properties in 1960.

But while the American government frequently has made diplomatic representations concerning the treatment of U.S. petroleum interests abroad, and has even employed diplomatic weapons at times, it is not always accurate to speak in terms of a country's basic oil policy or diplomacy. Because of the kaleidoscopic changes that sometimes occur in the governments of many nations, one perhaps should cite instead a nation's oil policies and diplomacies. This is especially true in those Latin American countries torn by revolutions and coups every year or two, such as Bolivia. Even more stable nations such as Venezuela and Colombia have been known to enact a series of petroleum laws over a short period of years that differ significantly from each other. In contrast, until the establishment of such Arab Socialist republics as Libya and Iraq, the oil policy and diplomacy of such Middle Eastern countries as Iran and Saudi Arabia evolved in a somewhat less capricious manner. This generalization holds true even though U.S. support for the newly created state of Israel seriously disturbed the Arab nations of that region. Certainly King Ibn Saud and Shah Mohammed Reza Pahlavi did not change their attitudes toward the American and other foreign petroleum interests as drastically as some Latin American political leaders, such as Rómulo Betancourt of Venezuela, Fernando Belaúnde Terry of Peru, Arturo Frondizi of Argentina, and Getúlio Vargas of Brazil.

It is unfair, though, to speak in terms of a foreign nation having oil policies or diplomacies, without recognizing that the United States itself has displayed various policies and diplomacies in the area of petroleum. These become especially evident in the historical rivalries among various branches of the executive branch of the federal government, each with its own special set of attitudes and objectives toward oil matters. As early as World War I the Navy Department favored the withdrawal of federal petroleum lands from public sale, while the Interior Department did not. This was only the first of many conflicts between or among executive branch departments over oil policy and diplomacy which one might cite.

Throughout the twentieth century, the State Department usually has adhered to the theoretical posture that American business interests operating abroad, including the petroleum industry, should be accorded economic opportunities equal to those enjoyed by the nationals of other foreign nations. Since 1947 the Defense Department basically has been concerned with the availability of an oil supply adequate to meet America's needs, especially in time of war. More negative in its approach has been the Justice Department, which has intermittently attempted to prosecute, frequently with little if any success, the multinational oil corporations for violating U.S. antitrust laws through their activities overseas. During the Nixon administration the Justice Department, which wished to abolish quotas on petroleum imports, was at odds with the Interior Department, which wanted to keep them.

The Interior Department has played a key role in shaping petroleum diplomacy, especially during World War II and the Johnson administration, even though it generally is identified with domestic policy. On both of these occasions the president then in office enlarged the authority of the Secretary of the Interior over oil. Thus in 1941 Franklin Roosevelt chose Harold Ickes to serve as Petroleum Coordinator, and in 1942 made him Petroleum Administrator for War, while two decades later Lyndon Johnson deferred many oil decisions to Stewart Udall. Although the Commerce Department also is concerned with petroleum matters, it only played a dominant role during the 1920s, when Herbert Hoover as Secretary of Commerce helped to lay the foundations of American oil policy and diplomacy.

Unfortunately, overlapping administrative jurisdictions have undermined the implementation of U.S. oil policy and diplomacy at times, especially during World War II. Harold Ickes as Petroleum Coordinator and as Petroleum Administrator for War, for example, could only forbid petroleum exports or imports with the approval of the President or Secretary of State Cordell Hull. In 1941, moreover, FDR instructed Vice-president Henry Wallace to set up a policy subcommittee under the Board of Economic Warfare, while in 1943 Hull established an informal interdepartmental committee, composed of members of the State, War, and Navy departments, in addition to Ickes' office. Ickes, however, scored a triumph in the same year when Franklin Roosevelt made him the head of the Petroleum Reserves Corporation. Despite this game of administrative musical chairs, there was enough oil for the United States to prosecute the war successfully.

Administrative rivalry in itself by no means exhausts the list of conflicting forces at work when the American government has attempted to formulate and execute petroleum policy and diplomacy. The division of opinion has not always been along political lines, but sometimes along sectional ones. For years the oil-hungry New England states have sought liberal policies for petroleum quotas, while such oil-producing states as Texas have protested continually what they have regarded as a glut of foreign petroleum inundating America. Independent oil producers in particular have demanded ceilings on oil imports, as have the coal industry, the railroads, the United Mine Workers, the Railroad Brotherhoods, and the CIO Oil Workers Union. Because in part of pressure from these groups, the Eisenhower administration imposed mandatory restrictions on petroleum from abroad in 1959. The controls were to remain in effect for fourteen years, until 1973: the year of the great Arab oil embargo.

Governmental controls over foreign petroleum imports, though, have not been a recent phenomenon. As early as 1932 the U.S. Congress specifically imposed a tariff on imported oil, and between 1933 and 1935 the National Recovery Administration supervised a code that also restricted petroleum imports. Furthermore, in 1939 and 1943 respectively the United States signed agreements with Venezuela and Mexico that gave these nations a preferential advantage in exporting oil to America. The Mexican agreement remained in effect until the closing years of the Truman administration, in 1952, when the United States also signed a new trade agreement with Venezuela.

Over the years, moreover, American governmental officials have been concerned with various matters pertaining to petroleum other than the regulation of imports. This is demonstrated by an examination of the congressional investigations centering on petroleum. In 1919 Republican Senator Henry Cabot Lodge of Massachusetts obtained senatorial approval for his Foreign Relations Committee to investigate whether or not the Mexican government had violated the rights of the United States or its citizens; the final report of New Mexico Republican Senator Albert Fall's subcommittee concluded that it might be necessary for the American government to send a police force to Mexico. During World War II the Truman Committee of the U.S. Senate took a critical look at the Canol project in the Canadian Northwest, and it also criticized the 1926 agreement between Jersey Standard and the German firm I.G. Farben providing for an exchange of patents and information.

Following World War II, a committee headed by Republican Senator Owen Brewster of Maine examined charges that the U.S. Navy had paid too much for the oil it had obtained from the joint California Standard-Texas Company firm operating in Saudi Arabia, Aramco. Its investigation, however, proved to be a "tempest in a teapot." Thus when Democratic Senator Frank Church's Subcommittee on Multinational Corporations of the Foreign Relations Committee began to investigate the petroleum firms a quarter century later, it was by no means the first Senate or House committee or subcommittee to examine their operations.

The Justice Department, too, has taken a great interest in the activities of the oil companies at times, including those with international operations. This has been especially true of the period since World War II. In 1942, for example, the federal government pressured Jersey Standard into signing a consent decree, under which Jersey Standard pledged not to violate U.S. antitrust laws abroad. A decade later during the Truman administration there was widespread talk of bringing a criminal suit against several of the petroleum giants; by the following year, though, the more pro-business Eisenhower administration decided to settle for a civil suit. After years of litigation, in 1960 both Jersey Standard and Gulf Oil agreed to consent decrees that enjoined them from entering into any sort of price fixing and marketing arrangements that might hinder competition. It was during the same year that a consent decree struck down Standard-Vacuum.

At the time of the European petroleum lift in connection with the second Arab-Israeli war in 1956, however, the Justice Department approved a plan that allowed fifteen American oil firms to join in a single marketing combine, which might supply Europe with petroleum without facing possible U.S. antitrust prosecution. But this schizophrenic attitude of tolerating, or even encouraging, various business activities outside the United States that might lead to indictments inside the country is hardly something new in American history. It dates back at least as far as the Progressive Era, when many domestic reformers were imperialists overseas. Philander Chase Knox, who as Theodore Roosevelt's Attorney General had prosecuted the Northern Securities case, as Secretary of State helped to implement President William Howard Taft's program of dollar diplomacy in the Caribbean.

While cases in the U.S. courts frequently have centered on this schizophrenic distinction between operations in America and operations abroad, cases in foreign courts, especially those of Latin America, often have focused on the so-called Calvo Clause. According to the underlying principle of the Clause, a business firm active outside its home country should not call upon its own government to intervene in any way, should the business become involved in a dispute that will go before a foreign court. Laboring under this disadvantage, it would seem that American oil firms involved in litigation abroad invariably would lose, but this has not always been the case. In Mexico, for example, the Supreme Court decreed in the *amparo* case of the Texas Company in 1921 that the fourth paragraph of Article 27 in the 1917 Constitution was not intended to be retroactive. Then in 1924 the Supreme Court of Rumania ruled that embatic (perpetual lease) holders also possessed the subsoil of embatic lands, a ruling that was a major setback for the state. Nevertheless, in the majority of the cases decided by foreign courts, American as well as European oil firms have emerged as losers. In 1925, for example, a Russian court upheld the cancellation of the Sinclair concession, and there have been a number of hostile court rulings from such Latin American countries as Mexico, Venezuela, Bolivia, and Colombia.

Yet many foreign nations are hesitant to permit cases to go before a higher tribunal over which they may have little, if any, control. Thus in 1951 Iran rejected Truman's suggestion that it allow the International Court of Justice to find a *modus vivendi* in the Anglo-Iranian dispute. Nor have these countries always been eager to submit a petroleum dispute to arbitration, the examples of Peru and Costa Rica during the early 1920s notwithstanding. Thus in 1938 when Under Secretary of State Sumner Welles recommended arbitrating the Bolivian expropriation of Standard Oil properties, he received a negative response. In 1940, when Secretary of State Cordell Hull suggested that Mexico arbitrate the controversy surrounding the expropriation of U.S. oil properties, the Cárdenas regime also declined on the grounds that this was purely a domestic matter.

Although various legal questions relating to petroleum matters are often in dispute during times of peace, the outbreak of war invariably adds to their number. There is, for example, the matter of neutral rights. During World War I the British seized the *John D. Rockefeller*, releasing it when they discovered that the oil it was carrying was bound for Denmark. A quarter of a century later, with the world at war again, the United States protested the shipment of American oil to Mussolini's Italy and Hitler's Germany via officially neutral Spain, Portugal, and Vichy France.

Another question of major consequence during World War II was the destruction—either with or without U.S. consent—of American oil properties in foreign countries as a war measure. In 1914 Belgium destroyed the Jersey Standard facilities there, rather than allow them to fall into the hands of the Germans. Two years later in oil rich Rumania the British persuaded the Americans to plug their wells, dismantle their tank "farms," and destroy their refining machinery for similar reasons. Then after the outbreak of World War II American planners considered the possible demolition of the Saudi Arabian fields to keep them out of the hands of the Axis powers, but this step proved unnecessary. Military authorities, though, did destroy the Standard-Vacuum refinery in Sumatra during 1942 to render it useless to the Japanese.

These observations and conclusions thus far have concentrated on the more negative or hostile responses by the American government to other nations. But there is another, more positive side. On occasion the American government not only has extended financial assistance to other nations for oil projects, but also has helped to make petroleum itself available to foreign countries during times of crisis and periods of shortage. After a period of delay the U.S. government made a $10 million loan available to Mexico in 1943 for the construction of a new, high octane refinery. Although Pemex chief Antonio Bermúdez failed to obtain a petroleum development loan in 1949 when Mexico was planning a $470 million program, by 1972 the Export-Import Bank had become willing to extend a loan that would cover most of the cost of exporting $46 million in equipment and services to Pemex.

On the other hand, the American government has played a more obstructionist role on those occasions when U.S. oil firms have endeavored to make processes, technical assistance, and materials available to other countries. In 1935, when Standard-Vacuum and Royal Dutch-Shell offered data on hydrogenation to the Japanese, the State Department assumed a hands-off position, but by 1938 the department was advising Stanvac to stall. A year later, at the urging of the State Department, the Universal Oil Products Company announced that it would not honor a contract under which the Japanese would receive data on its polymerization process for the production of iso-octanes. In 1941, after World War II had begun, the American government used its influence to encourage the Phillips Petroleum Company not to deliver information and technical assistance for the manufacture of gasoline to the Argentine state oil firm, the YPF, as there was considerable pro-Nazi sentiment in Argentina at that time. In addition, highly placed individuals in the State Department during World War II also obstructed efforts to supply the Soviet Union with U.S. oil processes and patents, despite several presidential directives. Even before the onset of the cold war a number of governmental officials were suspicious of the Russians.

At the same time, during World War II, the U.S. government set up an American Tanker Committee to facilitate the distribution of petroleum products throughout Latin America. After the formulation of the Marshall Plan in 1948, America made oil available to Western Europe, along with various other items, by means of the Economic Cooperation Administration. The U.S. government also set up the Foreign Petroleum Supply Committee in 1951 after the nationalization of Anglo-Iranian; the FPSC attempted to increase crude oil production in as many as eleven nations, as well as to stimulate the manufacture of refined products in as many as twenty-seven countries. Five years later, when Egypt seized the Suez Canal, the American government saw to it that both Great Britain and France received adequate oil supplies.

Examining the activities of the representatives of the U.S. government and those of the American petroleum firms, it is not always easy to distinguish between the two. Such diplomats and governmental officials as Albert Fall, Charles Evans Hughes, J. Reuben Clark, Donald Richberg, and John Connally have sometimes acted as agents, or even employees, of the oil companies. Conversely, a number of petroleum firm executives and advisers have held important governmental jobs in such departments as State, Defense, Commerce, and Interior. These include Max Thornburg, Charles Rayner, John McCone, Robert B. Anderson, George McGhee, Ralph K. Davies, Edwin Pauley, and Herbert Hoover. Richard Barnet once observed in this connection that:

> The extraordinary place of the oil companies in the world economy and the controlling influence they have in many producing countries could not have been accomplished without a close and continuing partnership with the U.S. government. The "public" operations of the State Department and the "private" operations of the oil companies blend in what the petroleum analyst Michael Tanzer, himself a former employee of an oil company, calls a "symbiosis." It is a symbiosis based on mutual dependency.[5]

As a rule oil firm executives have enjoyed great freedom in conducting their operations in foreign countries, but in periods of crisis and in times of war the American government has attempted to restrict their maneuverings. Two examples of the latter phenomenon—both from the beginning of World War II—were the independent petroleum operator William Rhodes Davis, and Torkild Rieber of Texaco. In both cases their firms had been shipping oil to Germany, and in both cases the Germans requested that they personally approach President Franklin Roosevelt and present to him certain diplomatic proposals. Significantly, Davis as well as Rieber failed in their roles as would-be diplomats.

There remains for consideration the question of the positive or negative influence of American petroleum firms on those foreign countries where they have been active. Critics have charged that self-interest rather than altruism usually has motivated the oil firms' acts of generosity. They point out that sometimes the host country has even passed legislation requiring the petroleum firms to undertake certain programs. Thus in the case of Venezuela the Congress enacted a law in 1922 that held the employers accountable for the health and well-being of their employees, and required them to build hospitals, supply medicine, and provide for accident compensation. It then passed a similar law fourteen years later, in 1936.

According to one estimate, by the end of World War II Creole and Mene Grande had expended as much as $36 million on housing, hospital and medical care, schools, recreational facilities, lighting, and sanitation in Venezuela. The company then launched a program of community integration, under which religious orders and other organizations would take over their social and educational programs. By 1957, according to another estimate, Creole had spent a quarter of a billion dollars in Venezuela on physical facilities for employee housing, schools, and hospitals. In the following decade Creole inaugurated a new "Dividend for the Community" program under which the firm dedicated several percent of its profits for new hospitals and social projects. There are similar examples from Colombia, Peru, Argentina, and the Netherlands East Indies.

One, though, must ask how much of its profits Creole in Venezuela—or any other American firm in any foreign country—devoted to altruistic causes, and how much of its budget the government of Venezuela—or any other nation—spent on these social programs. Unfortunately, year by year figures in dollars and cents, and percentages, are not readily available. It must be remembered, moreover, that had more of the profits of any U.S. oil company operating in any foreign nation gone to the treasury of the host country in royalties or taxes, there is no guarantee that the government of the country would have spent all, or most, of this additional income for the benefit of its people. Prime examples, but by no means the only ones, of the ruling class lavishing these moneys on itself rather than its people are the ornate palaces the royal family has maintained in Saudi Arabia. This, however, is not to deny that the Saudi Arabian government has spent petroleum revenues on badly needed railroads, highways, ports, airfields, schools, hospitals, electric power plants, and irrigation systems. In addition, since coming to Saudi Arabia in the 1930s Aramco has constructed houses for its laborers, undertaken an irrigation project, and furnished to the Saudi government technical assistance, agricultural aid, medical services, educational programs, trachoma and malaria research, and telecommunications service.

Programs such as these may represent the U.S. petroleum companies operating abroad at their most admirable, but the political payoffs and bribes they have made to foreign governments over the years have significantly tarnished their image. Some of these only came to light as the result of investigations by Senator Frank Church's Subcommittee on Multinational Corporations. It should be realized, however, that other American firms operating abroad also have been guilty of political payoffs and bribes to foreign governments and so have the non-American enterprises that compete with them in other countries. Moreover, such practices appear to be a way of life in certain nations; those firms not willing to participate in them soon find themselves without contracts or concessions. Thus to single out U.S. oil firms for criticism in this connection is decidedly unfair.

It must be remembered, too, that political payoffs and bribes have occurred in Europe, as well as in the Third World. In Italy, the Italian subsidiary of Exxon made $20 million in political payments between 1963 and 1972, while Mobil Oil Italiano gave financial contributions to the major political parties between 1970 and 1973. But data on these types of practices unfortunately are difficult to substantiate, and therefore must be discussed with caution.

American oil money also has been used to finance revolutions and counterrevolutions abroad. Thus in the case of World War I Mexico, Standard Oil apparently backed Francisco Madero by extending a loan to him when Porfirio Díaz continued to favor the British Cowdray interests. Several years later Standard Oil allegedly financed the revolution that Venustiano Carranza launched against Victoriano Huerta. It has been said, too, that Edward L. Doheny gave Carranza and the Constitutionalists both cash and fuel oil. In any case, it is highly significant that Lázaro Cárdenas used the real or alleged support of the rebels by both the American and British oil firms during the post-Díaz decade as one of the reasons for expropriating their holdings in 1938.

Examples from other Latin American countries could also be cited, including the role of the International Petroleum Company in Peru, and that of Gulf Oil in Bolivia. At times the State Department has felt it necessary to deny that American petroleum companies have supported the overthrow of a regime; this happened in 1948, when an Army coup drove Rómulo Gallegos from power in Venezuela. There have been instances, too, where a government's favorable treatment of U.S. oil firms has led to its fall, for example the Perón regime in Argentina during 1955.

Elsewhere in the world, it was rumored that Standard Oil had supported Sun Yat-sen in China at the time he came to power in 1911, in return for a promise of oil concessions. After World War I Jersey Standard opposed the recognition of the counterrevolutionary regime in the Caucasus because of the alleged ties between it and Sir Henri Deterding. By 1927, however, Jersey Standard and Royal Dutch-Shell had agreed not to deal with the Communist government then entrenched in Moscow, until it formally recognized their property rights in Russia.

American oil firms also have been involved, or allegedly involved, in foreign wars, as when Bolivia and Paraguay fought over the Gran Chaco between 1932 and 1935. Here one finds conflicting reports. According to one account, Standard Oil supported Bolivia, and Royal Dutch-Shell, Paraguay; another version maintains that, to the contrary, Standard Oil refused to lend Bolivia several million dollars, and to refine aviation gasoline for it during its war with Paraguay. One rumor even charged that Standard Oil had purchased sixty bombers in Italy, which it then turned over to the Paraguayan army. In any event, Bolivia used the supposedly uncooperative attitude of Standard Oil as one of the reasons for its expropriation of the American firm in 1937.

Conversely, there is the widespread interpretation that American business in general opposed the Spanish-American War until the eve of its outbreak. Joseph Pulitzer in his New York *World* charged that Standard Oil was a "peacemonger." Stung by this attack, Standard Oil both sold and leased some of its small coastal ships to the American government. Two generations later, during the Vietnam War, there were reports about the activities of U.S. oil companies in Southeast Asia. So widespread were these rumors in 1971 that the State Department felt compelled not only to minimize the importance of Vietnamese petroleum, but also to play down the official American role in developing it.

Although much more could be said about the activities of U.S. oil firms in other countries, perhaps the basic standard by which to determine whether or not the U.S. petroleum companies have treated the host nations unfairly is the share of the oil profits they have demanded and obtained over the years. Since the 50-50 profit sharing formula did not go into effect anywhere in the world until the 1940s, the firms probably profited unfairly, at least for several decades. As if to compensate for this long-standing imbalance, in recent years many countries have demanded so large a percentage of the profits that American petroleum companies often have found it unprofitable to enter into contracts with them.

A truly definitive assessment of the role the U.S. oil firms have played in foreign countries would necessitate a thorough investigation of the business archives of the "Seven Sisters," as well as those of dozens of independent companies, in addition to the diplomatic and other governmental archives that presently are open to researchers. While Irvine H. Anderson did examine some of these business archives from several decades ago in writing his *The Standard-Vacuum Company and United States East Asian Policy 1933-1941*, the work only represents an eight-year segment of one petroleum company's activities. It seems quite possible that the more sensitive files (especially the more current ones) of many oil firms may be withheld or destroyed, rather than made available to historians or other scholars. In addition, it is quite probable that some of their more compromising actions have been based upon oral agreements or understandings.

Despite the many negative actions by U.S. petroleum companies abroad, there is no question but that many Third World countries have lacked the capital with which to find, extract, refine, and distribute the oil reserves they may possess. Libya, for example, was little more than a large pile of sand of interest only to frustrated Italian imperialists until American petroleum entrepreneurs uncovered its oil bonanza during the 1950s. And many of those countries that have expropriated foreign controlled petroleum companies—such as Mexico in 1938 or Iran in 1951—have quickly discovered that they have placed themselves at an extreme disadvantage when they attempt to operate the petroleum industry as a state enterprise. Thus, until recent years when some of the state oil companies have become successful operations, the foreign petroleum firms did play an essential role in developing the oil reserves of many nations. Had it not been for the foreign petroleum companies, much if not all of the oil might remain in the ground today, yielding no profit to anyone.

Because of these circumstances, it would seem that the governments of petroleum-producing countries would be at the mercy of foreign oil firms, but far from meekly capitulating to them, nations have often defied and harassed them in a number of ways. Expropriation has been the most extreme weapon which they have employed, but by no means the only one. Certainly, had the petroleum-producing countries capitulated to the wishes of foreign oil firms every time, it would not have been necessary for governments like that of the United States to make diplomatic representations, and even to employ diplomatic weapons, because of challenges to the operations of American petroleum companies in these nations.

Obviously, the decisions made by foreign governments on oil matters, and the activities of foreign oil entrepreneurs operating there, are not always based on petroleum considerations in the narrow sense, or economic ones in the broader sense. Political, social, and cultural factors sometimes play a more important role in the shaping of governmental policy than economic factors. Aside from the actual officials of foreign governments, groups such as labor, the military, and the press may exert significant pressure on petroleum questions.

The role of labor has been especially pronounced in Mexico. The reluctance of Jersey Standard and other foreign oil firms there to accept a wage increase for their petroleum workers was one of the reasons behind President Lázaro Cárdenas' decision in 1938 to expropriate their holdings. Certainly American labor on more than one occasion has taken a sympathetic attitude toward its Mexican counterpart. The Perón Administration in Argentina also irritated Jersey Standard nine years later by stating that it would have to extend wage increases to its petroleum workers before it and the other foreign oil companies received compensatory price increases. A year earlier, in 1946, President Rómulo Betancourt of Venezuela had pointed out to the foreign petroleum firms that improving the lot of the workers was a political rather than a military issue, which led Jersey Standard and Royal Dutch-Shell to sign quite generous contracts with the unions.

Yet foreign governments have not always been slaves to the labor movements in their countries. When 13,000 Aramco workers went out on strike in Saudi Arabia during 1953, Crown Prince Ibn Saud told them either to go back to work or else. At this time highly conservative Saudi Arabia had no labor law of any kind.

Traditionally the Latin American military has been associated with a more conservative ideological stance than has Latin American labor. Nevertheless, in the case of Brazil it was the Brazilian army that pointed out to the National Petroleum Council during World War II that Brazil should nationalize its oil industry. Twenty-five years later, in Libya, a group of army officers led by Colonel Muammar el-Qaddafi set up a new revolutionary government that began to take over the foreign petroleum holdings there. Thus it would be a mistake to consider the military as the inevitable ally of foreign capitalists.

The press of the world has manifested all shadings of opinion, and certain publications have been quite vocal on oil matters. In Rumania during 1924 the *Romana Petrolifera* led the attack against Jersey Standard by publishing a series of articles dealing with "The American Peril." A generation later, in Peru, *El Comercio* (the best-known daily of that nation) took part in the successful drive to replace Fernando Belaúnde Terry as President at the time of the furor over the new International Petroleum Company contract.

Thus, while the "Seven Sisters" and other independent oil firms may have enjoyed many advantages in their dealings with foreign nations, they have had little, if any, control over the roles of various interest and pressure groups within these countries. There have been months, or even years, of stalemate in their dealings with foreign nations over petroleum matters. Even if they enlist the support of the most powerful government in the world—that of the United States—that government is in no position to dictate its wishes to other countries on a take-it-or-leave-it basis.

Writing at the end of World War II, Democratic Senator Joseph O'Mahoney of Wyoming observed about oil: "There is nothing that men and nations will not do to gain control of it [petroleum]. They have been known to bribe kings and potentates, to foment revolutions, to overthrow governments. Purely individual rights and interests have frequently been of very little moment in the struggle for petroleum."[6] If anything, O'Mahoney's observations are more valid today than ever before, as the global quest for oil reaches a fever pitch of intensity. This search, moreover, continues in an era when it is no longer possible for American petroleum entrepreneurs to approach conservative regimes in oil-producing nations and obtain a guaranteed supply of "black gold." Today the old liberal/conservative dichotomy is a thing of the past; no longer is it Lázaro Cárdenas versus Porfirio Díaz in Mexico, or Colonel Muammar el-Qaddafi versus King Idris in Libya. Instead, we are now in an age in which every petroleum-producing country places its own national interest first, and Americans in search of oil must play the role of the client rather than that of the patron if they are to gain access to any of this petroleum.

# 8. Epilogue: Developments Since 1973

Having devoted hundreds of pages to an examination of U.S. oil policy and diplomacy through 1973, one could easily write several hundred additional pages surveying this vast topic from 1973 down to the present, but limitations of space necessarily restrict us to a brief summary. An examination of the New York *Times* index for the years since 1973 reveals a staggering number of developments relating to petroleum; in fact, there are so many trees, that one frequently loses sight of the forest. Unfortunately, however, much of the published information takes the form of ephemeral journalistic accounts. Thus a companion scholarly monograph covering the events of the last decade is out of the question at present, and may remain an impossibility for many years.

Perhaps the best approach to the events since 1973 is to summarize the developments centering on OPEC, which today as before continues to play a major role in the global arena. Although the Arab oil embargo lasted from late 1973 to early 1974, its termination did not lead to a return to cheaper petroleum. In fact, during late 1973 the price of oil *quadrupled* within the space of two months, even though Venezuela and Iran did not participate in the boycott and Saudi Arabia followed a course of moderation and restraint. While in the years that followed the boycott Saudi Arabia continued to advocate caution, especially in price hikes, other OPEC members did not hesitate to push the cost of petroleum as high as possible. In the postembargo years, however, even in "friendly" Saudi Arabia the long dominant U.S. oil company, Aramco, was forced to execute policies that more reflected the wishes of the Saudi government than those of its American counterpart.

In 1974 the Nixon administration attempted to counter the increasing OPEC threat posed to U.S. oil companies by escalating its military aid to both Saudi Arabia and Iran, two countries that had long been rivals, not allies. Nevertheless, during the summer of that year the OPEC nations raised the price of oil again and increased the tax and royalty rates, as well as their share of participation in the foreign petroleum firms. There even was talk of creating other OPEC-type cartels throughout the Third World to monopolize control over such raw materials as copper, iron ore, bauxite, rubber, coffee, peanuts, and cocoa. President Houari Boumedienne of Algeria set forth such a proposal before the United Nations. Fortunately, this potential nightmare to the Western world has yet to materialize in concrete form.

The OPEC nations were divided in the first six months of 1977 over how much they should raise oil prices. Saudi Arabia held to a 5 percent increase, while other OPEC countries agreed upon a 10 percent increase, but that summer, a single price pattern had been reestablished. Seven months later, in January 1979, an even more important development occurred: the Shah of Iran fell from power after occupying the throne of that country for more than a generation. The new revolutionary government headed by the Ayatollah Ruhollah Khomeini spent the remainder of the year consolidating its position, before seizing the American Embassy hostages on November 4 and triggering a year-long international *cause célèbre*.

The new Iranian government, moreover, closed down the country's oil fields, leading to a gasoline shortage in the United States during the summer of 1979. Saudi Arabia failed to step up its own production to meet the increased American demand, but it charged only $18 a barrel for oil in contrast to the ceiling price of $23.50 set by OPEC. The war between Iran and Iraq during the following year led to the widespread destruction of each other's oil properties.

Yet, not every important petroleum development since 1973 has taken place in an OPEC nation. Oil discoveries are a case in point. In 1974 alone three major petroleum finds were made outside the Middle East, in the Chapas-Tabasco region of Mexico, the offshore Campos field of Brazil, and the offshore Bombay High region of India. Mexico, however, has declined to offer a bargain price to the United States for this oil, and Canada has announced plans eventually to terminate petroleum exports to this country.

While the movement toward the nationalization of foreign petroleum companies has been most evident in such Middle Eastern countries as Saudi Arabia, Kuwait, Libya, and Iraq, it is also to be found in such non-Arab nations as India. Furthermore, both Canada and Great Britain set up state oil firms in 1975. It is interesting to note, though, that even revolutionary Libya refrained from fully nationalizing Occidental in 1975, which may be an indication that those countries with large petroleum reserves are beginning to realize that foreign oil firms provide them with vital marketing functions. In fact, a number of nations—both Middle Eastern and non-Middle Eastern—have adopted polices since 1973 that are designed to encourage foreign oil firms to explore for oil in their countries. Among these nations are Algeria, Peru, Brazil, and Spain. Consortiums, of course, continue to be active on the world petroleum scene, as are individual companies.

To the average American, petroleum means Middle Eastern oil, even though such non-Middle Eastern states as Nigeria have become significant producers of petroleum in recent years. Considering the developments in Iran since the fall of the Shah and other events in that part of the world that have also been unfavorable to the United States, one might expect that most Americans would place the major share of the blame for any oil shortage on the Middle Eastern members of OPEC. But this has not been the case; on the contrary, the giant petroleum firms have been a more frequent target. This attitude is by no means irrational, since the profits Exxon earned in 1973 set an all-time record for any corporation at any time in any nation. By 1975 Exxon had become the world's largest business firm in terms of sales, dethroning General Motors. These two companies together with Ford and four other American "sisters" were now the seven largest U.S. corporations.

Still another factor working against the big petroleum firms was the public disclosures of their various unsavory activities. North Carolina Democratic Senator Sam Ervin's Committee, which was investigating secret contributions to political campaigns, examined the charge that both Gulf and Ashland had committed legal violations in this area. Even more damaging were the questionable practices uncovered by the subcommittee of the Senate Foreign Relations Committee headed by Idaho Democrat Frank Church. Although its original focus had been on International Telephone and Telegraph's attempts to undermine the Marxist Allende regime in Chile, it turned its investigation to a careful scrutiny of the "Seven Sisters." Elsewhere in the world at this time, antitrust probes were being conducted in West Germany, Japan, and France.

Even in these circumstances it was not easy to pass remedial legislation through Congress pertaining to energy in general and petroleum in particular. The historical rivalry between the oil-rich and the oil-poor states persists today. During the postembargo era Democratic Senator Henry Jackson of Washington emerged as one of the leading critics of the oil companies, while Democratic Senator Russell Long of Louisiana stood forth as one of their main defenders. At the same time Congress also was pressured by ecologists, who were extremely displeased by the Nixon Administration's decisions to construct the Alaskan pipeline, permit more stripmining for coal, and allow the leasing of foreign lands for oil exploration.

During November 1973—with his administration increasingly caught up in the Watergate scandal that was to force his resignation in August 1974—President Nixon unveiled his "Project Independence." Its goal was energy self-sufficiency for the United States by 1980. Two months later, in February 1974, Secretary of State Henry Kissinger presided over a gathering of thirteen leading oil-consuming nations in Washington, who were to draw up an energy policy acceptable to all parties involved. Such an agreement, though, was not reached. By the time Nixon left office in August, Congress had passed only eight pieces of energy legislation; the most important of these authorized the construction of the Alaskan pipeline, mandated federal price controls and the allocation of crude oil and oil products, established a new Federal Energy Agency and an Energy Research and Development Administration, and suspended various measures designed to control pollution.

Nixon's successor as chief executive, Gerald Ford, recommended a comprehensive energy program in January 1975 that among other things featured higher taxes on imported oil, domestic petroleum, and natural gas. A month later, in February, he issued a proclamation under existing laws that imposed a $1 tariff on oil imports, which would rise to $2 within a short period. The proclamation encountered opposition in both Congress and a federal court, but Ford kept it in effect until the end of the year, when he lifted it in response to the congressional enactment of the Energy Policy and Conservation Act. Among other things, this measure authorized the establishment of a strategic petroleum reserve, and called for a gradual phasing out over forty months of those price controls the government had placed on domestic oil. The cost of domestic crude oil was to be reduced by 12 cents a barrel and then allowed to rise slowly. Ford also came out in favor of a windfall profits tax on domestic petroleum, the decentralization of oil prices, the stockpiling of 1 billion barrels of petroleum, and an increased reliance on coal, electricity, and nuclear power.

By the spring of 1976 the average daily output of domestic crude oil had fallen to 8.1 million barrels per day, the lowest level in more than ten years. An indication of the mounting hostility against the oil companies was a bill, approved by the Senate Judiciary Committee, that would have dismantled the country's eighteen largest petroleum firms. While Congress did pass measures permitting oil production on three Naval petroleum reserves, and setting aside $1.2 billion for the coastal states to cushion the impact of oil and gas development offshore, it also rejected a high priority Ford administration bill providing for the deregulation of the price of natural gas.

Even more complex was the energy program of Jimmy Carter, who called for the conservation of petroleum leading to a reduction in demand, and thus invoked the spirit of self-sacrifice the American people had last willingly embraced during World War II. In attempting to encourage domestic production Carter endorsed both higher prices and higher taxes; Congress, however, felt that higher prices were unfair to the consumers, while higher taxes incurred the displeasure of the oil firms. Congress debated, but failed to approve, bills introduced by various members that called for both horizontal divestiture, which would have forced the oil firms to sell their nonpetroleum investments, and vertical divestiture, which would have fragmented the petroleum giants into smaller units, each performing a key function: refining, exploration, marketing, etc.

Carter, moreover, recommended the creation of a Department of Energy. This was initially presided over by James Schlesinger, whom Carter later fired in a midterm purge of Cabinet members. There was to be a gradual switch from petroleum to other types of energy; a high priority item was the insulation of inadequately protected houses and buildings. The Carter administration also advocated taxes on crude oil, the industrial use of petroleum, and gas-guzzling automobiles, as well as the creation of a six months' supply of oil reserves.

Although his energy proposals cleared the House of Representatives on August 5, 1977, with only minor changes, no legislation emerged from the Senate in that year. By this time the Alaskan pipeline was in operation, providing America with much needed oil. When Congress finally did pass an energy bill on

October 15, 1978—the last day of the 95th Congress—the finished product was very different from the original Carter proposals. Gone were the taxes on crude oil, gasoline, and the industrial use of petroleum and energy insufficient automobiles. Since the preceding eighteen months had witnessed one of the most extensive lobbying campaigns in the history of Congress, it is not surprising that such changes were made. What did survive was an emphasis on the conservation of oil and gas, a growing reliance on coal and other fuels, and an attempt to increase the domestic production of gas and oil. The final package had as its centerpiece the Natural Gas Policy Act, which provided for the gradual decontrol of natural gas.

In the spring of 1979 President Carter again directed his attention to the energy question and after the Three Mile Island nuclear power plant incident reaffirmed his support of nuclear power. The Nuclear Regulatory Commission, though, shut down six operating reactors in the East in March of 1979 because it feared potential earthquakes.

The world price of oil nearly doubled in 1979 from $16 a barrel to $30. In April Carter ordered the gradual lifting of price controls on crude oil, which was to take place between June 1, 1979, and September 30, 1981. Such a step was permitted under the Energy Policy and Conservation Act, which had become law in December 1975. Economist Thomas F. Hogarty has suggested that the decontrol of crude oil boosted domestic production, thereby lowering oil imports and causing OPEC members to battle over prices. Still, one must remember that the quadrupling of oil prices in 1973-74 set in motion a worldwide search for oil that continues today.

The President supplemented his order for decontrol by proposing the imposition of a two-level windfall profits tax on the petroleum firms and requesting stand-by authority to ration gasoline. Carter, however, encountered strong opposition from Congress in implementing both deregulation and rationing, but this did not stop him from offering still more energy proposals in August. This time he recommended a limit on oil imports, the creation of a synthetic fuels corporation, and the establishment of an energy mobilization unit to circumvent bureaucratic red tape. He also replaced Schlesinger as Secretary of Energy with Charles P. Duncan, after Schlesinger had exhibited certain deficiencies as a manager, and had offended Congress by defending the President's energy policy.

By the end of 1979, Congress had passed a bill designed to restrict the heating in winter and cooling in summer of nonresidential buildings. Despite widespread opposition to it in the House, the stand-by gasoline rationing scheme eventually won congressional approval. Only the Senate considered the synthetic fuels corporation bill; it agreed to an initial authorization of a mere $20 billion, rather than the $88 billion the Carter administration originally had requested.

Petroleum imports fell in 1980, but still constituted 40 percent of all oil used. Petroleum company profits were also up, despite the implementation of the so-called windfall profits tax, which Congress finally had agreed to levy in March; this act placed a 30 percent excise tax on crude oil from newly discovered wells. That body also set up a U.S. Synthetic Fuels Corporation in 1980, while rejecting the proposal to establish an Energy Mobilization Board. By this time, though, presidential election politics had begun to intrude into congressional debates and votes. Not only did the Republican members of Congress demonstrate their support for Ronald Reagan by opposing the EMB, they also blocked the Carter administration's appointments to the new Synthetic Fuels Corporation.

Ronald Reagan won the presidency that November, easily defeating Jimmy Carter. The Republican platform for 1980 had emphasized the need for additional energy production, the deregulation of oil and gas prices at the well head, a reduced governmental authority to allocate petroleum supplies, and an increased use of coal and nuclear power. It also called for the restricted application of the windfall profits tax and an accelerated phase-out of the tax on "old oil."

The new chief executive accelerated the Carter decontrol program, ending controls on January 28 instead of September 30, while Congress (with the Republicans now in control of the Senate) authorized the gradual reduction in the excise tax on crude oil from 30 percent to progressively lower levels. In striving for an economy of plenty, as contrasted to the economy of scarcity implicit in Jimmy Carter's stress on conservation, Ronald Reagan has ushered in a bold new era in government. Only time will tell how successful Reagan's new departure will prove in making America self-sufficient from the standpoint of energy.

# Chronology of Key Events Pertaining to Petroleum

1857 Rumania—beginnings of its oil industry
1859 United States—oil discovered at Titusville, Pennsylvania
1870 United States—establishment of Standard Oil Company of Ohio
1882 United States—Standard Oil devises the trust form of business organization
1888 Great Britain—Standard Oil founds the Anglo-American Oil Company
1890 United States—passage of the Sherman Antitrust Act
1890 Netherlands—setting up of the Royal Dutch Company
1896 Great Britain—creation of the Shell Transport and Trading Company
1900 United States—John Hay pronounces the "Open Door" policy
1901 Iran—William D'Arcy obtains an oil concession in the South
1902 Turkey—Germans negotiate petroleum contract with the Civil List
1902 Burma—Indian government rejects the bids by the Colonial Oil Company and by the Anglo-American Oil Company for oil prospecting licenses
1904 Rumania—establishment of Romano-Americana
1905 Colombia—granting of Barco and de Mares concessions
1905 Burma—Standard Oil fails in its attempt to obtain permission to operate a refinery at Rangoon
1906 Europe—Royal Dutch merges with Shell as Royal Dutch-Shell
1907 Iran—Anglo-Russian settlement establishes a British sphere of interest in the South, and a Russian one in the North
1908 Turkey—Colby Chester obtains an oil concession from the Sultan
1909 Great Britain—creation of Anglo-Persian as a result of an agreement between William D'Arcy and the Burmah Oil Company
1911 United States—U.S. Supreme Court breaks up Standard Oil
1911-14 United States—creation of Elk Hills, Buena Vista, and Teapot Dome petroleum reserves
1911 China—Standard and Royal Dutch-Shell agree to divide up the Chinese market between themselves
1913 Kuwait—Sheikh gives the British veto power over future oil deals
1913 Netherlands East Indies—Dutch government suspends the issuance of new concessions there

1913 Turkey—formation of the Turkish Petroleum Company
1913 Peru—Jersey Standard leases the La Brea y Pariñas holdings
1914 Great Britain—British government purchases a controlling interest in Anglo-Persian
1914 Turkey—Anglo-Persian obtains a 50 percent interest in the Turkish Petroleum Company, to which Turkey grants a concession
1914 Bahrain—Sheikh gives the British veto power over future oil deals
1914 China—Standard obtains a petroleum monopoly in the North
1914-18 World War I
1915 Arabian Peninsula—Ibn Saud agrees not to grant foreigners any economic concessions without British approval
1916 Rumania—destruction of Jersey Standard properties there
1916 Iran—Russian entrepreneur Khostaria obtains a concession in the North
1917 Netherlands—passage of mining law providing for at least a 50 percent government ownership in a future oil concession
1917 Mexico—Article 27 of the new constitution causes furor because of its potential effect on the petroleum industry
1917 China—Standard withdraws from its development company deal
1918 Netherlands—passage of another mining law giving the government the right to exploit oil concessions itself, or to sublet contracts
1918 Costa Rica—Tinoco regime grants Amory concession
1919 United States—founding of the American Petroleum Institute
1919 Colombia—Henry L. Doherty obtains the de Mares concession
1920 United States—Minerals Leasing Act attacks foreign discrimination
1920 Soviet Union—Communists take over the Baku oil fields, and seize foreign holdings there
1920 Europe—Great Britain and France draw up San Remo Agreement vis-à-vis Middle Eastern petroleum
1920 Bolivia—signing of Richmond Levering contract
1920 Colombia—Jersey Standard buys into the de Mares concession
1920 Iran—Anglo-Persian obtains the Khostaria concession
1920 Netherlands East Indies—Dutch Parliament grants a concession to a Royal Dutch-Shell subsidiary
1921 United States—secret oil conference held in Washington
1921 Rumania—agrarian law gives embatic owners full ownership of the land covered by their leases
1921 Mexico—Supreme Court ruling in the *amparo* case of the Texas Co.
1921 Costa Rica—Congress cancels the Amory concession
1921 Iran—signs a treaty of friendship with Russia, cancelling Iranian debts to Russia, and Russian petroleum concessions in Iran
1922 Europe—Genoa Conference, London Memo, and Paris Conference on Russian expropriation of foreign oil holdings
1922 Argentina—creation of the YPF
1922 Peru—international arbitration award vis-à-vis La Brea y Pariñas
1922 Bolivia—Jersey Standard obtains the Richmond Levering concession
1922 Sakhalin—government signs contract with Sinclair
1923 Mexico—Warren and Payne negotiate a petroleum settlement
1923 Costa Rica—Chief Justice Taft hands down ruling in the Amory Concession dispute
1923 Arabian Peninsula—Ibn Saud grants a concession to the Eastern and General Syndicate

1923 Turkey—government grants the Chester interests another concession, only to withdraw its support later in the year
1924 United States—exposure of Teapot Dome scandal
1924 Rumania—mining law provides in effect for the nationalization of the subsoil of that nation, including oil
1924 Peru—the International Petroleum Company purchases the La Brea y Pariñas petroleum fields outright
1925 Russia—Sinclair loses law suit in the courts there, with respect to the cancellation of the Sakhalin concession
1925 Bahrain and Kuwait—Eastern and General Syndicate obtains separate concession
1925 Iraq—government approves a new concession for the Turkish Petroleum Company
1925 Sahkalin—Japanese and Russians reach an agreement over its oil
1926 Russia—signs contract with both Vacuum and New York Standard
1926 Germany—Jersey Standard and I. G. Farben agree to exchange patents and information
1926 Rumania—law reserves embatic lands' subsoil rights to the state
1926 Colombia—Gulf Oil purchases the Barco concession, following which the government voids the latter
1927 Spain—establishes petroleum monopoly (Campsa)
1927 Mexico—Supreme Court hands down ruling in the Mexican Petroleum Company case
1927 Mexico—Calles-Morrow settlement
1927 Kuwait—Gulf Oil purchases concession from Frank Holmes
1927 India—Royal Dutch-Shell and New York Standard end their oil war
1928 world—Jersey Standard, Royal Dutch-Shell, and Anglo-Persian enter into the Achnacarry ("As Is") Agreement
1928 France—Law Poincaré imposes tariff on refined oil products
1928 Middle East—Red Line Agreement monopolizes petroleum throughout area, excluding Iran and Kuwait
1928 Netherlands East Indies—NKPM (Jersey Standard) obtains concession
1929 United States—establishment of the Independent Petroleum Association of America
1930 world—Memorandum for European Markets of 1930
1930 Bahrain—British approve the transfer of oil prospecting rights to California Standard
1931 Colombia—revalidates the Barco concession
1931 Iraq—Atlantic Refining, Gulf, and Pan American withdraw from the Iraq Refining Company
1931 Kuwait—Anglo-Persian obtains a concession
1932 United States—international oil conference in New York City
1932 United States—Revenue Act imposes duties on crude and residual oil
1932 Japan—oil cartel obtains the right to fix petroleum prices
1933 United States—formation of Standard-Vacuum Oil Company
1933 Saudi Arabia—California Standard obtains concession
1934 world—Draft Memorandum of Principles of 1934
1934 Great Britain—Petroleum Act empowers the Board of Trade to grant licenses to look for and obtain petroleum
1934 Germany—signs contract with Jersey Standard, Royal Dutch-Shell, and Anglo-Persian
1934 Italy—Licensing Act encourages both Italian and foreign oil companies to build refineries there

1934 Kuwait—Anglo-Persian and Gulf jointly obtain a concession
1934 Japan—Diet passes Petroleum Control Bill incorporating price controls, importing and refining quotas, and stockpiling requirements
1935 United States—Connally "Hot Oil" Act
1935 United States—Interstate Oil Compact
1935 Ethiopia—State Department forces cancellation of Rickett concession
1935 Italy—U.S. government imposes a moral embargo on petroleum exports
1935 Russia—Socony-Vacuum signs contract
1935 Bahrain—California Standard sells a half interest in Bapco to the Texas Company
1935 Manchukuo—oil monopoly officially begins operations
1936 Colombia—Texaco and Socony-Vacuum acquire the Barco concession
1936 Bolivia—sets up the YPFB
1936 Saudi Arabia—Texaco purchases a half-share in Casoc
1937 Brazil—new constitution bars foreigners from oil exploration
1937 Bolivia—expropriates the holdings of Jersey Standard
1937 China—Japanese attack three Stanvac vessels during the bombing of the *Panay*
1938 Brazil—establishes the National Petroleum Council
1938 Mexico—government expropriates U.S. and British oil properties
1939 Venezuela—agreement with the U.S. halves the 1932 oil import tax
1939 Japan—Universal Oil Products Company decides not to supply the Japanese with its polymerization process
1939-45 World War II
1940 Austria—Germany seizes Socony-Vacuum and Royal Dutch-Shell holdings
1940 Japan—Roosevelt Administration embargoes the shipment of aviation gas to countries outside the Western Hemisphere
1941 United States—FDR appoints Harold Ickes as Petroleum Coordinator
1941 Japan—U.S. places a virtual embargo on the exporting of motor fuel and lubricating gas from America
1942 United States—Jersey Standard agrees to a consent decree
1942 Canada—signing of Canol contract
1942 Spain—Spanish, Americans, and British enter into a joint regulatory system for petroleum
1942 Iceland—the United States replaces Great Britain as the supplier of oil
1942 Bolivia—settlement of expropriation dispute with Jersey Standard
1942 Mexico—Cooke and Zevada resolve the expropriation controversy
1943 United States—President Roosevelt establishes the Petroleum Reserves Corporation
1943 Mexico—agreement with the U.S. permits unlimited Mexican oil exports to America
1943 Mexico—U.S. makes $10 million loan for a petroleum refinery
1944 Rumania—Russia seizes American and British oil properties while occupying the country
1944 Great Britain—Draft Memorandum of Understanding with the United States on Petroleum
1944 Great Britain—first Anglo-American oil agreement (not ratified)
1944 Spain—U.S. suspends oil shipments because of trade with the Axis
1944 Iran—suspends the awarding of petroleum concessions for the remainder of World War II
1945 Great Britain—second Anglo-American oil agreement (not ratified)
1945 Venezuela—imposes a multimillion dollar extraordinary tax on the petroleum firms

1946 United States—creation of the National Petroleum Council
1946 Saudi Arabia—Aramco contract with Jersey Standard and New York Standard
1947 Canada—major oil discoveries in Alberta
1947 Iran—*Majlis* rejects petroleum concession which the Russians sought in the North
1948 Western Europe—inauguration of petroleum shipments under the Marshall Plan
1948 Hungary—government arrests two Maort employees and then seizes this oil firm
1948 Venezuela—writes into law the 50-50 profit splitting formula
1948 Colombia—establishes Ecopetrol
1948 Kuwait-Saudi Arabia Neutral Zone—Aramco surrenders its rights, acquiring rights to offshore petroleum elsewhere
1948 Kuwait sector of the Neutral Zone—American Independent Oil Company obtains a concession
1949 Great Britain—sterling-dollar dispute with the United States involving petroleum
1949 Argentina—new constitution declares that all underground petroleum belongs to the nation
1949 Mexico—U.S. rejects bid for a $470 million oil development loan
1949 Saudi sector of the Neutral Zone—Pacific Western Oil Corporation obtains a concession
1950 Great Britain—Standard-Vacuum and Caltex reach a petroleum agreement with the British
1950 Saudi Arabia—Aramco begins paying Saudi Arabia in taxes, not royalties
1951 United States—establishment of the Foreign Petroleum Supply Committee
1951 Colombia—de Mares concession reverts to the nation
1951 Iran—government expropriates Anglo-Iranian and sets up the National Iranian Oil Company
1952 United States—Federal Trade Commission and Mutual Security Administration reports on the international petroleum cartel
1952 Western Europe—creation of European Coal and Steel Community
1952 Venezuela—trade agreement with the United States reduces the duty on oil imports
1952 Mexico—termination of petroleum agreement with the United States
1952 Saudi Arabia—occupies the Buraimi Oasis by force
1953 United States—Eisenhower Administration brings civil suit against various big petroleum companies
1953 Brazil—establishment of Petrobrás
1954 Iran—creation of oil consortium to replace Anglo-Iranian
1955 Argentina—Perón signs a contract with a California Standard subsidiary, leading to his fall from power
1955 Libya—begins awarding concessions to foreign oil firms
1956 United States—creation of Middle East Emergency Committee
1956 Egypt—seizure of the Suez Canal by President Nasser, followed by a massive drive by the United States and other nations to supply Western Europe with petroleum
1956 Middle East—Second Arab-Israeli War
1957 United States—inauguration of voluntary oil import program
1957 Western Europe—setting up of the European Economic Community and the European Atomic Energy Community
1957 Peru—the International Petroleum Company obtains control of the Lobitos holdings
1958 Iran—Indiana Standard enters into a partnership with the National Iranian Oil Company

1959 United States—inauguration of mandatory oil import program, with Canada and Mexico exempted
1959 Western Europe—establishment of the European Free Trade Association
1960 United States—dissolution of the Standard-Vacuum Oil Company
1960 United States—Jersey Standard and Gulf Oil agree to consent decrees
1960 Venezuela—sets up the Venezuelan Petroleum Corporation
1960 Kuwait—establishes the Kuwait National Petroleum Company
1960 Indonesia—declares that the government controlled the oil of that nation, and alters the role of the oil firms to that of contractors
1960 world—founding of the Organization of Petroleum Exporting Countries at Baghdad
1961 Canada—formulates a National Oil Policy and sets up a National Energy Board
1961 Western Europe—creation of the Organization for Economic Cooperation and Development
1961 United States—the Kennedy administration tries to halt Russian penetration of free world petroleum markets
1961 Iraq—seizes 99 percent of the concession territory of the Iraq Petroleum Company under Public Law 80
1962 Mexico—Pemex obtains a $50 million loan from American investors
1962 Saudi Arabia—sets up Petromin
1963 United States—Texaco agrees to consent decree
1963 Argentina—President Illia cancels the Frondizi contracts
1963 Indonesia—foreign oil firms reach an agreement with Sukarno
1964 Iraq—establishment of the Iraq National Oil Company
1966 Argentina—President Onganía agrees to compensate the American oil firms for their cancelled contracts
1967 United States—Emergency Petroleum Supply Committee replaces FPSC
1967 Canada—secret agreement with the U.S. vis-à-vis oil imports
1967 Algeria—expropriates various foreign petroleum companies
1967 Middle East—Third Arab-Israeli War
1968 Peru—government takes over the properties of the International Petroleum Company after the two parties had agreed to the Act of Talara
1968 Indonesia—sets up Pertamina
1969 Western Europe—Phillips strikes oil in the Norwegian sector of the North Sea
1969 Bolivia—nationalizes the holdings of Gulf
1970 Mexico—ending of "Brownsville Loop" arrangement with the United States
1970 Libya—Occidental agrees to increase its taxes and royalties
1970 South Vietnam—new petroleum law with low tax rates
1971 Canada—sets up licensing procedures for imports of gasoline
1971 Iraq—Russia makes a $224 million loan for the construction of petroleum pipelines
1971 South Vietnam—foreign oil firms begin applying for concessions
1971 South Vietnam—U.S. government denies that the petroleum reserves there had affected American policy towards that country
1972 Rumania—Export-Import Bank extends loan to the governmental agency Impexmin
1972 Mexico—Export-Import Bank agrees to Pemex loan
1972 Bolivia—Hydrocarbons Act outlaws foreign concessions
1972 Saudi Arabia—government obtains a share of Aramco
1972 Iraq—nationalization of the Iraq Petroleum Company
1973 United States—termination of mandatory oil import program

1973 Canada—government imposes export controls on shipments of crude oil and gasoline to America, and sets charge on petroleum exports to the United States
1973 Iran—petroleum consortium becomes the Oil Service Company of Iran
1973 Iran—National Iranian Oil Company contract with Ashland Oil
1973 Middle East—Fourth Arab-Israeli War
1973 Middle East—Arab petroleum embargo
1974 United States—creation of the Federal Energy Agency and the Energy Research and Development Administration
1974 United States—authorizes the building of the Alaska pipeline
1974 Canada—announces a program to limit progressively crude oil exports to the United States
1974 Mexico—new petroleum fields discovered in Chapas-Tabasco region
1974 Brazil—Petrobrás makes its first major strike in the off shore Campos Field
1974 Saudi Arabia—Aramco grants the Saudi government a 60 percent participation, paving the way for an eventual complete takeover
1974 India—significant petroleum discoveries made in the offshore Bombay High region
1974 India—Indian government signs an agreement with Esso providing for its eventual ownership of the latter's holdings
1975 United States—Congress enacts the Energy Policy and Development Act
1975 United States—Exxon becomes the world's largest corporation in terms of sales
1975 Great Britain—creates the British National Oil Corporation
1975 Spain—enacts legislation providing a tax incentive for both native and foreign companies
1975 Canada—establishes Petro-Canada as a state oil firm
1975 Brazil—the government allows foreign petroleum firms to explore on the basis of a service contract
1975 Iraq—nationalizes all remaining foreign petroleum holdings
1976 Peru—the government reopens the door to foreign capital, including the oil companies
1976 Venezuela—the state officially finalizes its takeover of the petroleum industry
1976 Kuwait—the government assumes the ownership of the Kuwait Oil Company
1977 United States—the Carter administration shifts American energy policy towards an emphasis on conservation
1977 United States—setting up of the Department of Energy
1977 Saudi Arabia—temporarily splits from OPEC on raising oil prices
1978 United States—Congress passes a comprehensive energy package, featuring the Natural Gas Policy Act
1979 United States—firing of Energy Secretary James Schlesinger
1979 United States—gasoline rationing implemented during the Summer
1979 Iran—the Shah falls from power, interrupting oil exports
1979 OPEC—announces a two-tier system for pricing petroleum
1980 United States—Congress sets up a Synthetic Fuels Corporation
1980 Iran and Iraq—begin a war with each other, which damages oil properties
1981 United States—Reagan administration shifts focus of energy policy from scarcity to plenty

# Notes

Note: Some page references are missing to newspaper and magazine articles that were examined in clipping form at the various presidential libraries.

## 1. An Overview of the Historical Evolution of American Oil Policy

1. Ralph W. Hidy and Muriel E. Hidy, *History of Standard Oil Company (New Jersey): Pioneering in Big Business 1882-1911* (New York: Harper and Brothers, 1955), pp. 122-23.
2. Gerald Nash, *United States Oil Policy 1890-1964* (Pittsburgh: University of Pittsburgh Press, 1968), p. 10.
3. Sidney B. Fay, "Oil and the Middle East," *Current History*, April 1945, p. 336.
4. Norman Nordhauser, "Origins of Federal Oil Regulation in the 1920's," *Business History Review*, Spring 1973, p. 55.
5. Cordell Hull, *Memoirs*, 2 vols. (New York: Macmillan Company, 1948), Vol. 2, p. 1519.
6. Lloyd C. Gardner, *Economic Aspects of New Deal Diplomacy* (Madison: University of Wisconsin Press, 1964), p. 217.
7. John A. Loftus, "Oil in United States Foreign Policy," *Vital Speeches*, October 1, 1946, p. 755.
8. Nordhauser, "Origins of Federal Oil Regulation in the 1920's," p. 53 and p. 68.
9. John A. DeNovo, "The Movement for an Aggressive American Oil Policy Abroad, 1918-1920," *American Historical Review*, July 1956, p. 871.
10. Edward Mead Earle, "Oil and American Foreign Policy," *New Republic*, August 20, 1924, p. 357.
11. Ivan McKinley Stone, "The Relation of Petroleum to American Foreign Policy" (M. A. thesis, University of Nebraska, 1926), p. 40.
12. Ibid., p. 151.
13. Stanley K. Hornbeck, "The Struggle for Petroleum," *Annals of the American Academy*, March 1924, p. 170.
14. Arthur C. Veatch, "Oil, Great Britain and the United States," *Foreign Affairs*, July 1931, p. 668.
15. U.S., Congress, Senate, Special Committee Investigating Petroleum Resources, *Diplomatic Protection of American Petroleum Interests in Mesopotamia, Netherlands*

*East Indies, and Mexico*, Document No. 43, 79th Cong., 1st sess. (Washington, D.C.: Government Printing Office, 1945), p. 49.

16. DeNovo, "The Movement for an Aggressive American Oil Policy," p. 868.

17. "Memorandum Regarding the Foreign Oil Policy of the United States," New York City, April 15, 1921, Commerce Papers—Oil, Herbert Hoover Presidential Library, West Branch, Ia.

18. Ibid.

19. "Memorandum Relating to a Corporation for Acquisition of Foreign Petroleum," Commerce Papers—Oil, Hoover Library.

20. John E. Nelson to Singapore Office, Bureau of Foreign and Domestic Commerce, Department of Commerce, Washington, D.C., June 26, 1928, Bureau of Foreign and Domestic Commerce Record Group No. 151. No. 312, National Archives, Washington, D.C.

21. Samuel Guy Inman, "The Oil Dollar Loses Its Power," *Christian Century*, September 25, 1935, p. 1205.

22. Ray Lyman Wilbur to Hiram W. Johnson, Washington, D.C., January 29, 1931, Presidential Papers, Hoover Library.

23. Chairman, Advisory Committee, to Ray Lyman Wilbur, Washington, D.C., February 9, 1931, Presidential Papers, Hoover Library.

24. William D. Mitchell to Herbert Hoover, Washington, D.C., July 23, 1931, Presidential Papers, Hoover Library.

25. Herbert Hoover to C. C. Teague, Washington, D.C., October 18, 1932, Public Statements File, Hoover Library.

26. Franklin D. Roosevelt to Harold Ickes, Washington, D.C., June 18, 1941, Morgenthau Diaries, Vol. 410, Franklin D. Roosevelt Presidential Library, Hyde Park, N.Y.

27. Harold Ickes to Franklin D. Roosevelt, Washington, D.C., December 8, 1941, Official File—Oil, Roosevelt Library.

28. Blair Bolles, "Oil: An Economic Key to Peace," *Foreign Policy Reports*, July 1, 1944, p. 92.

29. "In Search of a Policy," *Time*, December 27, 1943, p. 77.

30. Unsigned memorandum, "The World Oil Cartel," Washington, D.C., n.d., Oscar Chapman Papers, Harry S. Truman Presidential Library, Independence, Mo.

31. "In Search of a Policy," p. 77.

32. U.S., Congress, House of Representatives, Committee on Interior and Insular Affairs, *Report on Mandatory Oil Import Control Program, Its Impact on the Domestic Minerals Industry*, Committee Print No. 11. 90th Congr., 2d sess. (Washington, D.C.: Government Printing Office, 1968), p. 4.

33. *1947 Britannica Book of the Year* (Chicago: Encyclopaedia Britannica, 1947), p. 593.

34. Ibid.

35. Wallace E. Pratt, "The United States and Foreign Oil," *Yale Review*, Autumn 1951, p. 112.

36. Ibid.

37. Wright Patman, report, "The Effect of Oil Imports on Independent Domestic Producers," Washington, D.C., January 26, 1950, Truman Papers—Official File, Truman Library.

38. Harry Truman to Oscar Chapman, Washington, D.C., May 2, 1950, President's Secretary's File, Truman Library.

39. Unsigned memorandum, "The World Oil Cartel."

40. Thomas C. Hennings, Jr. to Harry Truman, Washington, D.C., July 18, 1952, Sumner Pike Papers, Truman Library.

41. William Harlan Hale, "Troubled Oil in the Middle East," *Reporter*, January 23, 1958, p. 12.

42. Oscar Chapman, "A Report to the National Security Council by the Secretary of the Interior and Petroleum Administrator for Defense on National Security Problems Concerning Free World Petroleum Demands and Potential Supplies," Washington, D.C., December 8, 1952, *Declassified Documents Reference System* 59C (Arlington, Va.: The Carrollton Press, 1978); see p. 43 of this 47 page report on microfiche. This post World War II series consists both of a retrospective collection and annual supplements.

43. Stephen J. Spingarn to Harry Truman, Washington, D.C., October 29, 1952, Spingarn Papers, Truman Library.

44. "The FTC 'Cartel' Report—A Major Scandal," *Oil Forum*, November 1952, p. 370.

45. "Sparkman and the Big Bad Wolf," Washington *Post*, November 3, 1952, p. 10.

46. "Oil and National Security," Washington *Post*, November 10, 1952, p. 8.

47. Stephen J. Spingarn to Harry Truman, Washington, D.C., November 7, 1952, Spingarn Papers, Truman Library.

48. "Oil Companies Cry 'Blackmail' at Truman Plan," New York *Herald Tribune*, January 13, 1953.

49. Stephen J. Spingarn to Guy Gillette, Washington, D.C., January 21, 1953, Spingarn Papers, Truman Library.

50. "Civil Suit is Filed Against 5 Oil Firms," Washington *Post*, April 22, 1953, p. 10.

51. Charles F. Edmundson, "Iraq's Explosive Oil," *Nation*, August 2, 1958, p. 52.

52. Ibid., p. 50.

53. Ibid., p. 52.

54. Dwight Eisenhower to Dillon Anderson, Washington, D.C., July 30, 1957, Material Relating to Oral History—Dillon Anderson, Dwight Eisenhower Presidential Library, Abilene, Kans.

55. Mira Wilkins, *The Maturing of Multinational Enterprise: American Business Abroad from 1914 to 1970* (Cambridge: Harvard University Press, 1974), p. 388.

56. "Recommendations Concerning Energy Supplies and Resources," Department of State *Bulletin*, March 21, 1955, p. 489.

57. Robert Murphy to Percival Brundage, Washington, D.C., May ?, 1957, White House Central Files, Eisenhower Library.

58. Special Committee to Investigate Crude Oil Imports, report, "Petroleum Imports," Washington, D.C., July 1957, White House Central Files, Eisenhower Library.

59. "Eisenhower Administration Project: Dillon Anderson," Oral History Office, Columbia University, N.Y., 1972, p. 119. There is a copy in the Eisenhower Library.

60. Frederick Payne to Dwight Eisenhower, Washington, D.C., July 31, 1957, White House Central Files, Eisenhower Library.

61. Special Committee to Investigate Crude Oil Imports, "Supplementary Report," Washington, D.C., March 24, 1958, White House Central Files, Eisenhower Library.

62. Dwight Eisenhower to Milward Simpson, Washington, D.C., April 3, 1958, White House Central Files, Eisenhower Library.

63. Minutes of Cabinet Meeting, Washington, D.C., March 6, 1959, Ann Whitman File—Cabinet Series, Eisenhower Library.

64. Dwight Eisenhower to Rómulo Betancourt, Washington, D.C., April 28, 1959, White House Central Files, Eisenhower Library.

65. Recently declassified document (October 12, 1979), Eisenhower Library.

66. Douglas Dillon to Maurice Stans, Washington, D.C., December 1, 1960, White House Central Files, Eisenhower Library.

67. "Why the Oil Giants are Under the Gun," *Business Week*, October 25, 1969, p. 88.

68. Anthony Sampson, *The Seven Sisters: The Great Oil Companies and the World They Shaped* (New York: Bantam Books, 1976), p. 91.

69. Office of Emergency Planning, untitled release, Washington, D.C., February 16, 1962, Robinson-Ross Office Files, Lyndon Johnson Presidential Library, Austin, Tex.

70. "LBJ Should Take Back Interior's Oil Policy Authority, Patman Says," *Platt's Oilgram News Service*, June 18, 1965.

71. John Ricca to J. Cordell Moore, Washington, D.C., October 6, 1967, White House Central Files, Johnson Library.

72. Stewart Udall to Meyer Feldman, Washington, D.C., April 28, 1964, Confidential File, Johnson Library.

73. John Tower to Lyndon Johnson, Washington, D.C, June 8, 1964, White House Central Files, Johnson Library.

74. Task Force on Natural Resources Report, "Resource Policies for a Great Society," Washington, D.C., November 11, 1964, 1964 Outside Task Force on Natural Resources File, Johnson Library.

75. Stewart Udall to Marvin Watson, Washington, D.C., March 30, 1965, White House Central Files, Johnson Library.

76. "Residual Oil Quota End Urged," Washington *Post*, April 3, 1965.

77. "New England Loses Again in the Residual Oil Fight," Providence *Journal*, April 2, 1965.

78. Stewart Udall to Joseph Califano, Washington, D.C., December 23, 1965, Confidential File, Johnson Library.

79. Presidential Proclamation, July 17, 1967, White House Central Files, Johnson Library.

80. De Vier Pierson to Lyndon Johnson, Washington, D.C., May 1, 1968, Pierson Aide File, Johnson Library.

81. U.S., Congress, House of Representatives, Committee on Interior and Insular Affairs, *Report on Mandatory Oil Import Control Program, Its Impact on the Domestic Minerals Industry*, Committee Print No. 11, 90th Cong., 2d sess. (Washington, D.C.: Government Printing Office, 1968), pp. 24-25.

82. Ibid., p. 26.

83. Congressional Quarterly Service, *Congress and the Nation*, Volume III, *1969-1972* (Washington: Congressional Quarterly Service, 1973), p. 845.

84. Ibid.

85. Naiem A. Sherbiny and Mark A. Tessler, eds., *Arab Oil: Impact on the Arab Countries and Global Implications*, Praeger Special Studies in International Business, Finance and Trade (New York: Praeger Publishers, 1976), p. 281.

86. National Petroleum Council, *Petroleum Resources Under the Ocean Floor* (Washington, D.C.: National Petroleum Council, 1969), p. 13.

87. Dankwart A. Rustow and John F. Mungo, *OPEC: Success and Prospects*, Council on Foreign Relations (New York: New York University Press, 1976), p. 8.

88. Sampson, *The Seven Sisters*, p. 260.

89. Ibid., p. 273n.

90. "Old Land Offers Key to Middle East Oil Access," *Oil and Gas Journal*, May 21, 1973, p. 77.

91. *Newsweek*, September 17, 1973.

## 2. Western and Eastern Europe

1. Ludwell Denny, *America Conquers Britain: A Record of Economic War* (New York: Alfred A. Knopf, 1930), p. 229.

2. John A. DeNovo, "The Movement for an Aggressive American Oil Policy Abroad, 1918-1920," *American Historical Review*, July 1956, p. 859.

3. Ibid., p. 860.

4. Herbert Feis, *Petroleum and American Foreign Policy* (Palo Alto: Stanford University Food Research Institute, 1944), p. 4.

5. U.S., Department of State, *Foreign Relations of the United States 1936* (Washington, D.C.: Government Printing Office, 1953), Vol. 1, p. 738.

6. U.S., Department of State, *Foreign Relations of the United States 1928* (Washington, D.C.: Government Printing Office, 1943), vol. 3, p. 382.

7. Ibid., p. 401.

8. U.S., Congress, *Message from the President of the United States, Restrictions on American Petroleum Prospectors in Certain Foreign Countries*, Document No. 272. 66th Cong., 2d sess. (Washington, D.C.: Government Printing Office, 1921), p. 3.

9. Edwin B. George to John H. Nelson, Philadelphia, Pennsylvania, July 28, 1927, Bureau of Foreign and Domestic Commerce, Record Group No. 151, File No. 312, National Archives.

10. "Germany ousting Standard Oil," *Literary Digest*, November 23, 1912, p. 950.

11. U.S., Department of State, *Foreign Relations of the United States 1934* (Washington, D.C.: Government Printing Office, 1951), Vol. 2, p. 327.

12. "Odor of Oil," *Time*, September 16, 1935, p. 17.

13. Edward W. Chester, *Clash of Titans: Africa and U.S. Foreign Policy* (Maryknoll: Orbis Books, 1974), p. 207.

14. "Sanctions: League Cheers U.S. Check, Plans Oil Drought to Leave Duce High and Dry," *Newsweek*, December 7, 1935, p. 7.

15. Marvin H. McIntyre to F. J. Sisson, Washington, D.C., n.d., Official File, Roosevelt Library.

16. Wilkins, *The Maturing of Multinational Enterprise*, p. 221.

17. Emil Lengyel, "Rumanian Oil and Foreign Money," *Nation*, September 17, 1924, p. 296.

18. U.S., Department of State, *Foreign Relations of the United States 1924* (Washington, D.C.: Government Printing Office, 1939), Vol. 2, p. 636.

19. U.S., Department of State, *Foreign Relations of the United States 1926* (Washington, D.C.: Government Printing Office, 1941), Vol. 2, p. 324.

20. Ibid. 324-25.

21. H. C. MacLean to the Director of the Bureau of Foreign and Domestic Commerce, Rome, Italy, May 22, 1922, Bureau of Foreign and Domestic Commerce, Record Group No. 151, File No. 312, National Archives.

22. U.S., Department of State, *Foreign Relations of the United States 1925* (Washington: Government Printing Office, 1940), Vol. 1, p. 502.

23. Hidy and Hidy, *History of Standard Oil Company (New Jersey): Pioneering in Big Business 1882-1911*, p. 135.

24. Ibid., p. 511.

25. Philip S. Gillette, "American Capital in the Contest for Soviet Oil, 1920-23," *Soviet Studies*, April 1973, p. 482.

26. George Sweet Gibb and Evelyn H. Knowlton, *History of Standard Oil Company (New Jersey): The Resurgent Years 1911-1927* (New York: Harper and Brothers, 1956), p. 341.

27. Louis Fischer, *Oil Imperialism: The International Struggle* (New York: International Publishers, 1926), p. 193.

28. Robert James Maddox, *William E. Borah and American Foreign Policy* (Baton Rouge: Louisiana State University Press, 1969), p. 209.

29. Karl E. Ettinger, "What Does the Federal Trade Commission Cartel Report Mean," address to Empire State Petroleum Association, Lake Placid, N.Y.: September 29, 1952, p. 15. Copy in Spingarn Papers, Truman Library.

30. Fischer, *Oil Imperialism*, p. 145.

31. Hans Heymann, "Oil in Soviet-Western Relations in the Interwar Years," *American Slavic and East European Review*, December 1948, p. 311.

32. E. S. Land to Harry Hopkins, Confidential Memorandum, Washington, D.C.: April 28, 1941, Morgenthau Diaries, Vol. 394, Roosevelt Library.

33. "Oil on War's Waters," *Newsweek*, April 24, 1944, p. 68.

34. "Oil Agreement with Britain Places U.S. In First Cartel," *Newsweek*, August 21, 1944, p. 62 (quotation from New York *Times*).

35. "The Ambiguous Oil Agreement," *Nation*, August 19, 1944, p. 201.

36. "New Oil Agreement Drafted," *Business Week*, October 6, 1945, pp. 26 and 30.

37. "Oil, Blood and Ickes," *New Republic*, January 21, 1946, p. 94.

38. Oscar L. Chapman, Statement for Senate Foreign Relations Committee, Washington, D.C., March 1946, Warren Gardner Papers, Truman Library.

39. U.S., Department of State, *Foreign Relations of the United States 1941* (Washington, D.C.: Government Printing Office, 1959), Vol. 2, p. 916.

40. U.S., Department of State, *Foreign Relations of the United States 1943* (Washington, D.C.: Government Printing Office, 1964), Vol. 2, p. 680.

41. Ibid., p. 687.

42. "Suspension of Oil Shipments to Spain," U.S. Department of State *Bulletin*, January 29, 1944, p. 116.

43. Thomas A. Bailey, *A Diplomatic History of the American People*, 8th ed. (New York: Appleton Century-Crofts, 1969), p. 752.

44. U.S. Department of State, *Foreign Relations of the United States, 1943*, Vol. 2, p. 592.

45. See pp. 16-18, Vol. 587, of the Morgenthau Diaries at the Roosevelt Library for this undated report which was presented to the President along with other documents on November 16, 1942.

46. "Export of Petroleum Products to the Soviet Union," letter from Secretary of State Cordell Hull to Representative Frank E. Hook, Department of State *Bulletin*, February 24, 1940, p. 195.

47. Oscar Cox to Harry Hopkins, Washington, D.C., September 29, 1942, Cox Papers, Roosevelt Library.

48. U.S., Department of State, *Foreign Relations of the United States 1943*, Vol. 3, p. 754.

49. U.S., Department of State, *Foreign Relations of the United States 1945* (Washington, D.C.: Government Printing Office, 1967), Vol. 5, p. 653.
50. Wilkins, *The Maturing of Multinational Enterprise*, p. 259.
51. *Standard Oil Company (New Jersey) and Oil Production in Hungary by MAORT 1931-1948* (n.p.: European Gas and Electric Company, 1949), p. v.
52. Edward H. Shaffer, *The Oil Import Program of the United States* (New York: Frederick A. Praeger, 1968), p. 13.
53. Joyce Kolko and Gabriel Kolko, *The Limits of Power: The World and United States Foreign Policy, 1945-1954* (New York: Harper and Row, 1972), p. 447.
54. L. A. Minnich, Jr., Minutes, Bipartisan Legislative Meeting, November 9, 1956, Ann Whitman File, Eisenhower Library.
55. "The Pinch," *Newsweek*, December 10, 1956, p. 33.
56. James E. Akins, "The Oil Crisis: This Time the Wolf is Here," *Foreign Affairs*, April 1973, p. 472.

## 3. Canada and Mexico

1. Harold Ickes to Franklin Roosevelt, Washington, D.C., May 29, 1942, Official File, Roosevelt Library.
2. Wilkins, *The Maturing of Multinational Enterprise*, p. 273.
3. "Gas for the Planes to Asia," *Time*, October 4, 1943, p. 68.
4. $134,000,000," *Time*, January 17, 1944, p. 15.
5. "Canol, the War's Epic Blunder," *Nation*, May 5, 1945, p. 513.
6. "The Great Canol Fiasco," *American Mercury*, April 1948, p. 416.
7. "The Epic of Canol," *Canadian Geographical Journal*, March 1947, p. 137.
8. "Quota for the West," *Time*, January 6, 1958, p. 64.
9. *Oil and Gas Journal*, May 5, 1969, p. 100.
10. Samuel Guy Inman, "The Oil Dollar Loses Its Power," p. 1205.
11. Henry Lane Wilson, *Diplomatic Episodes in Mexico, Belgium, and Chile* (Garden City, N.Y.: Doubleday, Page and Company, 1927), pp. 238-39.
12. Scott Nearing and Joseph Freeman, *Dollar Diplomacy: A Study in American Imperialism* (New York: Viking Press, 1925), p. 92.
13. "Oil and Ideals in Latin Lands," *Literary Digest*, December 6, 1913, p. 1099 (quotation from New York *Times*).
14. Josephus Daniels, *Shirt-Sleeve Diplomat* (Chapel Hill: University of North Carolina Press, 1947), p. 211.
15. Samuel J. Astorino, "Senator Albert B. Fall and Wilson's Last Crisis with Mexico," *Duquesne Review*, Spring 1968, p. 7.
16. Graham H. Stuart, *Latin America and the United States* (New York: Century Company, 1928), p. 140.
17. Ibid.
18. Nearing and Freeman, *Dollar Diplomacy*, p. 270.
19. Astorino, "Senator Albert B. Fall and Wilson's Last Crisis with Mexico," p. 11.
20. Ibid.
21. Stone, "The Relation of Petroleum to American Foreign Policy," p. 82.
22. Ibid.
23. U.S., Congress, Senate, Special Committee Investigating Petroleum Resources, *Diplomatic Protection of American Petroleum Interests in Mesopotamia, Netherlands East Indies, and Mexico*, p. 68.

24. David Y. Thomas, *One Hundred Years of the Monroe Doctrine 1823-1923* (New York: Macmillan Company, 1923), p. 319.

25. Ibid., p. 322.

26. Gibb and Knowlton, *History of Standard Oil Company (New Jersey): The Resurgent Years 1911-1927*, p. 363.

27. James Morton Callahan, *American Foreign Policy in Mexican Relations* (New York: Macmillan Company, 1932), p. 596.

28. Ibid.

29. Ibid.

30. Harry Stegmaier, "Delaying the Crisis: Oil, Mexico and Dwight Morrow, 1925-1928," unpublished paper, p. 6.

31. Ibid., p. 9.

32. E. David Cronon, *Josephus Daniels in Mexico* (Madison: University of Wisconsin Press, 1960), p. 12.

33. Thomas A. Bailey, *A Diplomatic History of the American People*, 8th ed., p. 680.

34. L. Ethan Ellis, *Frank B. Kellogg and American Foreign Relations 1925-1929* (New Brunswick, N.J.: Rutgers University Press, 1961), p. 48.

35. Ibid., p. 51.

36. Callahan, *American Foreign Policy in Mexican Relations*, p. 612.

37. John W. Frey to Julius Klein, Washington, D.C., May 16, 1931, Bureau of Foreign and Domestic Commerce, Record Group No. 151, File No. 312, National Archives.

38. Cronon, *Josephus Daniels*, p. 156.

39. Ibid., p. 125.

40. Ibid., p. 185.

41. Peter R. Odell, "Oil and State in Latin America," *International Affairs*, October 1964, p. 660.

42. J. Richard Powell, *The Mexican Petroleum Industry 1938-1950* (Berkeley: University of California Press, 1956), p. 197.

43. William O. Scroggs, "Mexican Oil in World Politics," *Foreign Affairs*, October 1938, p. 175.

44. Cronon, *Josephus Daniels*, p. 232.

45. George C. Coleman, "The 'Good Neighbor' Tested: 1938," *Southwestern Social Science Quarterly*, December 1952, p. 219.

46. Eliot Janeway, "Mexico, Tokyo, Berlin and the Oil Axis," *Asia*, September 1938, p. 518.

47. Cronon, *Josephus Daniels*, p. 232.

48. Ibid., p. 233.

49. Donald Richberg, *The Mexican Oil Seizure* (New York: Arrow Press, 1939), p. 53.

50. Cronon, *Josephus Daniels*, p. 250.

51. Ibid., p. 252.

52. Franklin Roosevelt to Sumner Welles, Washington, D.C., December 18, 1941, Official File, Roosevelt Library.

53. Cronon, *Josephus Daniels*, p. 270.

54. Manuel Ávila Camacho to Franklin Roosevelt, Mexico City, Mexico, April ?, 1942, Official File, Roosevelt Library.

55. Jesse Jones to Franklin Roosevelt, Washington, D.C., April 7, 1942, Official File, Roosevelt Library.

56. U.S., Department of State, *Foreign Relations of the United States 1944* (Washington: Government Printing Office, 1967), Vol. 7, p. 1337.
57. Powell, *The Mexican Petroleum Industry*, p. 194.
58. U.S., Congress, House of Representatives, Committee on Interstate and Foreign Commerce, *Fuel Investigation: Mexican Petroleum*, 80th Cong., 2nd Sess. (Washington, D.C.: Government Printing Office, 1949), p. 16.
59. "Oil and Dollars," *Newsweek*, August 1, 1949, p. 32.
60. Harry Truman to Wayne Morse, Washington, D.C., March 8, 1950, Truman Papers—Official File, Truman Library.
61. Harry S. Truman, handwritten note on a letter from Secretary of State Dean Acheson, Washington, D.C., January 23, 1950, *Declassified Documents Reference System* 196F (Arlington, Va.: The Carrollton Press, 1975).
62. Antonio J. Bermúdez, *The Mexican National Petroleum Industry*, Institute of Hispanic American and Luso-Brazilian Studies (Stanford: Stanford University Press, 1963), p. viii.
63. "Pemex at Twenty-Five," *Fortune*, August 1963, p. 87.

## 4. The Noncontiguous Western Hemisphere

1. U.S., Department of State, *Foreign Relations of the United States 1921* (Washington, D.C.: Government Printing Office, 1936), Vol. 2, p. 937.
2. Gibb and Knowlton, *History of Standard Oil Company (New Jersey): The Resurgent Years 1911-1927*, p. 391.
3. U.S., Department of State, *Foreign Relations of the United States 1950* (Washington, D.C.: Government Printing Office, 1976), Vol. 2, p. 1025.
4. Wilkins, *The Maturing of Multinational Enterprise*, p. 224.
5. Ibid.
6. Bryce Wood, *The Making of the Good Neighbor Policy* (New York: Columbia University Press, 1961), p. 261.
7. "Letter from the Secretary of State to the Governor of Kansas," U.S. Department of State *Bulletin*, December 9, 1939, p. 671.
8. Wood, *The Making of the Good Neighbor Policy*, pp. 277-78.
9. U.S., Department of State, *Foreign Relations of the United States 1946* (Washington, D.C.: Government Printing Office, 1960), Vol. 9, p. 1331.
10. Ibid., p. 1346.
11. U.S., Department of State, *Foreign Relations of the United States, 1950*, Vol. 2, p. 1040.
12. Ibid., p. 1025.
13. "U.S. Signs New Trade Agreement with Venezuela," Department of State *Bulletin*, September 15, 1952, p. 403.
14. "The Case for Free Trade," *Time*, April 26, 1954, p. 46.
15. "Venezuelan Oil Squeeze," *New Republic*, January 5, 1959, p. 5.
16. "More—or Less," *Newsweek*, January 5, 1959, p. 52.
17. "Venezuela Asks for U. S. Market Quota," *Petroleum Week*, May 1, 1959, p. 19.
18. Philip A. Kay, *South Wind Red: Our Hemispheric Crisis* (Chicago: Henry Regnery Company, 1962), p. 68 (quotation from New York *Times*).
19. Ibid., p. 69.
20. "Government Intervention in Oil Plagues Creole," *Oil and Gas Journal*, May 24, 1971, p. 46.

21. Rómulo Betancourt, *Venezuela: Oil and Politics*, trans. Everett Bauman (Boston: Houghton Mifflin Company, 1979), p. 403.

22. E. Taylor Parks, *Colombia and the United States 1765-1934* (Durham: Duke University Press, 1935), p. 445.

23. Wilkins, *The Maturing of Multinational Enterprise*, p. 14.

24. Benjamin Williams, *Economic Foreign Policy of the United States* (New York: McGraw Hill Book Company, 1929), p. 76.

25. Stone, "The Relation of Petroleum to American Foreign Policy," p. 98.

26. Ibid., p. 99.

27. C. Reed Hill to Director of the Bureau of Foreign and Domestic Commerce, Washington, D.C., November 16, 1925, Bureau of Foreign and Domestic Commerce, Record Group No. 151, File No. 312, National Archives.

28. Stephen J. Randall, "The International Corporation and American Foreign Policy: The United States and Colombian Petroleum, 1920-1940," *Canadian Journal of History*, August 1974, p. 181.

29. George J. Eder to Mr. Domeratzy, Washington, D.C., October 21, 1927, Bureau of Foreign and Domestic Commerce, Record Group No. 151, File No. 312, National Archives.

30. Ludwell Denny, "British Intrigue at the Panama Canal," *Nation*, July 11, 1928, p. 36.

31. J. Fred Rippy, *The Capitalists and Colombia* (New York: Vanguard Press, 1931), p. 138.

32. Denny, *America Conquers Britain*, pp. 263-64.

33. Randall, "The International Corporation and American Foreign Policy," p. 180.

34. "Priced Out," *Time*, April 4, 1949, p. 38.

35. Ibid.

36. Hubert Herring, *A History of Latin America from the Beginnings to the Present* (New York: Alfred A. Knopf, 1972), p. 638.

37. Herbert S. Klein, "American Oil Companies in Latin America: The Bolivian Experience," *Inter-American Economic Affairs,* Autumn 1964, p. 50.

38. Ibid., p. 54.

39. Mira Wilkins, "Multinational Oil Companies in South America in the 1920's: Argentina, Bolivia, Chile, Colombia, Ecuador, and Peru," *Business History Review*, Autumn 1974, p. 440.

40. Louis Turner, "The Oil Majors in World Politics," *International Affairs*, July 1976, p. 369.

41. Klein, "American Oil Companies in Latin America," p. 57.

42. Wood, *The Making of the Good Neighbor Policy*, p. 168.

43. U.S., Department of State, *Foreign Relations of the United States 1937* (Washington, D.C.: Government Printing Office, 1954), Vol. 5, p. 287.

44. Ibid., p. 297.

45. Wood, *The Making of the Good Neighbor Policy*, pp. 177-78.

46. Ibid., p. 181.

47. Samuel Flagg Bemis, *The United States as a World Power: A Diplomatic History 1900-1950* (New York: Henry Holt and Company, 1950), pp. 298-99.

48. U.S., Department of State, *Foreign Relations of the United States 1942* (Washington, D.C.: Government Printing Office, 1962), Vol. 5, p. 525.

49. Henry Ozanne, "Bolivia Plans New Oil Policy," *Oil Weekly*, June 3, 1946, p. 4.

50. Kolko and Kolko, *The Limits of Power*, pp. 416-17.
51. George M. Ingram, *Expropriation of U. S. Property in South America: Nationalization of Oil and Copper Companies in Peru, Bolivia, and Chile* (New York: Praeger Publishers, 1974), p. 341.
52. Selden Rodman, "The New-Left Rightists in Bolivia," *National Review*, July 7, 1972, p. 740.
53. Wilkins, "Multinational Oil Companies in South America in the 1920's," p. 437.
54. Ibid.
55. Jessica P. Einhorn, *Expropriation Politics* (Lexington: Lexington Books, 1974), p. 13.
56. Julian D. Smith to Director, Bureau of Foreign and Domestic Commerce, Lima, Peru, July 9, 1931, Bureau of Foreign and Domestic Commerce, Record Group No. 151, File No. 312, National Archives.
57. Herring, *A History of Latin America*, p. 604.
58. Rieck B. Hannifin, *Expropriation by Peru of the International Petroleum Company* (Washington: Library of Congress Legislative Reference Service, 1969), p. 1.
59. Adalberto J. Pinelo, *The Multinational Corporation as a Force in Latin American Politics: A Case Study of the International Petroleum Company in Peru* (New York: Praeger Publishers, 1973), p. 149.
60. Peru: Dirección General de Informaciónes, *Petroleum in Peru* (Lima: Petróleos del Peru: Public Relations Department, 1969), p. 44.
61. Ingram, *Expropriation of U.S. Property in South America*, p. 369.
62. "Oxy Deal May Spark Wider Peru Search," *Oil and Gas Journal*, July 12, 1971, p. 28.
63. Harold F. Peterson, *Argentina and the United States 1810-1960* (New York: State University of New York, 1964), p. 349.
64. U.S., Department of State, *Foreign Relations of the United States 1936* (Washington: Government Printing Office, 1954), Vol. 5, p. 184.
65. U.S., Department of State, *Foreign Relations of the United States 1943* (Washington: Government Printing Office, 1965), Vol. 5, pp. 395-96.
66. George Messersmith to George Marshall, Buenos Aires, Argentina, May 9, 1947, Office Files of the Assistant Secretary of State for Economic Affairs, Truman Library.
67. Laura Randall, *An Economic History of Argentina in the Twentieth Century* (New York: Columbia University Press, 1978), p. 204.
68. U.S. Embassy to State Department, Telegram, Buenos Aires, Argentina, December 6, 1963, *Declassified Documents Reference System* 46C (Arlington, Va.: Carrollton Press, 1976).
69. "Go Home Yankee Trader," *New Republic*, November 23, 1963, p. 9.
70. Walter La Feber, *The New Empire: An Interpretation of American Expansion 1860-1898* (Ithaca: Cornell University Press, 1963), p. 215.
71. Wilkins, "Multinational Oil Companies in South America in the 1920's," p. 443.
72. Walter J. Donnelly to Jefferson Caffery, Rio de Janeiro, Brazil, April 27, 1938, Bureau of Foreign and Domestic Commerce, Record Group No. 151, File No. 312, National Archives.
73. U.S., Department of State, *Foreign Relations of the United States 1947* (Washington, D.C.: Government Printing Office, 1972), Vol. 8, p. 467.
74. Herring, *A History of Latin America*, p. 871.

75. Peter Seaborn Smith, "Petrobrás: The Politics of a State Company, 1953-1964," *Business History Review*, Summer 1972, p. 190.

76. Odell, "Oil and State in Latin America," p. 669.

77. John D. Wirth, *The Politics of Brazilian Development 1930-1954* (Stanford: Stanford University Press, 1970), p. 214.

78. Dana G. Munro, *International and Dollar Diplomacy in the Caribbean 1900-1921* (Princeton: Princeton University Press, 1964), p. 431.

79. U.S., Department of State, *Foreign Relations of the United States 1921* (Washington: Government Printing Office, 1936), Vol. 1, p. 647.

80. John H. Lind, "U. S. Oil Companies under Transition in South America," *Magazine of Wall Street*, May 21, 1960, p. 261.

81. "Red Oil for the Lamps of Cuba," *Newsweek*, July 11, 1960, p. 65.

82. Raymond Hare to Clarence Randall, Washington, D.C., August 16, 1960, Joseph Rand Records, Eisenhower Library.

## 5. The Middle East and North Africa

1. Thomas A. Bryson, *American Diplomatic Relations with the Middle East, 1784-1975: A Survey* (Metuchen: Scarecrow Press, 1977), p. 133.

2. U.S., Department of State, *Foreign Relations of the United States 1946* (Washington: Government Printing Office, 1969), Vol. 7, p. 23.

3. Henry Cattan, *The Evolution of Oil Concessions in the Middle East and North Africa* (Dobbs Ferry: Oceana Publications, 1967), p. 4.

4. "Six Kingdoms of Oil," *Time*, March 3, 1952, p. 26.

5. "Secretary Dulles Discusses U.S. Foreign Policy for British Television Broadcast," in Department of State *Bulletin*, November 10, 1958, p. 737.

6. Victor Perlo, "American Oil Companies in the Middle East," *International Affairs*, December 1967, p. 40.

7. Sherbiny and Tessler, eds., *Arab Oil*, p. 269.

8. "Nixon Sees Big Arab Spending in U.S.," *Oil and Gas Journal*, April 2, 1973, p. 46.

9. Lee H. Hamilton, "The Energy Crisis: The Persian Gulf," *Vital Speeches*, February 1, 1973, p. 229.

10. James A. Field, *America and the Mediterranean World 1776-1882* (Princeton: Princeton University Press, 1969), p. 311.

11. U.S., Department of State, *Foreign Relations of the United States 1883* (Washington, D.C.: Government Printing Office, 1884), p. 823.

12. Allan Nevins, *Study in Power: John D. Rockefeller, Industrialist and Philanthropist*, 2 vols. (New York: Charles Scribners' Sons, 1953), Vol. 2, p. 124.

13. C. M. Chester to Edwin Denby, Washington, D.C., March 30, 1921, Commerce Papers—Oil, Hoover Library.

14. John Carter, "The Bitter Conflict over Turkish Oilfields," *Current History*, January 1926, p. 495.

15. Henry Woodhouse, "American Oil Claims in Turkey," New York *Times Current History*, March 1922, p. 957.

16. Ibid.

17. Carter, "The Bitter Conflict over Turkish Oilfields," p. 494.

18. U.S., Department of State, *Foreign Relations of the United States 1921*, Vol. 2, p. 90.

19. Ibid., p. 91.
20. U.S., Department of State, *Foreign Relations of the United States 1920* (Washington, D.C.: Government Printing Office, 1936), Vol. 2, p. 650.
21. W. Y. Elliot, "The Turkish Petroleum Company—A Study in Oleaginous Diplomacy," *Political Science Quarterly*, June 1924, p. 272.
22. H. C. Morris to Herbert Hoover, Washington, D.C., April 24, 1922, Commerce Papers—Oil, Hoover Library.
23. "Oil Troubling the Diplomatic Waters," *Literary Digest*, December 11, 1920, p. 18 (quotation from the Boston *Globe*).
24. Ibid., p. 19 (quotation from the Indianapolis *News*).
25. Michael J. Hogan, *Informal Entente: The Private Structure of Cooperation in Anglo-American Economic Diplomacy 1918-1928* (Columbia: University of Missouri Press, 1947), p. 178.
26. U.S., Department of State, *Foreign Relations of the United States 1921*, Vol. 2, p. 920.
27. Chester to Denby, Commerce Papers—Oil, Hoover Library.
28. Levantine Section, Eastern European Division, Bureau of Foreign and Domestic Commerce, to C. A. Herter, Washington, D.C., April 16, 1923, Commerce Papers—Oil, Hoover Library.
29. Herbert Hoover to Charles Evans Hughes, Washington, D.C., August 19, 1922, Commerce Personal—Charles Evans Hughes, Hoover Library.
30. U.S., Department of State, *Foreign Relations of the United States 1926*, Vol. 2, p. 989.
31. Stone, "The Relation of Petroleum to American Foreign Policy," p. 30.
32. U.S., Department of State, *Foreign Relations of the United States 1923* (Washington, D.C.: Government Printing Office, 1938), Vol. 2, p. 1247.
33. Bryson, *American Diplomatic Relations with the Middle East*, p. 83.
34. Edward M. Earle, "Oil and American Foreign Policy," *Political Science Quarterly*, Summer 1924, p. 277.
35. Shoshana Klebanoff, *Middle East Oil and U. S. Foreign Policy with Special Reference to the U. S. Energy Crisis* (New York: Praeger Publishers, 1974), p. 9.
36. Lloyd C. Gardner, *Economic Aspects of New Deal Diplomacy* (Madison: University of Wisconsin Press, 1964), p. 231.
37. Dean Rusk to Japanese Embassy in Iraq, Washington, D.C., September 2, 1967, *Declassified Documents Reference System* 42H (Arlington, Va.: The Carrollton Press, 1975).
38. Assistant Secretary of State to Herbert Hoover, Washington, D.C., November 20, 1922, Bureau of Foreign and Domestic Commerce, Record Group No. 151, File No. 312, National Archives.
39. Wilkins, *The Maturing of Multinational Enterprise*, p. 215.
40. Joseph Walt, "Saudi Arabia and the Americans 1926-1951" (Ph. D. diss., Northwestern University, 1960), p. 105.
41. Gardner, *Economic Aspects of New Deal Diplomacy*, p. 232.
42. Anonymous State Department Memorandum, Washington, D.C., April 21, 1941, PSFD Diplomatic, Roosevelt Library.
43. Gardner, *Economic Aspects of New Deal Diplomacy*, p. 233.
44. Wilkins, *The Maturing of Multinational Enterprise*, p. 276.
45. Patrick Hurley to Franklin Roosevelt, Saudi Arabia, June 9, 1943, PSFD Diplomatic, Roosevelt Library.

46. U.S., Department of State, *Foreign Relations of the United States 1943* (Washington: Government Printing Office, 1964), Vol. 4, p. 936.

47. Gardner, *Economic Aspects of New Deal Diplomacy*, p. 234.

48. Franklin Roosevelt to Cordell Hull and Harold Ickes, Washington, D.C., January 10, 1944, PSFD Diplomatic, Roosevelt Library.

49. "Pipeline in Arabia," *New Republic*, February 14, 1944, p. 196.

50. "Oil Leads to Tie in Near East," *Christian Century*, February 16, 1944, p. 196.

51. "Well Chosen Words," *Time*, April 3, 1944, p. 77.

52. U.S., Department of State, *Foreign Relations of the United States 1944* (Washington: Government Printing Office, 1965), Vol. 3, p. 103.

53. U.S., Department of State, *Foreign Relations of the United States 1945* (Washington: Government Printing Office, 1969), Vol. 8, p. 852.

54. Joseph W. Walt, "Saudi Arabia and the Americans," unpublished paper, p. 14.

55. "Politics Has a Part in International Oil," *Life*, March 28, 1949, p. 79.

56. "Oil War: Will It Split the West?" *Newsweek*, January 9, 1956, p. 28.

57. Simon G. Siksek, *The Legal Framework for Oil Concessions in the Arab World* (Beirut: Middle East Research and Publishing Center, 1960), pp. 47-48.

58. Ibid., p. 60.

59. "Saudi Arabia Seeking Share of Aramco," *Oil and Gas Journal*, June 17, 1968, p. 60.

60. "Saudi Arabia Awaits U.S., European Oil Moves," *Oil and Gas Journal*, April 9, 1973, p. 34.

61. Walt, "Saudi Arabia and the Americans," p. 18.

62. Wallace E. Pratt, "The Value of Business History in the Search for Oil," unpublished paper, p. 13.

63. Ibid., pp. 12-13.

64. William Appleman Williams, *American Russian Relations 1781-1947* (New York: Reinhart and Company, 1952), p. 195.

65. Fischer, *Oil Imperialism*, p. 138.

66. Ibid.

67. Hogan, *Informal Entente*, p. 172.

68. U.S., Department of State, *Foreign Relations of the United States 1924*, Vol. 2, p. 544.

69. Denny, *America Conquers Britain*, p. 287.

70. Henry P. Crawford, Confidential Report, Laws Governing the Petroleum Industry in Foreign Countries—Iran, Tehran, November 29, 1932, Record Group No. 40, File No. 71744, National Archives.

71. Wilkins, *The Maturing of Multinational Enterprise*, p. 219.

72. Justus D. Doenecke, "Revisionists, Oil and Cold War Diplomacy," *Iranian Studies*, Winter 1970, p. 28.

73. Ibid., p. 29.

74. "Private Deal," *New Republic*, January 13, 1947, p. 7.

75. U.S., Department of State, *Foreign Relations of the United States 1949* (Washington, D.C.: Government Printing Office, 1977), Vol. 6, p. 126.

76. Sampson, *The Seven Sisters*, p. 145.

77. Mohammed Mossadegh to Harry Truman, Tehran, Iran, June 11, 1951, President's Secretary's File, Truman Papers, Truman Library.

78. Mohammed Mossadegh to Harry Truman, Tehran, Iran, June 28, 1951, President's Secretary's File, Truman Papers, Truman Library.

79. Harlan Cleveland, "Oil, Blood and Politics: Our Next Move in Iran," in *Reporter*, November 10, 1953, p. 16.

80. John F. Simmons, Memorandum of Conversation, "Call of the newly appointed Ambassador of Iran on the President," Washington, D.C., September 24, 1952, Truman Papers—Confidential File, Truman Library.

81. "U. S. Attitude towards Purchase of Oil from Iran," Department of State *Bulletin*, December 15, 1952, p. 946.

82. Cleveland, "Oil, Blood and Politics: Our Next Move in Iran," p. 18.

83. Edmundson, "Iraq's Explosive Oil," p. 52.

84. Wilkins, *The Maturing of Multinational Enterprise*, p. 322.

85. Herbert Brownell to Dwight Eisenhower, Washington, D.C., September 15, 1954, Areeda Papers, Eisenhower Library.

86. Michael K. Sheehan, *Iran: The Impact of United States Interests and Policies 1941-1954* (Brooklyn: Theo Gaus' Sons, 1968), p. 66.

87. Ibid.

88. "Standard Indiana's Entrance Fee," *Economist*, May 3, 1958, p. 433.

89. State Department to U.S. Embassy (Tehran, Iran), Washington, D.C., June 8, 1964. Confidential File, Johnson Library.

90. James Akins to Robert L. Dowell, Jr., Washington, D.C., March 18, 1968, *Declassified Documents Reference System* 38E (Arlington, Va.: The Carrollton Press, 1975).

91. Daniel Yergin, "The One Man Flying Multinational, Part II: Armand Hammer Tries Harder," *Atlantic Monthly*, July 1975, p. 59.

## 6. The Far East and South Asia

1. U.S., Department of State, *Foreign Relations of the United States 1887* (Washington, D.C.: Government Printing Office, 1888), p. 48.

2. Howard K. Beale, *Theodore Roosevelt and the Rise of America to World Power* (Baltimore: John Hopkins Press, 1956), p. 228.

3. Noel H. Pugach, "Standard Oil and Petroleum Development in Early Republican China," *Business History Review*, Winter 1971, p. 453.

4. U.S., Department of State, *Foreign Relations of the United States 1938* (Washington, D.C.: Government Printing Office, 1955), Vol. 4, p. 25.

5. Hidy and Hidy, *History of Standard Oil Company (New Jersey): Pioneering in Big Business 1882-1911*, p. 266.

6. U.S., Congress, Message from the President of the United States, *Restrictions on American Petroleum Prospectors in Certain Foreign Countries*, Document No. 11. 67th Cong., 1st sess. (Washington, D.C.: Government Printing Office, 1921), p. 27.

7. U.S., Department of State, *Foreign Relations of the United States 1920* (Washington: Government Printing Office, 1936), Vol. 3, p. 290.

8. U.S., Congress, *Restrictions on American Petroleum Prospectors*, p. 22.

9. Peter Mellish Reed, "Standard Oil in Indonesia, 1898-1928," *Business History Review*, Autumn 1958, p. 311.

10. Caldwell Johnston to Director, Bureau of Foreign and Domestic Commerce, The

Hague, the Netherlands, May 25, 1921, Bureau of Foreign and Domestic Commerce, Record Group No. 151, File No. 312, National Archives.

11. "Dutch Oil, Lubricant and Irritant," *Literary Digest*, May 21, 1921, p. 14 (quotation from New York *Journal of Commerce*).

12. "Holland-American Oil Friction," *Literary Digest*, July 16, 1921, p. 17.

13. Ibid., p. 18 (quotation from *Rotterdammer*).

14. Reed, "Standard Oil in Indonesia, 1898-1928," p. 311.

15. Louis Fischer, "The Greased Wheel of Diplomacy," *Nation*, October 8, 1924, p. 357.

16. Floyd J. Fithian, "Dollars without the Flag: The Case of Sinclair and Sakhalin Oil," *Pacific Historical Review*, May 1970, p. 209.

17. Ibid., p. 207.

18. Ibid., p. 216.

19. Ibid., p. 222.

20. Edward W. Chester, *Sectionalism, Politics, and American Diplomacy* (Metuchen, N.J.: Scarecrow Press, 1975), p. 200.

21. Cordell Hull, *Memoirs*, 2 vols. (New York: Macmillan Company, 1948), Vol. 1, p. 275.

22. U.S., Department of State, *Foreign Relations of the United States 1935* (Washington, D.C.: Government Printing Office, 1953), Vol. 3, p. 931.

23. U.S., Department of State, *Foreign Relations of the United States 1937* (Washington: Government Printing Office, 1954), Vol. 4, p. 728.

24. Karl E. Ettinger, "What Does the Federal Trade Commission Cartel Report Mean."

25. U.S., Department of State, *Foreign Relations of the United States 1935*, Vol. 3, p. 899.

26. Joseph C. Grew, *Turbulent Era: A Diplomatic Record of Forty Years 1904-1945*, 2 vols. (Boston: Houghton Mifflin Company, 1952), Vol. 2, p. 980.

27. Charles Callan Tansill, *Back Door to War: The Roosevelt Foreign Policy 1933-1941* (Chicago: Henry Regnery Company, 1952), p. 143.

28. U.S., Department of State, *Foreign Relations of the United States: Japan 1931-1941* (Washington, D.C.: Government Printing Office, 1943), Vol. 2, p. 210.

29. Herbert Feis, *The Road to Pearl Harbor: The Coming of the War Between the United States and Japan* (New York: Atheneum, 1965), p. 97.

30. Harry Dexter White to Henry Morganthau, memorandum, "Effectiveness of Export Control on Aviation Gasoline, Aviation Lubricating Oil and Tetraethyl Lead," Washington, D.C., August 6, 1940, Morgenthau Diaries, Vol. 289, Roosevelt Library.

31. Verbatim transcript of telephone conversation, July 19, 1940, Morgenthau Diaries, Vol. 284, Roosevelt Library.

32. Ibid.

33. Tansill, *Back Door to War*, pp. 624-25.

34. U.S., Department of State, *Foreign Relations of the United States: Japan 1931-1941*, Vol. 2, p. 219.

35. Feis, *The Road to Pearl Harbor*, p. 158.

36. K. M., "Washington-Tokyo," *Amerasia*, May 1942, p. 100.

37. Feis, *The Road to Pearl Harbor*, pp. 237-38.

38. Clayton D. Carus and Charles Longstreth McNichols, *Japan: Its Resources and Industries* (New York: Harper and Brothers, 1944), p. 198.

39. Ibid.

40. U.S. Embassy to State Department, Djakarta, Indonesia, May 18, 1963, *Declassified Documents Reference System*, 301B Retrospective Collection (Arlington, Va.: The Carrollton Press, 1975), p. 3.
41. "A Complete Draw," *Fortune*, August 1963, p. 79.
42. "Oil and Nationalism Mix Beautifully in Indonesia," *Fortune*, June 1973, p. 99.
43. Gabriel Kolko, "An Economic Incentive for Winning the War?: Oiling the Escalator," *New Republic*, March 13, 1971, p. 19.
44. Frederick Jaeger, "Sea of Oil," *Nation*, July 6, 1970, p. 2.
45. "The Development of Vietnam's Petroleum Resources," *Asian Survey*, June 1976, p. 569.
46. U.S., Department of State, *Foreign Relations of the United States 1922* (Washington, D.C.: Government Printing Office, 1938), Vol. 2, p. 353.
47. U.S., Department of State, *Foreign Relations of the United States 1921*, Vol. 2, p. 80.
48. U.S., Department of State, *Foreign Relations of the United States 1941* (Washington: Government Printing Office, 1959), Vol. 3, p. 192.
49. Norman D. Palmer, *South Asia and United States Policy* (Boston: Houghton Mifflin, 1966), p. 123.
50. "Esso Plan to Sell Assets Shocks India," *Oil and Gas Journal*, November 13, 1972, p. 114.

## 7. Observations and Conclusions

1. Allan Nevins, *Study in Power: John D. Rockefeller, Industrialist and Philanthropist*, Vol. 2, p. 707.
2. Ibid.
3. Michael Hudson, *Global Fracture: The New International Economic Order* (New York: Harper and Row, 1977), p. 41.
4. Gibb and Knowlton, *History of Standard Oil Company (New Jersey): The Resurgent Years 1911-1927*, p. 381.
5. Richard J. Barnet, *Roots of War* (Baltimore: Penguin Books, 1973), p. 199.
6. Inge Kaiser, "Oil, Blood and Sand," *New Republic*, December 17, 1945, p. 831.

# Bibliographical Essay

## Unpublished Materials and Governmental Publications

It is impossible to write a comprehensive account of U.S. oil policy and diplomacy in the twentieth century without visiting the various presidential libraries. The collections in them that are open to researchers, though, are more valuable from the standpoint of oil policy than diplomacy, with the Johnson library in Austin at present being a case in point. The author refrained from visiting the Kennedy Library, because archivists informed him that the materials dealing with U.S. oil policy and diplomacy were minimal.

Let us assess the holdings pertinent to this volume at five of these presidential libraries in chronological order. The Hoover Library possesses the Commerce Papers-Oil, the Presidential Papers-Oil, Commerce Personal-Charles Evans Hughes, and the Ray Lyman Wilbur Papers. The holdings at the Hoover Library cover Hoover's tenure as Secretary of Commerce (1921-28), as well as his presidential years. The most important collections in the Roosevelt Library are concentrated in the Official File, the Morgenthau Diaries, and the President's Secretary's File. The Truman Library offers a large amount of material dealing with the antitrust campaign against the oil companies during the Truman administration in the Stephen Spingarn Papers. The President Secretary's File, the Official File, the Sumner Pike Papers, and the Oscar Chapman Papers also proved useful.

Aside from the White House Central Files, the Eisenhower Library also includes a most informative Legislative Meetings Series and the Cabinet Series as a part of the Ann Whitman Files. Other collections of interest are the Philip Areeda Papers and the Fred Seaton Papers: Ewald Research File. The Johnson Library houses the White House Central Files, the Confidential File, the Robson-Ross Office Files, and the Gaither Files. To date the Nixon Papers are not catalogued and open to researchers.

Much less material dealing with U.S. oil policy and diplomacy is available at the National Archives in Washington, D.C. The files of the Bureau of Foreign and Domestic Commerce (Record Group No. 151) do contain a wide scattering of pertinent data, and a card index to the Department of Interior records yielded a number of promising leads, but almost all of these documents were missing from the National Archives. The most valuable collection in Record Group No. 80 of the old Navy records was a folder in the Forrestal Papers dealing with Middle Eastern petroleum during the 1940s. There were only a few items in the War Department files (Record Group No. 165) pertaining to this subject; the correspondence concerning the proposed destruction of the Netherlands East

Indies oil wells at the beginning of World War II was the most illuminating group of documents.

One does find an immense quantity of published material on petroleum in the U.S. Department of State's annual *Foreign Relations* series. Unfortunately, as of this writing the series only covers the years through the early 1950s; and despite the passage of congressional legislation facilitating access by private individuals to public records, it is extremely difficult to obtain documents from State Department files for the last three decades.

The first *Foreign Relations* volume to refer to oil is that for the year 1872, in the section on Italy. Beginning in 1882 there are mentions of this topic every year through 1890, following which there is no mention of oil until 1905. After World War I references to petroleum become more and more common, and by the time of World War II the growing demand for oil had generated a huge quantity of diplomatic correspondence, which continued into the postwar period.

Investigating the question of which countries have appeared the most times in the *Foreign Relations* series, one discovers that there were more references from the standpoint of petroleum to Turkey than any other nation prior to World War I and more to Great Britain during the interwar years. For the entire seventy-eight-year span from 1872 to 1950 no less than twenty European countries appear; the countries most frequently cited after Great Britain are Russia, Rumania, France, and Spain. Mexico dominates the correspondence and other documents on Latin America, especially between the two world wars, but the United States also was involved in petroleum diplomacy with at least fifteen other Latin American countries during this period. Those most often mentioned are Venezuela, Argentina, and Brazil.

As for the Middle or Near East, the focus has been on Turkey and more recently on Iran, but Saudi Arabia also has appeared frequently. On a number of occasions the editors have treated the countries of the Middle East and the Far East as units. Japan individually has dominated the Far Eastern scene, with China the runner-up in number of citations. References to South Asia have been minimal; India only is to be found in the 1941 volumes. In the case of Africa, there also is limited data, since there are only two references to Ethiopia and one to Morocco. The *Foreign Relations* series also contains documents dealing with such topics as neutral rights and oil pollution.

Still another regular State Department publication is the weekly Department of State *Bulletin*, which offers the reactions of the State Department as events unfold around the world. Unfortunately, for the generation between the outbreak of World War II and the Arab oil embargo of 1973 there are only fifty references to U.S. petroleum policy and diplomacy. Taken as a group, probably the most important collection is that dealing with international efforts to prevent oil pollution. There are also occasional references to various countries not generally considered in discussions of petroleum, such as Rhodesia, North Vietnam, and Ceylon, but taken collectively the *Bulletin* articles are of secondary or even tertiary importance compared to the *Foreign Relations* series.

Turning to other useful governmental publications which did not originate at the Department of State, one might cite U.S. Department of Commerce, Bureau of the Census, *Historical Statistics, Colonial Times to 1957* (1960). The annual *Statistical Abstract of the United States* volumes also contain some helpful information. *Historical Statistics* includes several tables on petroleum production in the United States, imports and exports, and American investment overseas. Assessments of the oil situation in a number of countries can be found in U.S. Congress, Message from the President of the United

States, *Restrictions on American Petroleum Prospectors in Certain Foreign Countries*, two editions, 1920 and 1921.

Most publications dealing with the subject of American oil policy and diplomacy, however, date from the post-World War II period. Two such publications are U.S., Congress, Senate Special Committee Investigating Petroleum Resources, *Diplomatic Protection of American Petroleum Interests in Mesopotamia, Netherlands East Indies, and Mexico* (1945), and *American Petroleum Interests in Foreign Countries* (1946). Chapter 7 of *American Petroleum Interests in Foreign Countries* presents a history of political and diplomatic policies and activities pertaining to oil, while Chapter 12, the summary and conclusions, offers a wealth of information on the subject. More controversial is U.S. Congress, Senate Select Committee on Small Business, Subcommittee on Monopoly, *The International Petroleum Cartel Staff Report to the Federal Trade Commission* (1952). Each chapter has a valuable summary.

Oil imports have become an increasingly critical issue in recent years. A brief study that summarizes developments since the end of World War II, and makes a half-dozen recommendations, is U.S., Congress, House of Representatives, Committee on Interior and Insular Affairs, *Report on Mandatory Oil Import Control Program, Its Impact on the Domestic Minerals Industry* (1968). Much longer is the document drawn up under the chairmanship of George P. Shultz, which examines the relationship of petroleum imports to the national security: Cabinet Task Force on Oil Import Control, *The Oil Import Question* (1970). Far more extensive than this is U.S., Congress, House Committee on Banking and Currency, *Oil Imports and Energy Security: An Analysis of the Current Situation and Future Prospects* (1974).

A highly controversial publication is U.S., Congress, Senate Committee on Foreign Relations, Subcommittee on Multinational Corporations, *The International Petroleum Cartel, the Iranian Consortium, and U.S. National Security* (1974). Senator Frank Church provides a four page preface; his subcommittee made a number of startling revelations during the course of its hearings, throwing the oil companies on the defensive. Those wishing an up-to-date, nation-by-nation survey of petroleum developments in historical perspective should turn to U.S., Department of Energy, Office of International Affairs, *The Relationship of Oil Companies and Foreign Governments* (1975) and *The Role of Foreign Governments in the Energy Industries* (1977).

## Books, Pamphlets, Dissertations, Theses, and Unpublished Papers

Although the author extracted material from nearly 200 items falling into this category, those that yielded only a few scattered bits of data will not be discussed. Herbert Feis, *Petroleum and American Foreign Policy* (1944), offers a classic brief study that focuses on the Middle East during World War II. For the World War I era there is a 1974 University of Missouri doctoral dissertation by Dennis O'Brien, entitled "The Oil Crisis and the Foreign Policy of the Wilson Administration." Representative examples of the New Left perspective of the post-World War II era include Gabriel Kolko, *The Politics of War: The World and United States Foreign Policy, 1943-1945* (1968), and Joyce Kolko and Gabriel Kolko, *The Limits of Power: The World and United States Foreign Policy, 1945-1954* (1972). The former emphasizes Great Britain, the Soviet Union, and the Middle East, while the latter examines U.S. governmental support for the activities of petroleum firms and other American businesses abroad. Gerald D. Nash's *United States Oil Policy 1890-1964* (1968) treats both domestic and foreign developments, while Edward H. Shaffer's *The Oil Import Program of the United States* (1968) is a highly technical study. An important publication by the National Petroleum Council, *Petroleum Resources under the Ocean Floor* (1969), lays claim to off-shore petroleum on behalf of the United States. A supplementary report with this same title appeared two years later.

If one were to recommend the writings of a single author in the field of business history who has examined oil developments overseas, it should be Mira Wilkins. Wilkins has authored *American Business Abroad from the Colonial Era to 1914* (1970) and *The Maturing of Multinational Enterprise: American Business Abroad from 1914 to 1970* (1974). The present volume reflects many of Ms. Wilkins' insights. In contrast, Harold F. Williamson, Ralph L. Andreano, Arnold R. Daum, and Gilbert C. Klose emphasize the domestic scene in *The American Petroleum Industry: The Age of Energy 1899-1959* (1963), but they do have several excellent chapters on foreign oil. The definitive treatment of Jersey Standard's operations, including the overseas ones, is Ralph W. Hidy and Muriel E. Hidy, *History of Standard Oil Company (New Jersey): Pioneering in Big Business 1882-1911* (1955) and George Sweet Gibb and Evelyn H. Knowlton, *History of Standard Oil Company (New Jersey): The Resurgent Years 1911-1927* (1956). Gibb and Bennett Wall also wrote a biography of Walter Teagle, published in 1974. A journalistic rather than scholarly study that is nevertheless quite informative is Anthony Sampson, *The Seven Sisters: The Great Oil Companies and the World They Shaped* (1975, revised edition 1979).

It has only been in recent years that books and articles have begun to appear dealing with American interest in Canadian oil. One such work is Alan R. Plotnick, *Petroleum: Canadian Markets and United States Foreign Policy* (1964); another is the Canadian Ministry of Energy, Mines, and Resources, *An Energy Strategy for Canada: Policies for Self-Reliance* (1976). See chapter 5 of Plotnick's work for United States oil imports and chapter 3 of the Canadian Ministry's for Canadian-American energy relations.

Many of the first large-scale works dealing extensively with the foreign oil relations of the United States that appeared in the decade following World War I emphasized Great Britain. Two of these are E. H. Davenport and Sidney Russell Cooke, *The Oil Trusts and Anglo-American Relations* (1923), and Ludwell Denny, *America Conquers Britain: A Record of Economic War* (1930). A retrospective volume based on heavy research into archival materials—British as well as American—is Michael J. Hogan, *Informal Entente: The Private Structure of Cooperation in Anglo-American Economic Diplomacy, 1918-1928* (1947). There is also Sister Gertrude Mary Gray's 1950 University of California doctoral dissertation, "Oil in Anglo-American Diplomatic Relations, 1920-1928."

As for the continent of Europe, Brice Harris goes into the controversy over petroleum shipments from America to Italy in *The United States and the Italo-Ethiopian Crisis* (1964), and Robert Strausz-Hupé discusses the role of oil in Germany's defeat during World War II in *The Balance of Tomorrow: Power and Foreign Policy* (1945). Hungary's postwar nationalization of Standard Oil properties is examined from Standard Oil's viewpoint in European Gas and Electric Company, *Standard Oil Company (New Jersey) and Oil Production in Hungary by Maort* (1949).

Still other monographs treat Western Europe as an entity. Charles L. Robertson in *The Emergency Oil Lift to Europe in the Suez Crisis* (1965) takes the position that the U.S. government could not have done more than it did, while Walter J. Levy examines "The Suez Oil Crisis: Its Impact on the European Economy and Its Effects on International Oil Policy" in an unpublished paper written for release in 1957. A more recent study, chapter 8 of which deals with Western Europe, is Joseph A. Yager and Eleanor B. Steinberg, *Energy and U. S. Foreign Policy* (1974).

Russia also has attracted the attention of a number of writers interested in oil both before and after the Soviet takeover. An early title is J. D. Henry, *Baku: An Eventful History* (1956), which discusses American petroleum imports from the end of the U.S. Civil War on. The anticapitalist view is best exemplified by Louis Fischer, *Oil Imperialism: The International Struggle for Petroleum* (1926). A more recent volume of a more scholarly nature by a historian sympathetic to Marxist theory is William Appleman Williams, *American Russian Relations 1781-1947* (1952). Williams includes data on the struggle by Sinclair and Standard to obtain a Persian oil concession during the World War I decade.

Unlike the Middle East, no one volume adequately summarizes American interest and involvement in Latin American petroleum. As a point of departure there are the chapters written by Peter R. Odell in Edith T. Penrose, *The Large International Firm in Developing Countries: The International Petroleum Industry* (1968), and Claudio Véliz, *Latin America and the Caribbean: A Handbook* (1968). After that one must consult books on individual nations.

Not surprisingly, the most substantial body of literature deals with Mexico. A relatively early work is James Morton Callahan, *American Foreign Policy in Mexican Relations* (1932), which traces developments from the Díaz era. An in-depth, scholarly account from the Mexican standpoint that features extensive research into primary sources is Lorenzo Meyer, *Mexico and the United States in the Oil Controversy, 1917-1942* (1977), which originally appeared in Spanish. As for the 1938 expropriation, E. David Cronon, *Josephus Daniels in Mexico* (1960) is a key biography that carries the story through the Cooke-Zevada settlement; Donald R. Richberg, *The Mexican Oil Seizure* (1939?) sides with the foreign oil companies against the Cárdenas regime. More recent developments are treated by J. Richard Powell in his *The Mexican Petroleum Industry 1938-1950* (1956), which looks at the period from the 1938 expropriation through the early years of Antonio J. Bermúdez as the Director of Pemex. Bermúdez himself offers his views in *The Mexican National Petroleum Industry* (1963).

One might begin a nation-by-nation exploration of American oil activities in South America by examining Edwin Lieuwen's *Petroleum in Venezuela: A History* (1955), or the more recent book by Franklin Tugwell, *The Politics of Oil in Venezuela* (1975). An important account by a participant is Rómulo Betancourt, *Venezuela: Oil and Politics* (translated from the Spanish, 1979). Two older but still valuable monographs on Colombia are J. Fred Rippy, *The Capitalists and Colombia* (1931), and E. Taylor Parks, *Colombia and the United States 1765-1934* (1935). A much later study featuring an elaborate comparative analysis is George M. Ingram, *Expropriation of U. S. Property in South America: Nationalization of Oil and Copper Companies in Peru, Bolivia, and Chile* (1974).

Peru seems to have attracted the attention of a number of writers, especially its post-World War II years leading up to the expropriation of the International Petroleum Company in 1968. Four important volumes are James C. Carey, *Peru and the United States, 1900-1962* (1964); Daniel A. Sharp, ed., *U.S. Foreign Policy and Peru* (1972); Adalberto J. Pinelo, *The Multinational Corporation as a Force in Latin American Politics: A Case Study of the International Petroleum Company in Peru* (1973); and Jessica P. Einhorn, *Expropriation Politics* (1974). A nationalization from an earlier generation is viewed from the American standpoint in Standard Oil, *Confiscation: A History of the Oil Industry in Bolivia* (1939), while developments from 1921, including the 1937 expropriation, are treated from the Bolivian point of view by Victor Andrade in *My Missions for Revolutionary Bolivia, 1944-1962* (1976).

Michael Tanzer devotes a chapter of his *The Political Economy of International Oil and the Underdeveloped Countries* (1969) to Latin America, especially Argentina and Brazil. Works dealing with these two nations separately include Harold F. Peterson, *Argentina and the United States 1810-1960* (1964), which stresses the events of the 1920s and the 1950s, and Laura Randall, *An Economic History of Argentina in the Twentieth Century* (1978), which offers a balanced historical survey of petroleum developments. Peter Seaborn Smith in *Oil and Politics in Modern Brazil* (1976) covers the period before and after the creation of Petrobrás. One of the few works to treat events related to oil in the islands of the Caribbean is Philip W. Bonsal's *Cuba, Castro and the United States* (1971); Bonsal examines the seizure of the U.S. oil company refineries from the viewpoint of the American Embassy. Equally scarce are volumes that investigate events relating to petroleum in the republics of Central America other than Mexico. Dana G. Munro, *Intervention and Dollar Diplomacy in the Caribbean 1900-1921* (1964), does include material on Costa Rica.

There are a number of general summaries of the Middle East that contain information relating to petroleum. Among the best are Benjamin Schwadran, *The Middle East, Oil and the Great Powers* (1959, revised edition 1973); Shoshana Klebanoff, *Middle East Oil and U. S. Foreign Policy* (1974); Henry Cattan, *The Evolution of Oil Concessions in the Middle East and North Africa* (1967); and Simon G. Siksek, *The Legal Framework for Oil Concessions in the Arab World* (1960). Siksek's work is a highly technical but invaluable study that investigates a topic on which most books on Middle Eastern petroleum only touch.

A recently published survey that provides a good historical treatment of the subject is Thomas A. Bryson, *American Diplomatic Relations with the Middle East, 1784-1975: A Survey* (1977). Herbert Feis' *The Diplomacy of the Dollar: First Era 1919-1932* (1965) is more chronologically restricted, but it does examine the struggle to obtain control over the petroleum resources of Mesopotamia. Unfortunately, unlike Latin America, there are few books on individual countries dealing with Middle Eastern oil. One such work is Joseph Walt's 1960 Northwestern University doctoral dissertation, "Saudi Arabia and the Americans 1926-1951," which emphasizes the activities of Aramco. Also of interest are such works centering on the smaller Arab nations as Ragaei El Mallakh's *Economic Development and Regional Cooperation: Kuwait* (1968), and Donald Hawley's *The Trucial States* (1970). As for Iran, one might cite Mark H. Lytle's unpublished paper "Corporations and American Policy for Iran 1941-47" and Michael K. Sheehan's *Iran: The Impact of United States Interests and Policies 1941-1954* (1968). Similarly useful are Richard J. Barnet's *Roots of War* (1973), which studies the anti-Mossadegh coup, and Andrew F. Westwood, *Foreign Aid in a Foreign Policy Framework* (1966), which investigates Point Four aid and the Anglo-Iranian dispute. One of the more neglected countries and periods from the standpoint of American petroleum activity is Iraq since 1928, although writers have examined rather thoroughly developments prior to the signing of the Red Line Agreement.

Those books on the Far East that contain a significant amount of material on oil frequently center on the Japanese quest for petroleum during the decade prior to Pearl Harbor, and the American reaction to these maneuverings. An in-depth study using information from Exxon files is Irvine H. Anderson, *The Standard Vacuum Oil Company and United States East Asian Policy 1933-1941* (1975). Older established scholarly accounts include William L. Langer and S. Everett Gleason, *The Undeclared War 1940-1941* (1953) and Herbert Feis, *The Road to Pearl Harbor: The Coming of the War Between the United States and Japan* (1965). There is a great deal of invaluable data on earlier decades elsewhere in the Far East in two unpublished papers: Michael H. Hunt, "Americans in the China Market: Economic Opportunities and Economic Nationalism, 1890's-1931" and John A. Morgan, "The Exporting of American Petroleum Products to China." Dorothy Borg also deals with the protection of U.S. oil interests in China in *The United States and the Far Eastern Crisis of 1933-1938* (1964).

A number of studies covering the post-World War II period focus on the now independent nation of Indonesia. These include the Indonesia Project (Benjamin Higgins, Director), *Stanvac in Indonesia* (1957), and Roger Hillsman, *To Move a Nation: The Politics of Foreign Policy in the Administration of John F. Kennedy* (1967). James W. Gould's *Americans in Sumatra* (1961) offers a historical treatment of a single island.

The evolution of U.S. interest in South Asian oil can be traced only through a scattering of information in a few volumes, including Norman D. Palmer, *South Asia and United States Policy* (1966). Palmer examines recent developments in India and Ceylon. Two other studies of interest are Biplab Dasgupta, *The Oil Industry of India: Some Economic Aspects* (1971), and R. K. Pachauri, *Energy and Economic Development in India* (1977). The Dasgupta work has several pages on American involvement in Indian petroleum, while Pachauri treats the developments from the early 1950s to date.

One might mention at least four books that examine petroleum happenings in various nations on different continents. Herbert Feis in his *Seen from E. A.: Three International Episodes* (1947) discusses economic sanctions at the time of the Italo-Ethiopian War, and Saudi-American relations a decade later. More recently, Lloyd C. Gardner in his *Economic Aspects of New Deal Diplomacy* (1964) looked at Mexico, Venezuela, Iran, and Saudi Arabia and their oil relations with the United States during World War II. The 1964 Pennsylvania State University doctoral dissertation by William G. Prast, *The Role of Host Governments in the International Petroleum Industry* (1964) uses Iran, Kuwait, and Indonesia as case studies. Finally, a large number of countries are treated in Dankwart A. Rustow and John F. Mungo, *Opec: Success and Prospects* (1976), a volume that focuses on petroleum developments since 1960.

## Magazine and Journal Articles

Out of the thousands of articles dealing either directly or indirectly with U.S. oil policy and diplomacy, the author selected approximately 500 as part of that body of research materials which he used in the writing of the present volume. Although nearly 100 separate journals or magazines are represented in this collection, over half of the articles were drawn from only five publications: *Time, Business Week, Oil and Gas Journal, Newsweek*, and *Nation*. As for the other journals and magazines, only *Fortune, New Republic*, and *Literary Digest* furnished ten items or more. Most of these articles, however, are the products of contemporary journalism rather than retrospective scholarship. While some of the former are excellent, this discussion will be restricted to the scholarly articles. The majority of these retrospective studies are the product of the last decade, an indication that in the past, concern over petroleum was less intense than it has been since 1973.

There have been two important pronouncements on oil diplomacy by State Department officials: John A. Loftus, "Oil in United States Foreign Policy" (*Vital Speeches*, October 1, 1946) and James E. Akins, "The Oil Crisis: This Time the Wolf is Here" (*Foreign Affairs*, April 1973). The Loftus article generalizes about developments down to 1945, and makes some predictions for the future; the Akins one takes a pessimistic view of the years to come, and downplays the North Sea and Alaska fields. As for an earlier period,

John A. De Novo examines U.S. petroleum operations overseas in "The Movement for an Aggressive American Oil Policy Abroad 1918-1920" (*American Historical Review*, July 1956).

An article by Norman Nordhauser, "Origins of Federal Oil Regulation in the 1920's" (*Business History Review*, Spring 1973) argues that "oil executives rather than governmental officials led the way in the movement for regulation and that their dominant goals were the stabilization of prices and profits." Especially valuable for its treatment of the overland exemption is Kenneth W. Dam's "Implementation of Import Quotas: The Case of Oil" (*Journal of Law and Economics*, April 1971). As for the more recent years, Burton I. Kaufman shows that both "the Truman and Eisenhower administrations used American controlled multinational corporations operating in the Mideast as instruments of United States foreign policy" in his "Mideast Multinational Oil, U.S. Foreign Policy, and Antitrust: the 1950's" (*Journal of American History*, March 1977).

Unfortunately, it is only possible to recommend a handful of scholarly articles on American petroleum relations with Europe. These include: Michael J. Hogan, "Informal Entente: Public Policy and Private Management in Anglo-American Petroleum Affairs, 1918-1924" (*Business History Review*, Summer 1974); Herbert Feis, "Oil for Spain: A Critical Episode of the War" (*Foreign Affairs*, January 1948); Philip S. Gillette, "American Capital in the Contest for Soviet Oil, 1920-23" (*Soviet Studies*, April 1973); and Hans Heymann, "Oil in Soviet-Western Relations in the Interwar Years" (*American Slavic and East European Review*, December 1948). Hogan stresses developments in Mesopotamia, and points out how the State Department was pressured into redefining the "Open Door" policy, thus allowing the institutionalizing of cooperation between Great Britain and the United States at the private level. The Feis article, which deals with World War II, shows the problems of oil diplomacy from the standpoint of a leading neutral power. As for the two articles on the Soviet Union, Gillette focuses on the scramble for concessions in a Russia that recently had nationalized foreign oil holdings, and Heymann examines developments during the entire twelve years of Republican rule in the United States.

One is able to point to several good surveys of Latin America. D. M. Phelps in his "Petroleum Regulation in Temperate South America" (*American Economic Review*, March 1939) deals with a number of countries, including Brazil, Uruguay, and Chile, extending back to the beginning of the twentieth century. Two more recent articles are Peter R. Odell, "Oil and State in Latin America" (*International Affairs*, October 1964), and Mira Wilkins, "Multinational Oil Companies in South America in the 1920's: Argentina, Bolivia, Brazil, Chile, Colombia, Ecuador, and Peru" (*Business History Review*, Autumn 1974). While Odell examines recent developments in a number of nations, Wilkins' retrospective essay points out that responses to the foreign petroleum firms varied from country to country and according to their business functions.

Again there are more articles on Mexico than on any other Latin American nation, and many of them focus on the years immediately following the adoption of the Constitution of 1917. Kenneth J. Grieb, "Standard Oil and the Financing of the Mexican Revolution" (*California Historical Quarterly*, March 1971) deals with the fall of Porfirio Díaz. According to Grieb, fragmentary evidence suggests that there was probably some sort of an agreement between Standard Oil and the revolutionaries.

Turning to the years from 1917 on, Mark T. Gilderhus points out in his "Henry P. Fletcher in Mexico, 1917-1920" (*Rocky Mountain Social Science Journal*, October 1973) that Wilson and Fletcher disagreed on a policy for Mexico, with Fletcher advocating a more forceful approach than the President. Samuel J. Astorino in his "Senator Albert B.

Fall and Wilson's Last Crisis with Mexico" (*Duquesne Review*, Spring 1968) charges that Fall made a deliberate attempt to tarnish the Wilson administration. On the other hand, N. Stephen Kane plays down business influence on U.S. foreign policy in "American Businessmen and Foreign Policy: the Recognition of Mexico, 1920-1923" (*Political Science Quarterly*, Summer 1975). In an earlier article entitled "Bankers and Diplomats: The Diplomacy of the Dollar in Mexico, 1921-1924" (*Business History Review*, Autumn 1973), Kane concluded that "the resulting exercise in dollar diplomacy achieved some immediate ends but proved very short-sighted and self-defeating in the end." A highly technical study of events over a fifteen-year period is Harold E. Davis, "Mexican Petroleum Taxes" (*Hispanic American Historical Review*, November 1932).

Two decades of friction over foreign oil holdings in Mexico culminated in the expropriation of various American and British oil properties by President Lazaro Cárdenas in 1938. In praising U.S. Ambassador to Mexico Josephus Daniels, George C. Coleman in his "The 'Good Neighbor' Tested: 1938" (*Southwestern Social Science Quarterly*, December 1952) takes the position that the Roosevelt administration followed the correct policy toward Mexico after expropriation. Another article dealing with this period is Merrill Rippy, "The Economic Repercussions of Expropriation: Case Study: Mexican Oil" (*Inter-American Economic Affairs*, Summer 1951). Rippy discusses the legal cases arising from the expropriation, Mexican economic relations with countries other than the United States, and the American decision to suspend the silver purchase agreement.

Elsewhere in Latin America, Stephen J. Randall offers "The International Corporation and American Foreign Policy: The United States and Colombian Petroleum, 1920-1940" (*Canadian Journal of History*, August 1974). Randall attributes the virtual monopolistic position of the American oil companies in Colombia to both the U.S. government and the Colombian oligarchy. Herbert S. Klein gives a scholarly and detailed account of petroleum developments in Bolivia between 1920 and 1942 in his "American Oil Companies in Latin America: The Bolivian Experience" (*Inter-American Economic Affairs*, Autumn 1964), while Enrique A. Baloyra touches at times on U.S. oil relations with Venezuela in his "Oil Policies and Budgets in Venezuela, 1938-1968" (*Latin American Research Review*, Summer 1974). Peter Seaborn Smith's "Petrobrás: The Politicizing of a State Company, 1953-1964" (*Business History Review*, Summer 1972) demonstrates that "Petrobrás was born politicized and remained the focus of political agitation for most of the first ten years of its life."

As there are a number of books, or chapters or sections of books, that more than adequately trace American oil relations with the Middle East, we will not cite all the articles surveying this general area. There is, however, a need for more scholarly articles that focus on a single nation during various eras, although in recent years there have been a few on Iran. These include Justus D. Doenecke, "Revisionists, Oil and Cold War Diplomacy" (*Iranian Studies*, Winter 1970), and Jane Perry Clark Carey, "Iran and Control of Its Oil Resources" (*Political Science Quarterly*, March 1974). Doenecke endorses the findings of such revisionists as Lloyd Gardner and Gabriel Kolko; in contrast, Carey deals with the two decades following the nationalization of Anglo-Iranian in 1951.

Unfortunately there are few, if any, books or articles surveying U.S. petroleum diplomacy with the Far East, but there are some articles covering various periods in the histories of different countries. Among these is Peter Mellish Reed, "Standard Oil in Indonesia, 1898-1928" (*Business History Review*, Autumn 1958). According to Reed, "any success which Standard met in its quest for oil concessions in the Dutch colony prior to 1928 should partially be credited to an active interest in petroleum on the part of the Department of State." Another article appearing in this publication was Noel H. Pugach's "Standard Oil and Petroleum Development in Early Republican China" (Winter 1971). Pugach shows how the efforts to set up a joint Chinese-American company for the purpose of exploiting Chinese oil was unsuccessful in Wilson's first presidential term.

Sometimes U.S. oil relations with the Far East have involved more than one country, as was the case with the Japanese and Russian contest over Sakhalin Island following World War I. In his "Dollars without the Flag: The Case of Sinclair and Sakhalin Oil" (*Pacific Historical Review*, May 1970), Floyd J. Fithian concludes that the American government's reluctance to aid Sinclair was a significant deviation from its official policy. Another important article that deals with Japan prior to World War II is Jamie W. Moore's "Economic Interests and American-Japanese Relations: The Petroleum Monopoly Controversy" (*The Historian*, August 1973). Moore points out that the Roosevelt administration decided to treat the dispute over the Japanese petroleum monopoly "as a routine diplomatic matter rather than a burgeoning crisis." Elsewhere in the Orient, David G. Brown's "The Development of Vietnam's Petroleum Resources" (*Asian Survey*, June 1976) concerns the post-World War II period prior to the fall of the South Vietnamese government in the spring of 1975.

Any researcher investigating U.S. oil policy and diplomacy since 1973 will, of course, be forced to process a glut of materials. This is readily apparent from an examination of the recent listings under petroleum in the New York *Times* annual index. The overwhelming bulk of the current writing on this topic, though, is highly ephemeral; much of the analysis quickly goes out of date because of rapidly unfolding global developments. This, however, should not keep the researcher from turning to the more scholarly retrospective studies published since 1973, some of which are of lasting value.

# Index

## About the Author

EDWARD W. CHESTER is Professor of History at the University of Texas, Arlington. He has written *Clash of Titans: African and U.S. Foreign Policy* and *The U.S. and Six Atlantic Outposts: The Military and Economic Considerations*.